NATIONS

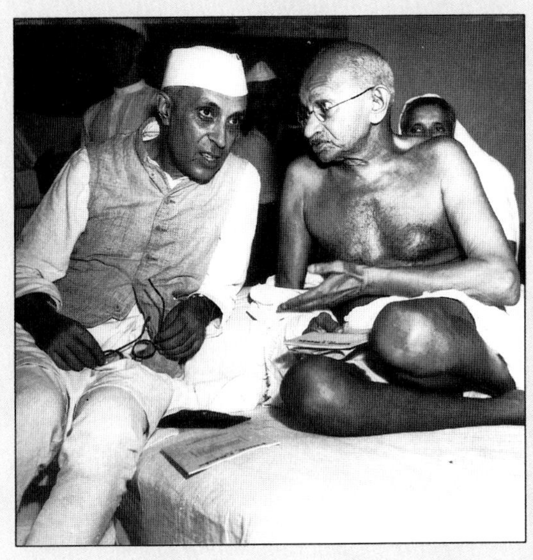

NATIONS

A SURVEY OF THE TWENTIETH CENTURY

EDITED BY
JACK SPENCE

Oxford • New York
OXFORD UNIVERSITY PRESS
1992

CONTENTS

Volume editor Peter Furtado
Art editor Ayala Kingsley
Assistant editors Louise Jones, Caroline Sheldrick, Elaine Welsh
Designers Dave Sumner, Frankie Wood
Cartographic manager Olive Pearson
Cartographic editor Clare Cuthbertson
Picture research manager Alison Renney
Picture researcher David Pratt

AN EQUINOX BOOK

Devised and produced by
Andromeda Oxford Ltd
11-15 The Vineyard, Abingdon,
Oxfordshire OX14 3PX, England

Published in the United States by
Oxford University Press, Inc.,
200 Madison Avenue,
New York, N.Y. 10016

Oxford is a registered trademark of Oxford University Press

Printed in Singpaore by
CS Graphics

Library of Congress Cataloging-in Publication data

Nations in the twentieth century /
edited by J.E. Spence.
p. cm.
Includes bibliographical references and index.
ISBN 0-19-520924-9
1. History, Modern–20th century–Handbooks, manuals etc.
I. Spence, J.E. (John Edward) II. Title: Nations in the 20th century
D421.N38 1992 91-27993
909.82–dc20 CIP

CONTRIBUTORS

Alan Angel
St Antony's College, Oxford, UK

David Arter
Leeds Polytechnic, UK

Jo Beall
London School of Economics, UK

Robert Bideleux
University College of Swansea, UK

Robert Borthwick
University of Leicester, UK

Steven Chapman
Freelance writer, Oxford, UK

Christopher Clapham
Lancaster University, UK

John Crabtree
Oxford Analytica, UK

Martin Dean
University of Keele, UK

Kenneth Dyson
University of Bradford, UK

Roberto Espindola
University of Bradford, UK

Henry Finch
University of Liverpool, UK

Robin Hallett
Freelance writer, Shropshire, UK

Galen Irwin
Leiden University, the Netherlands

Louise Jones
Freelance writer, Oxford, UK

Wilfrid Knapp
St Catherine's College, Oxford, UK

TC McCaskie
University of Birmingham, UK

Richard Nile
University of London, UK

David Renwick
Freelance writer, Trinidad

Gowher Rizvi
Nuffield College, Oxford, UK

Alisdair Rogers
Keble College, Oxford, UK

Peter Savigear
University of Leicester, UK

Franco Scabbiolo
Freelance writer, Oxford, UK

Richard Tames
Freelance writer, London, UK

Peter Woodward
University of Reading, UK

مونديال

PREFACE

Throughout the 20th century the nature and role of the nation state has provoked fierce debate: critics have vilified it as the primary source of war and insecurity in the international system; supporters have claimed that the state at the very least provides the benefit of order and stable government. Without its protective net, men would rapidly return to a state of nature where, in the evocative phrase of 17th-century philosopher Thomas Hobbes, life is "nasty, brutish, and short".

The persistence of the nation state is perhaps its most remarkable feature: those who are stateless because of alien rule or incorporation into multinational empires such as the Soviet Union have – with varying degrees of success – struggled to create nation states in which the principle of national self-determination can find expression.

In the aftermath of World War I, new states emerged in central and southern Europe and the Middle East as old empires disintegrated. A similar process occurred after 1945 as the great European empires conceded – in some cases after violent struggle – independence to their former colonies. Today some 150 states belong to the United Nations, but many of the newcomers are artificial creations, uneasily embracing a variety of ethnic groups and lack any profound sense of national identity. Thus, the dream of 19th-century liberals that state and new-born nation would happily fuse to produce democratic self-government has, in many instances, proved an illusion. Relatively few of the world's new states (and by no means all of the old) pass that test as polyglot empires jostle with Third World autocracies and the minority of mature democracies of Western Europe and North America in the international arena.

Indeed, today there is a challenge to the overarching sovereignty of the state from three very different sources. The Soviet Union, Yugoslavia and Ethiopia are examples of states containing suppressed nationalities seeking autonomy, if not independence, from repressive control by the center. This trend is reversed in Western Europe, however, where a gradual process of political and economic integration offers the longterm prospect of a continent united under supranational institutions of government. Similar developments are under way elsewhere – in West Africa and Southern Africa, for example, where transnational structures exist to promote common interests – especially in the realm of economic development. Finally, the inviolability of state sovereignty is under threat as international organizations such as the UN debate intervention in states where human rights are most flagrantly abused.

Both the professional student and the lay reader will – it is hoped – find this volume a helpful guide through the complex history of the nation state in the 20th century. It is meant to be more than a work of reference as each essay seeks to offer an interpretation of the rise and fall of the variety of states which constitute late 20th century international society.

J. E. Spence   Natal

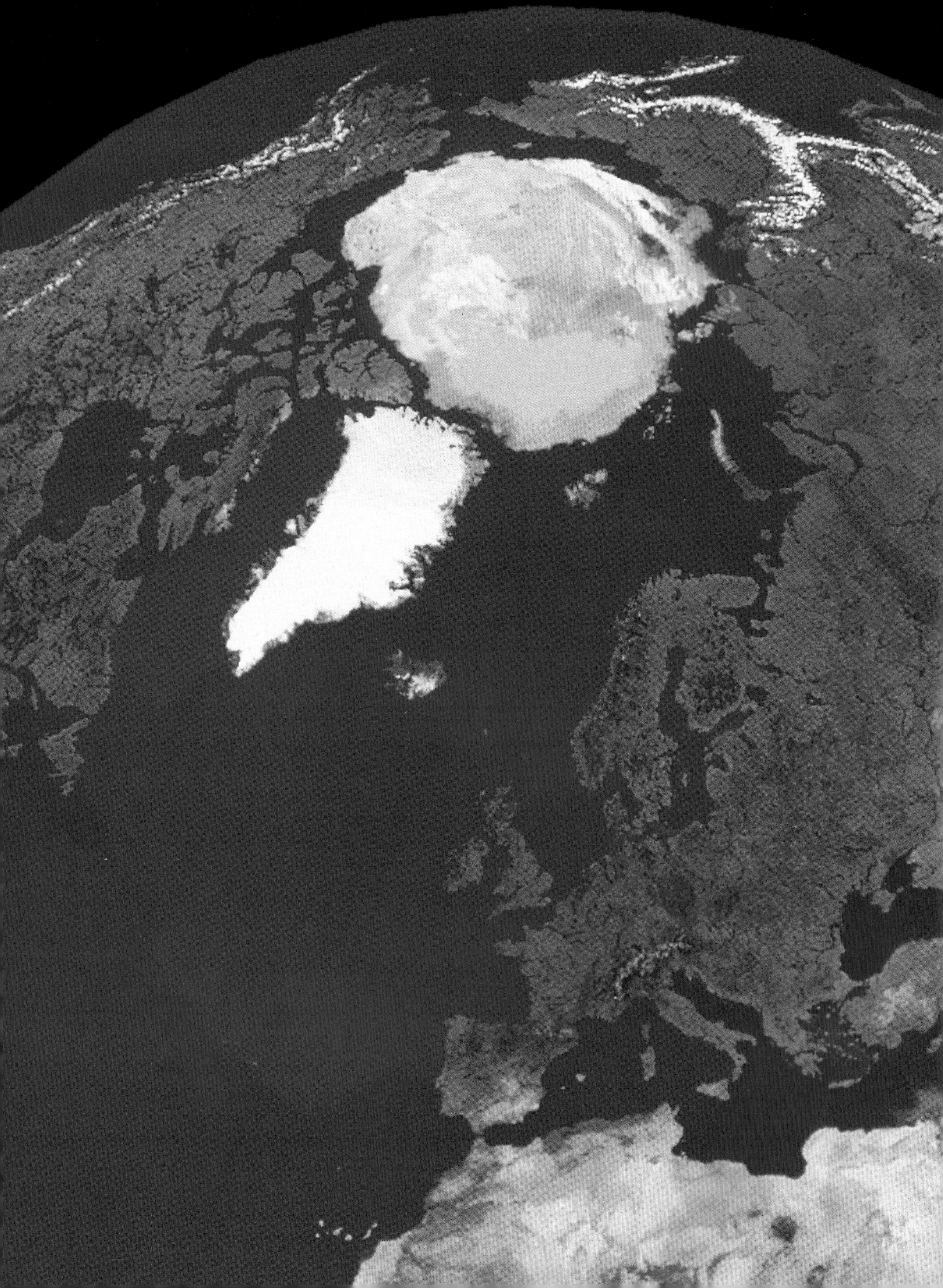

EUROPE

INDEX OF COUNTRIES

NORWAY

KINGDOM OF NORWAY

CHRONOLOGY

1940	Nazi puppet regime established under Vidkun Quisling
1949	Norway joins NATO
1963	Defeat of Labor government after 28 years

ESSENTIAL STATISTICS

Capital Oslo

Population (1989) 4,228,000

Area 323,878 sq km

Population per sq km (1989) 13.0

GNP per capita (1987) US$17,110

CONSTITUTIONAL DATA

Constitution Constitutional monarchy with one legislative house (Parliament)

Date of independence 7 June 1905

Major international organizations UN; I-ADB; NATO

Monetary unit 1 Norwegian krone (NKr) = 100 øre

Official language Norwegian

Major religion Evangelical Lutheran

Overseas territories Jan Mayen, Svalbard

Heads of state since independence (Prime minister) C. Michelsen (1905–07); J. G. Lövland (1907–08); G. Knudsen (1908–10); W. Konow (1910–12); J. Brathé (1912–13); G. Knudsen (1913–20); O. Halvorsen (1920–21); O. Blehr (1921–23); O. Halvorsen (Mar–May 1923); A. Berge (1923–24); J. Mowinckel (1924–26); I. Lykke (1926–28); J. Mowinckel (1928–31); N. Kolstad (1931–32); J. Hundseid (1932–33); J. Mowinckel (1933–35); J. Nygaardsvold (1935–40); E. Gerhardsen (1945–51); O. Torp (1951–55); E. Gerhardsen (1955–63); J. Lyng (Aug–Sep 1963); E. Gerhardsen (1963–65); P. Borten (1965–71); T. Bratteli (1971–72); L. Korvald (1972–73); T. Bratteli (1973–76); O. Nordli (1976–81); G. Brundtland (Feb–Oct 1981); K. Willoch (1981–86); G. Brundtland (1986–89); J. Syse (1989–)

Norway was the first of the new states of 20th-century Europe. Conjoined in a loose union with the Swedish Crown as a byproduct of the Napoleonic Wars, she engineered her full independence in 1905 under the adroit prime ministership of Christian Michelsen.

The distinctive single-chamber parliament, the 165-member Storting, constitutes itself into two internal divisions, the Odelsting and Lagting, for certain legislative purposes and cannot be dissolved ahead of the scheduled four-yearly general elections. The force of nationalism in the last quarter of the 19th century contributed significantly to the mobilization of political parties in Norway and in 1884 the principle of ministerial responsibility was accepted for the first time in the Nordic region. Universal male suffrage was achieved in 1898, and women won the vote by 1913.

During World War I Norway remained formally neutral, but suffered some economic dislocation owing to the war at sea and the decline in trade with Germany.

In the late 1930s the Labor party won its domination of government, which it has maintained ever since. It came to power in 1935 when its leader, Johan Nygaardsvold, did a "crisis deal" with the Agrarians. In foreign policy Norway again tried to maintain neutrality but on 9 April 1940 it was invaded and by June 10 the entire country was finally overrun by Germany, despite Allied support of the Norwegian army. The royal family went into exile and a "national government" was set up by Vidkun Quisling, leader of the tiny National Socialist party. An effective resistance movement developed, with some notable successes against industrial targets.

German forces surrendered in 1945, and in the ensuing election in October the Labor party won its first absolute majority in the Storting, which it maintained until 1961. In this period a number of parties emerged, reflecting a broadly left-right spectrum though with few extreme elements. It was an era of unprecedented political stability and, after the inevitable postwar controls had been lifted (rationing was only ended in 1952), one of sustained economic growth. This was the heyday of egalitarianism, the creation of a new form of state representing all sectors of society, and of an advanced program of reform to promote welfare.

The continuity of the Norwegian political system was facilitated by the absence of significant extremist challenges (except during the Nazi occupation) and radical leftism prospered only intermittently. Even so between 1919 and 1923 a majority of the Labor party adhered to the Third International (Comintern) and in 1945 the Communist party, a stalwart element in the resistance movement, gained a record 11 Storting seats. The Communist coup in Czechoslovakia in February 1948, and the fear of Russian intervention which led to Norway becoming a founder-member of NATO the following year, however dashed the Norwegian party's hopes and the Communists were not represented in the Storting after 1961. In 1961 a breakaway Socialist People's party became the mainstay of a Socialist Electoral Alliance which was formed to oppose Norwegian membership of the European Community; in 1975 this umbrella organization adopted the designation Socialist Left party. This latter made great strides forward at the polls in 1989, more than tripling its number of Storting seats.

On the nonsocialist side, there emerged three center-based parties plus the rightwing Conservatives. Of the former, the

Liberals, as the successors to the 19th-century, nonsocialist left, were a significant force for many years, but split irrevocably over the Common Market issue and both factors had lost their parliamentary representation by 1985. The Agrarian party, whose emergence was facilitated by the introduction of proportional representation in 1919, changed its name to Center party in 1959. Finally, the Christian People's party emerged in the strongly fundamentalist southwest region. Cooperation between four parties concerned to maintain their separate identities and electoral clienteles has never been easy, and this has led to a tendency towards the formation of a Labor minority as a last resort. None the less, between 1965 and 1990 there were three periods of four-party, non-socialist coalition.

Economic and social trends

Norway has developed this century into a welfare state affording citizens comprehensive social security. Benefiting economically from neutrality during World War I, Norway had accumulated substantial profits by the late 1920s from, among other things, the mercantile marine, Antarctic whaling and flourishing metal industries.

The need for reconstruction inspired economic planning in postwar Norway. External factors played their part – particularly the post-Korean War boom, and the discovery of substantial reserves of oil and gas in the Norwegian sector of the North Sea in the early 1970s. Until the collapse of oil prices in 1986, oil accounted for about 40 percent of Norway's foreign earnings.

Norway joined the European Free-Trade Association (EFTA) in 1959. Nevertheless, the basic strength of the economy, coupled with the paramount need for a small economy to trade to survive, prompted a majority in the government, corporate sector and Storting to favor Norwegian membership of the Community when Britain, Denmark and Ireland made applications to join in the early 1970s. The EC question aroused extraordinary passion among a normally undemonstrative populace. In the event, the Labor government proceeded with the application on condition that the ultimate decision to join or not lay with the people at a consultative (albeit effectively binding) referendum. This produced a narrow majority against membership, leaving a Free Trade Agreement (signed in 1973) as the only realistic alternative. The referendum cleared the Common Market question from the political agenda for a decade, although the prospect of the single European market in 1992 reopened the debate and created new pressures to join. In 1990 Norway took a step in this direction in linking the krone to the ECU.

The EC question mobilized citizens in an unprecedented manner and inspired other issue-based popular movements, especially those concerned with energy and environmental protection. As the 1970s advanced, moreover, there was growing evidence of a grassroots reaction against high income taxes and the bureaucratic welfare state to which they contributed. A new vein of radical rightism emerged in the form of the (Anti-tax, subsequently) Progress party, modeled on Danish lines, and led by the outspoken Carl I. Hagen. This combined a frontal attack on the welfare state with the type of exaggerated anti-immigration rhetoric associated with the National Front in France. In fact, the immigrant population remained small.

Over the course of the 20th century, Norway became one of the wealthiest countries in the world. Citizens enjoyed a high standard of living and affluent lifestyle. Values became largely materialistic and secularism advanced to a large degree, despite the fact that 92 percent of the population were formal members of the established church, the Evangelical Lutheran Church of Norway. Major tax reforms and a program of expenditure on the infrastructure were initiated in 1991 with the aim of invigorating the economy and cutting down unemployment.

Morality, or more precisely the search for moral authority, has been an underlying theme of Norwegian foreign policy, which has been based on the belief that small nations can play an active and not merely reactive role on the world stage. This has involved a vigorous commitment to the work of the United Nations (UNO). In addition Norway in recent years has met the UN target of assigning one percent of its GNP to Third World development aid. Such altruism has, however, been tempered by a spirit of realism and self-interest when defense and security interests have been at stake.

▲ An oil platform being towed from Norway into the North Sea.

DENMARK
KINGDOM OF DENMARK

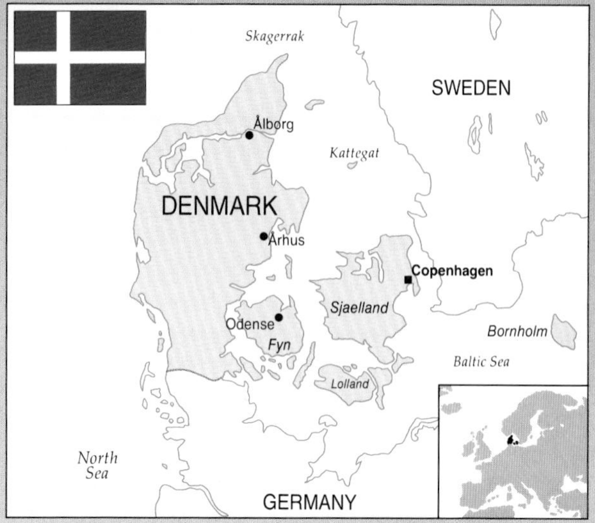

CHRONOLOGY

1901	The principle of a parliamentary government established
1919	Introduction of proportional representation
1920	Referendum on Schleswig leads to partition
1940	Nazi occupation of Denmark
1944	Iceland gains full independence from Denmark
1949	Denmark joins NATO
1958	Denmark becomes part of EFTA
1961	Denmark applies to join the Common Market
1982	Period of Conservative-led coalition under Poul Schlüter

ESSENTIAL STATISTICS

Capital Copenhagen

Population (1989) 5,135,000

Area 43,093 sq km

Population per sq km (1989) 119.2

GNP per capita (1987) US$15,010

CONSTITUTIONAL DATA

Constitution Parliamentary state and constitutional monarchy with one legislative house (Folketing)

Major international organizations UN; EC; EEC; I-ADB; NATO

Monetary unit 1 krone (Dkr) = 100 øre

Official language Danish

Major religion Evangelical Lutheran

Overseas territories Greenland; Faeroe Islands

Heads of government since 1900 (Prime minister) J. H. Deuntzer (1901–05); J. C. Christensen (1905–08); C. T. Zahle (1909–10); K. Berntsen (1910–13); C. T. Zahle (1913–20); N. Neergaard (1920–24); T. Stauning (1924–26); T. Madsen-Mygdal (1926–29); T. Stauning (1929–42); E. Scavenius (1942–45); V. Buhl (May–Oct 1945); K. Kristensen (1945–47); H. Hedtoft (1947–50); E. Eriksen (1950–53); H. Hedtoft (1953–55); H. C. Hansen (1955–60); V. Kampmann (1960–62); J. O. Krag (1962–68); H. Baunsgaard (1968–71); J. O. Krag (1971–72); A. Jørgensen (1972–73); P. Harling (1973–75); A. Jørgensen (1975–82); P. Schlüter (1982–)

In line with its Scandinavian neighbors, Denmark contrived, with no little skill, to maintain a neutral stance during World War I and yet to profit territorially from the defeat of Germany. As a result of a referendum in 1920, the northern part of Schleswig was incorporated into the Danish kingdom and, along with it, an ethnic German minority, which remained strong enough to elect its own representatives to parliament (Folketing) until the late 1960s. During the interwar period, however, Denmark lived in the shadow of renewed German imperialism and in April 1940 was summarily occupied by the Nazis as they advanced toward Norway. The resistance movement had by autumn 1943 spawned a Freedom Council and this subsequently coordinated the return to peacetime politics. Mindful of its vulnerable position astride the major Soviet naval corridor between the Baltic and the North Sea, Denmark, like Norway, opted for NATO membership in 1949 in preference to the Scandinavian Defense Union, for which Sweden was canvassing. Nonetheless, Denmark was the driving force in the creation of a consultative Nordic Council and in 1958 followed her Nordic neighbors into the European Free Trade Association (EFTA). Shortly afterward, however, the imperatives of the German and British export markets dictated an application to join the European Community (EC) and in 1973 Denmark became the first Scandinavian member.

Under the dominant influence of the Social Democrats there evolved an underlying consensus among the four "old parties" (Social Democrats, Radicals, Conservatives and Liberals) regarding the welfare state, management of a mixed economy and pragmatism in policy-making. While never possessing a parliamentary majority, the Social Democrats became the largest Folketing party in 1924 and, in 1933 at the height of the recession, they struck the historic Kanslergade agreement with the farm-based Liberals. This ensured that, in return for measures of agricultural support, the Social Democrats could proceed with wide-ranging welfare legislation.

The achievement in 1936 of a Social Democratic–Radical majority in both houses of parliament prompted the prime minister Stauning to initiate constitutional reform. This provided for a single-chamber assembly, a Swedish-style ombudsman and distinctive provisions for parliamentary control of the executive. The first leftwing majority in the assembly in 1966 seemed to portend a polarization of the party system, but in 1973 five new parties entered the Folketing, including the antitax Progress party of Mogens Glistrup. This proliferation of parties coincided with, and partly reflected, the rise in inflation and unemployment that were the byproducts of the 1973 oil crisis. It meant that it became more difficult to form effective governments, at precisely the time when the role of government in managing the economy had been rendered increasingly problematic. A unique left–right Social Democratic-Liberal coalition between 1978 and 1979 signified at once a reassertion of the "old parties" and an attempt to achieve broad-based direction of the economy. The protest parties of 1973, while mostly retaining a parliamentary toehold, were largely accommodated within the traditional framework of party politics. Two of them, the Center Democrats and Christian People's Party, participated in the Conservative-led coalitions which began in 1982 under Poul Schlüter, the first 20th-century Conservative premier, whose longevity in office marked a clear shift to the right in Danish politics.

Economic and social trends

Denmark in 1990 was an advanced welfare state. It offered citizens comprehensive cradle-to-grave protection. In terms of GDP per head, moreover, it had one of the highest living standards in the world. However, as a nation highly dependent on trade and possessing no raw materials and virtually no indigenous energy resources, Denmark has been more than usually subject to the vagaries of the international economy. A period of sustained growth and full employment beginning in the 1950s was thus ended by the first oil crisis in 1973, and subsequent oil price rises, plunged the country into a chronic balance of payments deficit. Recovery was hampered by a relatively narrow manufacturing base, which represented only some 20 percent of GDP.

Denmark entered the 20th century as a predominantly agricultural economy, which had adapted effectively from grain to dairying in response to the intensified competition in corn from the Americas, the Ukraine and Romania. Agricultural producer cooperatives became a distinctive feature of the economic landscape and by 1945 there were more than 1,400. Agriculture and industry vied for supremacy during the 1930s and light industry, such as food processing, furniture and high-technology goods, became the typical mode. Since World War II, there have been two main socio-economic developments. Agriculture declined to such a point that in 1986 it contributed only 4.8 percent of GDP. There was also a substantial expansion in the public sector's role in the economy: it contributed 21 percent of GDP and employed nearly 38 percent of the workforce in the same year. In short, although beginning later than some West European countries, Denmark shifted very rapidly to a preponderantly service economy.

A steady decline in the size of the ethnic German minority reinforced Denmark's traditionally homogenous culture. At the same time, an influx of political refugees entered the country on the back of a liberal government policy toward those seeking asylum from authoritarian systems, and this injected an element of cosmopolitanism into the culture, especially in the larger urban concentrations. Other citizens originated from the Home Rule territories of Greenland and Faeroes, as well as workers from EC countries (Greece, Turkey, Italy) and the other Scandinavian countries who benefited from agreements on the free movement of labor. Many Danes remained skeptical of the nation's external attachments. An anti-EC movement persisted, while an anti-NATO movement was vigorous and vociferous.

In the late 1980s there was also something of a moral backlash to the licentiousness of the previous 20 years and growing support for "green ideas", especially the need to clean up the Baltic Sea. Exposed to "cultural penetration" from a plethora of satellite television shows, the Danes evinced a growing concern to promote and protect the heritage of one of the oldest nations on earth.

▲ **Sights on the beaches explain the Danish reputation for broadmindedness.**

SWEDEN

KINGDOM OF SWEDEN

CHRONOLOGY

1905	Dissolution of union with Norway
1919	Universal voting introduced and recognition of the parliamentary principle
1933	Social Democratic–Agrarian agreement inaugurates period of welfare reform
1980	Nuclear power referendum
1986	Assassination of prime minister Olof Palme
1990	Swedish government apply to join the EC in 1991

ESSENTIAL STATISTICS

Capital Stockholm

Population (1989) 8,498,000

Area 449,964 sq km

Population per sq km (1989) 20.7

GNP per capita (1987) US$15,690

CONSTITUTIONAL DATA

Constitution Constitutional monarchy and parliamentary state with one legislative house (Parliament)

Major international organizations UN; I-ADB

Monetary unit 1 Swedish krona (SKr) = 100 ore

Official language Swedish

Major religion Evangelical Lutheran

Heads of government (Prime Minister) H. Hammarskjöld (1911–17); C. Swartz (Mar–Oct 1917); N. Eden (1917–20); H. Branting (1920–21); O. von Sydow (Feb–Oct 1921); H. Branting (1921–23); E. Trygger (1923–25); R. Sandler (1925–26); C. Ekman (1926–28); A. Lindman (1928–30); C. Ekman (1930–32); F. Hamrin (Aug–Sep 1932); P. Hansson (1932–36); M. Pehrsson (Jun–Sep 1936); P. Hansson (1936–46); T. Erlander (1946–69); O. Palme (1969–76); T. Fälldin (1976–78); O. Ullsten (1978–79); T. Fälldin (1979–82); O. Palme (1982–86); I. Carlsson (1986–)

Of all the smaller democracies in Western Europe, few have attracted so much attention and admiration abroad as Sweden. Whether in the field of labor relations, social policy or economic management, the "Swedish model" has been the envy of the outside world. Perhaps the crucial ingredient in Sweden's political culture has been a historic propensity for decision-making by consensus, in the spirit of *saklighet* or pragmatic realism. Whatever the case, in the 1980s Sweden could lay claim to possessing the most developed welfare state and one of the highest standards of living in the developed world or OECD developed bloc of countries. Indeed, as a nation it spent more as a proportion of gross domestic product (GDP) on health, education and related services than anywhere else in the world.

Political modernization in Sweden in the 20th century was characterized by considerable continuity, especially in terms of party politics. In April 1902 the central "blue-collar" trade union federation LO organized a general strike in support of universal and equal suffrage, but these goals were not attained until 1919 (for both local and parliamentary elections) and then as a result of pressure from the Liberal and Social Democratic parties. The size of the national territory was reduced in 1905 when the union with Norway, which had endured since 1814, was peacefully dissolved. In 1909 a major constitutional reform was undertaken of both houses of Parliament within the framework of a constitutional monarchy.

On the outbreak of European war in 1914 all parties backed the government's policy of neutrality. However, following the completion of mass democracy, the 1920s was a decade of government instability, characterized by a succession of short-lived, minority cabinets. An important turning point in Swedish history came in spring 1933 when another minority government, a Social Democratic cabinet under Per-Albin Hansson, reached agreement with the Agrarian party on a program of antirecession measures that were planned on the basis of "deficit spending", as proposed by the influential economist J. M. Keynes. In essence, the Social Democrats won measures to combat unemployment in return for the introduction of price supports for agricultural products. This "red-green" cooperation also gave rise to a social welfare program comparable to Franklin D. Roosevelt's New Deal legislation in the United States. Under Gustav Möller's influence, measures establishing an unemployment insurance system, raising basic pensions, regulating working hours for agricultural laborers and introducing a mandatory two-week vacation for all workers followed in short order. The costs of this embryonic welfare state were to be met by higher progressive taxes on income, inheritance and wealth.

Since 1932 there has been an exceptionally high level of partisan continuity in Swedish government. Cabinets have been dominated by a single party – the Social Democrats – more than in any country in Western Europe. The Social Democrats have been in power from 1932 to 1991, with only a six-year interlude of three-party nonsocialist cooperation between 1976 and 1982. In August 1940 the Social Democrats won their biggest-ever victory with 53.8 percent of the vote and in 1968, too, they polled a narrow absolute majority. In recent years they have governed as a minority cabinet and relied on support from the opposition-based radical left (the Left-Party Communists, which in 1990 became simply Leftist Party). Al-

though fluctuating between three parties, the Agrarians (since 1957 called the Center party), Liberals and Conservatives, the nonsocialist share of the poll remained relatively constant.

The late 1960s was a period of constitutional reform, resulting in the abolition of the old two-chambered Parliament that had been set up in 1909. In 1971 the first elections to the new single-chamber Riksdag were held with parties having to gain four percent of the vote nationally to gain entry to the national assembly. The king was reduced to performing purely ceremonial duties; the Speaker assumed responsibility for orchestrating the discussions leading to the formation of the new government; and the principle of parliamentarism was enshrined in the form of government, though it had been applied in practice since 1919. As the gap between the parliamentary ranks of the socialists and nonsocialists narrowed – in 1973–76 there was a deadheat in the Riksdag – the ideas of the New Left gained currency and environment issues, especially the future of atomic energy, were placed on the political agenda. At this stage, it was the Center party which was the main beneficiary of the new "green consciousness".

In foreign affairs the continuity of policy has been particularly strongly marked. For well over half a century Sweden pursued a policy of nonalignment in foreign affairs, backing a concern to remain outside Great Power conflict with the capacity to defend the nation if the need should arise. This was the strategy of so-called "armed neutrality". During World War I Sweden remained neutral, and asserted its right to trade with any belligerent nation. This caused the British to blockade Sweden in an attempt to reduce trade with Germany, leading to severe food shortages in 1916–17. In May 1918 an agreement was reached with the Allies in these matters.

In the later 1930s Sweden's defenses were strengthened to counter the Nazi threat, while the country retained a strictly neutral line. However, after the invasion of Norway and Denmark by Germany in 1940, Sweden was forced to permit German troops to pass through to Norway and Finland.

More recently, armed neutrality has not implied assuming a merely passive or reactive role on the world stage. Rather, Sweden has been active in a number of areas. First, Swedish governments have not been reluctant in speaking out against superpower imperialism in its many guises. Thus Sweden condemned the Soviet suppression of the Hungarian revolution of 1956 and the 1979 invasion of Afghanistan by the same country; US intervention in Vietnam in the late 1960s and more recently the American policy in Central America. In 1987 the Ingvar Carlsson government instigated a trade boycott against South Africa in protest against apartheid. On some occasions this tendency for Sweden to adopt a tone of apparent moral superiority has caused friction with the superpowers. One such example occurred in 1973, when prime minister Olof Palme criticized United States policy in Vietnam, an act that led to the withdrawal of the American ambassador in Stockholm.

Second, Sweden has played a prominent role in the mediation of international conflicts, especially through the agency of

▲ The Nobel Peace Prize ceremony, held annually in Stockholm.

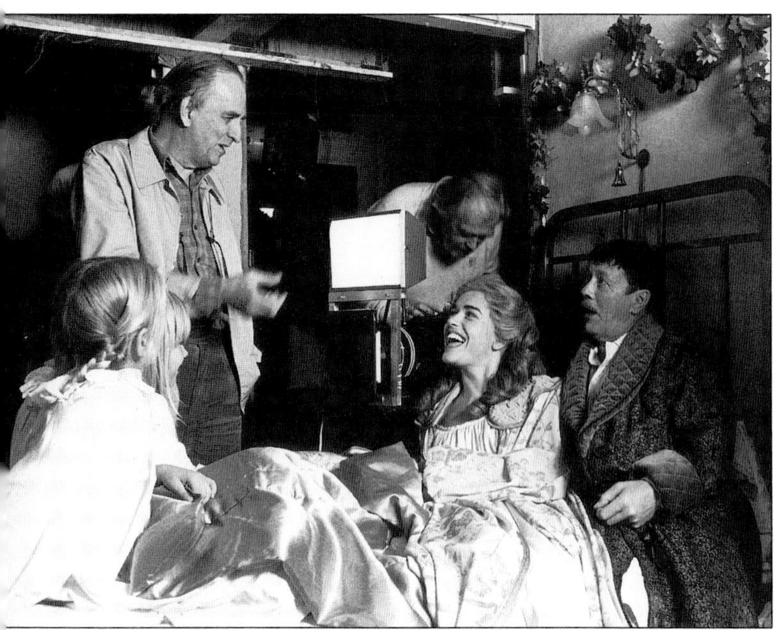

the United Nations and its peacekeeping forces. Prominent Swedish figures have acquired an international reputation as conciliators. Dag Hammarskjöld (who became UN Secretary-General in 1953) was active in the UN peacekeeping mission in the Congo (Zaire) where he met his death in 1961; Olof Palme was engaged in seeking a solution to the Iran–Iraq conflict in the 1980s.

Finally, Sweden has initiated proposals designed to further the cause of international detente and disarmament. The Unden Plan in the early 1960s (so-called after Östen Unden, the foreign secretary), for instance, envisaged an international agreement that countries with no nuclear weapons would pledge not to acquire such weapons on condition that the superpowers stopped developing their nuclear arsenals. Sweden has also participated in the process of Security and Cooperation in Europe and staged one of the rounds of the Conference. Throughout, a note of realism has been maintained. In 1981 longstanding fears of the Soviet Union were powerfully revived when a Soviet submarine ran aground in a restricted military area in the Karlskrona archipelago off the southeast coast of Sweden. Further incursions into Swedish waters prompted an official note of protest to Moscow.

Some events in the 1980s and early 1990 served somewhat to tarnish Sweden's reputation abroad and to challenge the essentials of the "Swedish model". After the shock of the assassination of prime minister Olof Palme on 28 February 1986 came an unsatisfactory private investigation into the murder. By 1991 nobody had been convicted of the crime. Then, despite a referendum in 1980 at which all three politicial groups advocated the phasing out of nuclear power, the government in 1990 decided to fudge the 1995–96 deadline for closing down the first two nuclear reactors. Industrial relations deteriorated, as did the rapport between the Social Democrats in office and the blue-collar LO. In February 1990, the Social Democrats imposed and then withdrew a temporary strike ban in connection with an austerity program and this led to wholesale defections from the party.

The problem of strengthening export competitiveness, reducing the size of the public sector and stimulating the economy generally prompted a radical tax reform, to be implemented in full in 1991, abolishing income tax for 85 percent of wage-earners and shifting the burden from direct to indirect taxation. In October 1990 Sweden, a founding member of EFTA, applied to join the European Community (EC). Running the state in hard economic times had taken its toll on the popularity of the ruling Social Democrats. For the first time ever the opinion polls showed the party standing below thirty percent, with the Greens (who broke the four-percent barrier in 1988) and a protest party called New Democracy picking up support. The September 1991 general election looked likely to produce the worst Social Democratic result ever.

Economic and social trends

Like its Nordic neighbors, Sweden experienced accelerated economic change in the 20th century. In 1900 the nation had reached the point of industrial take-off, though a majority of its economically active population still worked on the land. In 1900 half of Sweden's exports comprised textiles, pulp, paper and engineering products and the other half were raw materials such as iron, grain, metal ores and timber. Shortly after World War I, agriculture and forestry still employed more than forty-three percent of the labor force, but by 1940 this had fallen to just over thirty percent. Distributive trades and communications rose notably quickly in the interwar period from fourteen percent to twenty percent of the labor force. Despite the recession there was a strong growth in industrial production between the wars and, for example, in the period 1925–29 it rose by more than a third. The basic economic trend in the 1950s and 1960s, too, was growth. In the former decade GDP increased by an annual average of 3.5 percent while in the first half of the 1960s it topped five percent. By this time, the engineering industry had become an increasingly important contributor to exports and Swedish industry was undergoing a rapid structural transformation. A shortage of factory labor was met by a sizeable influx of immigrants. Between 1961 and 1965 170,000 immigrants, mainly from Finland, Germany, Italy, Greece and Yugoslavia, swelled the labor force and generated problems of social integration. Growth declined in the early 1970s although the real crisis followed in the wake of the dramatic rise in oil prices in 1973–74. In 1977 GDP declined for the first time in the postwar period. By the 1970s, however, Sweden had become an urban, service-based, postindustrial society. Some eighty-five percent of the population lived in urban areas and only four percent were dependent on agriculture.

As well as inaugurating the age of Social Democratic government, the 1930s saw the development of a system of collective wage bargaining which brought together the two sides of industry, while precluding interference from the state. The Saltsjöbaden agreement of 1938 represented a historic partnership between capital and labor which served considerably to stabilize conditions on the labor market. The institutional framework took the form of a Labor Market Council, which included delegates from LO and the central employers' organization SAF and functioned as a national negotiating body. It was the early 1970s before LO began to demand (in contravention of the "Saltsjöbaden spirit") that the government should intervene to limit the power of the employer in the workplace. In July 1974 employees were given legal protection against arbitrary dismissal and a Co-Determination Act followed in 1977. In December 1983 parliament approved a highly controversial scheme of employee funds, trade union-controlled funds deriving their income from a levy on corporate taxes which would buy shares in Swedish companies. The employee funds scheme soured Swedish industrial relations for many years.

Facilitated by the Social Democrats' control of government, economic management guidelines were formulated and implemented in the postwar period that sought to transcend the supposed contradiction between economic efficiency and social equality by giving pride of place to a solidarity wage policy. This was one of several strategic economic goals which were set out by Gösta Rehn and Rudolf Meidner at the LO Congress in 1951. Others included measures to combat unemployment, centralized pay negotiations and government responsibility for canceling out excessive increases in purchasing power. Costs would be met from increased corporate taxes. It was accepted that this pay policy meant that unprofitable companies would go out of business more quickly than if wage levels were determined by viability. Rehn and Meidner, however, considered this to be desirable, in order to remain competitive internationally, provided that there was "an active labor market policy" in which workers were helped to find jobs in profitable industries.

In its heyday in the 1960s, the "Swedish model" provided for high growth, low unemployment, cradle-to-grave protection, high living standards and harmonious industrial relations. The postwar years saw a new surge of social reforms. In 1950 a nine-year comprehensive school model was launched and this was put into practice over the following decade and a half. In 1958 a very controversial supplementary incomes-related pensions scheme was enacted after a popular referendum. In the early 1960s, the government introduced its "million program": 100,000 dwellings were to be built annually for 10 years; there was legislation, too, introducing a 40-hour working week and a minimum four-week paid vacation a year. By 1965 social welfare accounted for nearly thirty percent of the central government budget. The approach of the ministers of finance, Wigforss and later Strang, was to pay for these reforms through a sharply progressive tax system while instituting only modest increases in corporate taxes.

By the 1970s the industrial sector was becoming too small and traditional industries such as shipbuilding, steel and mining were plagued by structural problems brought about in part by the effects of global oil price rises and partly, too, by the impact of lower-cost producers in South Korea, Brazil and Hong Kong. Ironically, after 44 years in opposition, the nonsocialist governments of 1976–82 were forced to bail out ailing companies by partly nationalizing them. Faced by a mounting central budget deficit and current account deficit, the last acts of the nonsocialist government were a 10 percent devaluation of the Swedish crown in fall 1981, coupled with a far-reaching austerity program. The returning Social Democratic government devalued the crown a further 16 percent in fall 1982 and introduced a broad-ranging price freeze designed to counteract wage levels that had risen too high. The reality was dawning that governments of whatever complexion were faced with a vexed economic scenario: an outmoded industrial base, a very large public sector, strong labor unions and spiralling personal taxation.

◀ **Swedish filmmaker Ingmar Bergman on set.**

▼ **Swedes grieve at the site of Olof Palme's assassination.**

FINLAND

REPUBLIC OF FINLAND

CHRONOLOGY

1917	Finnish government's unilateral declaration of independence
1918	Outbreak of Civil War
1932	Unsuccessful coup by the Lapua Movement
1939	Soviet Union invade Finland
1941	Continuation War with the Soviet Union
1944	Armistice with Soviet Union
1955	Finland joins United Nations and Nordic Council
1973	Free Trade Agreement with the EC
1975	European Security Conference in Helsinki
1990	Gulf Crisis summit held in Helsinki

ESSENTIAL STATISTICS

Capital Helsinki

Population (1989) 4,960,000

Area 338,145 sq km

Population per sq km (1989) 16.3

GNP per capita (1987) US$14,370

CONSTITUTIONAL DATA

Constitution Multiparty republic with one legislative house (Parliament)

Date of independence 6 December 1917

Major international organizations UN; I-ADB

Monetary unit 1 markka (Fmk) = 100 pennia

Official language Finnish; Swedish

Major religion Evangelical Lutheran

Heads of government (President) K. J. Ståhlberg (1919–25); L. K. Relander(1925–31); P. E. Svinhufvud (1931–37); K. Kallio (1937–40); R. Ryti (1940–44); C. G. Mannerheim (1944–46); J. K. Paasikivi (1946–56); U. K. Kekkonen (1956–81); M. H. Koivisto (1982–)

Finland entered the 20th century as a Grand Duchy of the Russian Empire, but the collapse of the czarist regime and the Bolshevik revolution of 1917 in Russia emboldened the Finnish government unilaterally to declare national independence on 6 December 1917. There followed a struggle for the soul of the new state. The middle-class bloc, or Whites, which had proclaimed independence, prevailed over the socialist Reds in a bloody civil war the following year. Ironically, the victorious Whites then engaged in a parliamentary battle among themselves over the most appropriate forms of government for the new state: the middle-class left (Agrarians and Liberals) favored a republican system, whereas the bourgeois right (Conservatives and Swedish People's Party) advocated constitutional monarchy. Ultimately, a republican constitution was enacted which had the support of the Social Democrats who returned to parliament as the largest party in March 1919. Interwar Finland was dominated by the former Whites. They controlled the government, the army, the clergy and the civil service, while the Social Democrats were never fully reconciled to the republic, and the Communists were proscribed.

The Soviet Union under Stalin invaded Finland in November 1939. During the ensuing Winter War, the Finns attracted international acclaim for the heroic way in which they fought against impossible odds but substantial territorial

concessions followed defeat and there were further losses after the resumption of hostilities against the Soviet Union in the Continuation War (1941–44). Finland lost most of the Karelian isthmus and incurred a heavy indemnity.

At the outbreak of World War I, and again following defeat in the Winter War, there was a pro-German lobby in Finland. Accordingly, the destinies of the two countries became intertwined. During World War I, Germany trained what was designed to be a revolutionary force of volunteer Finnish infantrymen, while a German battalion landed in southern Finland and proceeded to "liberate" Helsinki from the Reds during the final phase of the civil war. Finally, Hitler's attack on the Soviet Union in summer 1941 afforded Finland the possibility of salvaging "conceded Karelia", in alliance with Nazi Germany. The gamble did not pay off and, as part of the armistice with Stalin, Finland was obliged to clear the retreating Nazis from its territory. Remarkably, despite these two defeats by the Soviet Union, Finland was not occupied and retained both her independence and her Western political and economic system.

Finland was, however, required to rent the strategic naval base at Porkkala near Helsinki to the Soviet Union. President Paasikivi nonetheless challenged popular anti-Soviet prejudice and pressed home the lesson of recent history, that only by remaining friendly with the Soviet Union could Finland hope to retain sovereignty. In 1948, the Soviet Union renounced its 50-year lease on Porkkala and in the same year Finland acceded to the Nordic Council and the United Nations. During his long tenure of office, President Urho Kekkonen engineered an active role for Finland as a broker between the two superpowers.

Economic and social trends

Finland remained a predominantly agricultural nation significantly longer than its Nordic neighbors. The export of forest products to the West yielded the currency with which to modernize, while the boom after the Korean War stimulated foreign demand. By the late 1950s Finland was a competitive industrial nation producing mainly for the British, German and Scandinavian markets. For much of the postwar period, about one-fifth of total Finnish exports (mostly luxury goods in return for oil) went to the Soviet Union and this buttressed Finland against the worst effects of recession. In the late 1980s, however, the Soviet economic restructuring led to a drop in Finnish exports to the Soviet Union. Understandably, Finland prepared the ground thoroughly in Moscow before becoming associated with the European Free Trade Association (EFTA) in 1961 and reaching a Free Trade Agreement with the European Community (EC) in 1973.

The election in 1966 brought the Social Democrats, People's Democrats (Communists), Center (formerly Agrarians), Liberals and Swedish People's party (the organ of the ethnic minority) together in a broad Popular Front government. There followed a thorough rationalization of agriculture, a belated surge of welfare legislation, including health and sickness insurance systems, the comprehensivization of secondary schooling and the expansion of higher education; and liberal reforms in moral questions such as the laws on alcohol.

▲ **Russian prisoners taken by the Finns in 1940.**

◀ **Postwar building by Finnish master architect Alvar Aalto.**

SOVIET UNION

UNION OF SOVIET SOCIALIST REPUBLICS

CHRONOLOGY

1905	"Bloody Sunday" triggers first Revolution
1917	Bread riots lead to revolution against Czarism – Bolsheviks seize power
1918	Red Army led by Trotsky begin civil war
1929	Forced collectivization of agriculture begins
1936	Stalin's "show trials" are launched
1949	Soviet Union tests its first atomic bomb
1953	Prisoners released from labor camps on Stalin's death
1955	Warsaw Pact established
1957	Soviet Union launches world's first space satellite
1985	Inauguration of *perestroika* and *glasnost*
1986	Nuclear reactor at Chernobyl explodes
1988	Emergence of nationalist "Popular Fronts" in non-Russian Soviet borderlands
1991	Coup by Communist hardliners thwarted

ESSENTIAL STATISTICS

Capital Moscow

Population (1989) 287,800,000

Area 22,403,000 sq km

Population per sq km (1989) 12.8

GNP per capita (1987) US$8,160

CONSTITUTIONAL DATA

Constitution Federal socialist republic with two legislative houses (Congress of People's Deputies; Supreme Soviet (two chambers: Soviet of the Union; Soviet of the Nationalities))

Major international organizations UN; COMECON; WTO

Monetary unit 1 ruble = 100 kopecks

Official language Russian

Major religion Russian Orthodox

Heads of government since 1900 (Czar) Nicholas II (1894–1917); (Chairman) V. Lenin (1917–22); (General secretary) J. Stalin (1922–53); N. Khrushchev (1953–64); L. Brezhnev (1964–82); Y. Andropov (1982–84); K. Chernenko (1984–85); M. Gorbachev (1985–)

Imperial Russia was a vast multinational peasant society, dominated by the czarist autocracy, a powerful state bureaucracy, a privileged landed nobility and the Russian Orthodox church. During the repressive reigns of Alexander III (1881–1894) and Nicholas II (1894–1917), the regime increasingly alienated the ethnic and religious minorities who together constituted over half the Empire's population. Simultaneously, the development of capitalist industry produced a proliferation of class–based organizations.

These social, ethnic and religious tensions were the major ingredients in the "1905 Revolution". From 1903 to 1905 a so-called "Liberation Movement" succeeded in building a coalition of the liberal bourgeoisie, socialists and ethnic-minority nationalists. Following the "Bloody Sunday" massacre of several hundred workers in St Petersburg in January 1905 as they attempted to petition the czar for social changes, liberal industrialists even provided strikers with strike pay, to put pressure on the government to make concessions. There was a huge wave of strikes, and during October-December 1905, workers' councils or soviets were formed in over twenty towns. Soviets, which initially arose as organs of the revolutionary strike movement, became "embryonic forms of popular state power", setting up workers' militias.

These events accelerated the emergence of Russia's major political parties. Bureaucrats, the professionals and landowners and industrialists who had been active in the 1903–05 liberation movement launched the liberal Kadet (Constitutional Democratic) party in October 1905, while wealthy conservative liberals founded the so–called "Octobrist" party, which sought an accommodation with the czarist government. A peasant-oriented Socialist Revolutionary (SR) party, which was Russia's largest political party in 1905, had been founded in 1901–02. The (Marxist) Russian Social Democratic Workers' party, founded in 1898, led a rather nominal existence until its second congress in 1903, when it split into rival factions: Vladimir Lenin's Bolsheviks (derived from *bolshinstvo*, meaning "majority"), who favored a tight-knit centralist organization; and the Mensheviks (from *menshinstvo* meaning "minority"), who favored a looser confederal workers' movement.

In 1904–05 Russia also suffered a series of humiliating military defeats at the hands of Japan in Manchuria. This disastrous war cost over 100,000 Russian lives, and demoralized Russia's armed forces. These circumstances forced the czar to issue a "Manifesto" in October 1905 conceding a limited form of constitutional monarchy, including a Duma (parliament) weighted heavily in favor of Russians and the propertied classes. In late 1905, amid escalating strikes and peasant unrest, the government resorted to repression and encouraged pogroms (violent attacks on Jews and intellectuals). A new minister of the interior, Pyotr Stolypin, initiated systematic repression of the peasantry and in December 1905 the government arrested over two hundred members of the St Petersburg Soviet, provoking the soviets in Moscow and several other cities to stage armed uprisings. These were suppressed bloodily.

In April 1906 the czar reasserted his control of foreign policy, military matters and ministerial appointments and claimed the right to veto Duma decisions, dissolve or suspend the Duma and legislate by decree. The liberal-dominated first Duma, which first met in April 1906, was dismissed within three months and when its members defiantly reassembled in

Finland, they were arrested and debarred from reelection.

The more leftwing second Duma lasted only from February to June 1907. The premiership of Stolypin was distinguished both by repression and by attempts to broaden the Russian nationalist base of the regime through discrimination in favor of Russians at the expense of non-Russians and vigorous promotion of the new class of "strong" peasant proprietors.

Repression, recession, and the liberals' partial accommodation with czarism sapped the strength of the workers' movement in 1908–10. However, socialism underwent a dramatic revival in 1911–14 in the face of rising food prices, tighter labor markets and the police crackdown on mass demonstrations. The increasing militancy of Russian workers was also reflected in dramatic Bolshevik gains, mainly at the expense of the more moderate Mensheviks. By July 1914 St Petersburg and several other cities were in the grip of mass strikes.

For two years, World War I rallied most Russians to the defense of "czar and Motherland". Social unrest quickly subsided. Russian's armed forces expanded from 1.4 million in early 1914 to nearly seven million by early 1917 and performed well against Austria-Hungary and the Ottoman Empire. Yet the war dragged on indecisively while over three million Russians perished in 1914–17, other casualties and refugees from the German-occupied western borderlands multiplied, shortages mounted, prices rocketed and real wages fell.

As the center of Russia's engineering and armaments industries, the industrial workforce of Petrograd (formerly St Petersburg) grew sharply during the war years while its food, fuel and raw material supplies were curtailed by the German naval blockade and wartime strains on the rail network. In January 1917 workers were antagonized by the arrest of workers' representatives on the central war industries committee. The czar's prolonged absences at the front made it hard for him to assess the situation in Petrograd and rally support in February 1917, when Petrograd strikers and bread riots unexpectedly escalated into a full-blown revolution. This led to the czar's abdication, the formation of a provisional government and the reemergence of workers' soviets.

The czar's abdication broke the bonds of social deference and obedience to state and ecclesiastical authority. The relatively impotent propertied classes and the unelected provisional government were still universally expected to govern Russia, even though it was primarily the workers and soviets that had overthrown czarism. The ensuing instability only

▼ **1906 news photo of massacre of Moscow rioters.**

ended when Lenin's Bolsheviks seized power in the name of the soviets and the working class in October/November 1917. The second All-Russia Congress of Soviets immediately ratified Lenin's government. Lenin temporarily persuaded many "left SRs" and anarchists to support the Bolsheviks, at least until the long-awaited constituent assembly had been democratically elected. Support for the Bolshevik regime rapidly shrank, as real wages and urban food supplies fell and as the armistice concluded with Germany in December 1917 encouraged unplanned demobilization and desertion of millions of soldiers, reduced employment in the war industries and utterly disrupted transport, distribution and labor discipline. Moreover, after the SRs won 58 percent of the votes cast in the elections to the constituent assembly, the Bolsheviks suppressed it in early 1918. During 1918 the Bolshevik regime increasingly resorted to armed requisitioning of peasant grain stocks and suppression of soviets on which Mensheviks and SRs were regaining majorities, and a "Red Terror" against "counter-revolutionaries" and rival socialist parties. This and the draconian peace terms accepted by the Bolsheviks in the treaty of Brest-Litovsk (which ceded large parts of Russia to Germany) led the SRs to mount the massive socialist opposition to the Bolshevik dictatorship. By proving himself unwilling to share power in a broad socialist coalition, Lenin plunged Russia into a civil war which cost over twelve million lives in 1918–21. Lenin's regime created the conditions in which a counter-revolution launched by the "whites" (supporters of the czarist white flag) and their European backers temporarily gained some popular support on Russia's peripheries. War with Poland in 1920 resulted in the loss of western Belorussia and western Ukraine, but in 1920–21 the regime reconquered the rest of the Ukraine and Belorussia, Georgia, Armenia, Azerbaijan and Muslim-Turkic central Asia, each of which had seceded from the Russian Empire in 1918.

In 1920–21 there were popular uprisings against the Bolshevik regime in the Ukraine, Tambor, the Volga basin, central Asia and Kronstadt, all of which were bloodily suppressed. But despite these revolts and the failure of the predicted international European socialist revolution, the Bolshevik victory in Russia's civil war ushered in Lenin's more relaxed and conciliatory New Economic Policy (NEP) in March 1921. NEP provided a breathing space, allowing Russia to recover and the Communist party to "regroup its forces" in preparation for the next "socialist offensive". But the restoration of a mixed market economy went hand-in-hand with a consolidation of one-party rule. Independent trade unions, professional associations, schools, publishers, opposition parties and even factions within the ruling party were banned completely in 1921. Yet the party tried to persuade Soviet citizens that this was still *their* revolution by promoting an unprecedented ferment of cultural experimentation, iconoclasm and festivity, mass organizations and the mobilization of hitherto underprivileged groups which had not received the endorsement of their society, including women and minority nationalities.

◀ **Soviet liberation after World War II.**

A Union of Socialist Republics

With the Treaty of Union, on 30 December 1922, the Soviet Union was constituted from the Russian Soviet Federal Socialist Republic (RSFSR) – which itself had been created in July 1918 – and the Ukrainian (25 December 1917), Belorussian (1 Jan 1919), and Transcaucasian (15 Dec 1922) Soviet Socialist Republics (SSRs). The huge RSFSR contained a number of Autonomous (that is, ethnically based) Soviet Socialist Republics (ASSRs) and autonomous regions of *oblasti* (AOs). Ind October 1924 Uzbekistan and Turkestan, and in December 1929 Tadzhikistan, joined the Union as SSRs.

In 1936 the Union adopted a new constitution, and Transcaucasia became three SSRs, Armemia, Azerbaijan and Georgia; and two ASSRs of the RSFSR, Kazakhstan and Kirghizia, became SSRs. In 1939 some of Poland was absorbed into the Ukraine and Belorussia. On 31 March 1940 the Karelo-Finnish SSR was created from the Karelian ASSR with some territory ceded by Finland. On 2 August 1940 the Moldavian SSR was created from the Moldavian ASSR and most of Bessaraboa; and on 3, 5 and 6 August 1940 respectively Lithuania, Latvia and Estonia became SSRs. On 16 July 1956 the Karelo-Finnish SSR became an ASSR in the RSFSR.

The political infrastructure of this union was based on a hierarchy of SSRs, then ASSRs, AOs, and a series of ethnic or geographical divisions down to village level; each level had an elected soviet, with the Supreme Soviet of the USSR, which elected the Council of Ministers, over all. The guiding force in this structure was the Communist party (CPSU), through its elected Central Committee, which in turn elected the Presidium or Politburo, to direct party affairs between the plenary sessions. The head of government was the CPSU general secretary. Four rules were held as integral to the system, styled "democratic centralism". They were: all executive CPSU organs to be made up of elected members; unconditional obedience from those lower to those higher in the hierarchy; majority rule; and periodic accountability.

In practice this "democratic centralism" was more centralized than democratic, with voting according to orders and accountability only for the lower echelons to their bosses. The Union was dominated by the RSFSR, which had over two thirds of its territory in 1991, when the CPSU was abolished and the Union structure dissolved in recognition of the sovereignty of the individual republics.

After Lenin's premature death in January 1924, the major defenders of NEP and a guided market economy were Nikolai Bukharin and Leon Trotsky. But Joseph Stalin, who rose to prominence as people's commissar for nationalities and (from 1922) as general secretary of the party, outmaneuvered both, while accumulating enormous power of patronage in his own hands. By repeatedly raising the production targets of Russia's First Five-Year Plan (1928–32), by staging "show trials" of "bourgeois specialists" and independent-minded intellectuals, by launching a brutal all-out assault on prospering peasants (*Kulaks*) and by forcibly collectivizing Soviet agriculture in 1929–36, Stalin brought NEP to a violent end.

Celebration of Stalin's 50th birthday in 1929 launched the Soviet "cult of personality", glorifying the Georgian dictator as the infallible party leader. Millions fervently believed in the

rightness of Stalin's programs of forced industrialization, coupled with collectivization and mechanization of agriculture. They were prepared to mobilize vast armies of labor to build dams, canals, railroads and giant industrial complexes, as well as to force 100 million peasants to join collective and state farms. The supreme tests of the Stalinist regime occurred in World War II when the Soviet Union bore the brunt of the war against Nazi Germany, in which nearly thirty million Soviet citizens perished. Hitler's failure to capture Moscow in late 1941 was a crucial setback and the Soviet victories in the battle of Stalingrad in late 1942/early 1943 and in the battle of Kursk in July 1943 turned the tide of the entire war. Techniques of resource-mobilization developed under the five-year plans of the 1930s came into their own in wartime. However, Stalin's policies in the 1930s had also depleted Soviet manpower, weakened the Red Army's officer corps, caused terrible setbacks in agriculture, antagonized the peasantry and induced many Ukrainians and Belorussians initially to welcome the Nazis as "liberators".

Victory in World War II gave the Soviet regime a new lease of popular support. When Soviet reconstruction was pronounced complete in 1950, the United States was still producing three to four times as much as the Soviet Union, and the burdens incurred as a result of the pursuit of "superpower" status were therefore far greater than they were for the Americans. This intrinsic inequality of the two "superpowers" was further magnified by the fact that, until 1949, the US alone possessed atomic weapons. Postwar Soviet military occupation of Eastern Europe partly held the region hostage against possible Western nuclear aggression, until a "balance of nuclear deterrence" was established by the testing of the USSR's first atomic and hydrogen bombs in 1949 and 1953, respectively. Stalin's death in March 1953 permitted a gradual "thaw". In the mid 1950s, under Malenkov and Khrushchev, millions of people were released from prisons and labor camps, the armed forces shrank, labor mobility increased, industrial growth-rates rose, minimum wages and pensions were raised, social expenditure was increased and censorship and police surveillance were relaxed.

Khrushchev's "Virgin Lands" program (1954–63) caught the imagination of the public and some 300,000 volunteers were mobilized to bring under cultivation nearly thirty million hectares of grassland (steppes) in North Kazakhstan and West Siberia. The optimism, vitality and experimentation fostered by the Khrushchev regime was slowly stifled under Leonid Brezhnev (1964–82). The new freeze involved a reassertion of bureaucratic regulation, ideological dogmatism and widespread violations of human rights. This climate contributed to the increasing demoralization, cynicism and corruption which, along with ever-growing defense burdens, helped to sap the strength of the Soviet Union. Brezhnev died in November 1982. His successor, Yuri Andropov, launched a crackdown on corruption, indiscipline and inefficiency, but he died after only 15 months in office. He was succeeded by a frail conservative nonentity, Konstantin Chernenko, who died after little more than a year. But his energetic successor, Mikhail Gorbachev, soon initiated an ambitious program of *perestroika* ("restructuring"), *glasnost* ("openness") and "democratization". The leading technocratic champions of *perestroika* advocated increased autonomy and financial accountability for state enterprises, a lifting of restrictions on small-scale cooperatives and private enterprise, freedom of expression, the rule of law and "socialist pluralism". Gorbachev won crucial military and KGB backing in the mid 1980s by promising to double the housing stock and health-care provision (by 2000) and greatly increased

salaries, education provisions and supplies of consumer goods. Gorbachev's commitments to overambitious goals overstretched the economy, and in 1990 contributed to the reassertion of central planning and controls. When challenged by striking workers and ethnic separatist movements who strove to go beyond "restructuring" of the Soviet system to a bolder dismantling of that system, the Gorbachev regime repeatedly floundered.

From 1989 to early 1991, however, there was a growing backlash amongst hardline Communists, bureaucrats and military personnel. The initial military and KGB support for *perestroika* and *glasnost* ended in 1990 as it became clear that the survival of the USSR and the Soviet armed forces in their existing forms had been placed in jeopardy and that Gorbachev and the president of the Russian Federation (RSFSR), Boris Yeltsin, were prepared to back a radical plan for a 500-day transition to a decentralized state and a full-blown market economy. Senior military personnel were also annoyed that Soviet withdrawal from Afghanistan and the ending of Soviet tutelage over Eastern Europe in 1989 meant that the Soviet Union had lost some of its "superpower" status. In 1989–90, moreover, reformers temporarily lost support among Soviet workers, for whom *perestroika* came to mean economic chaos and political indecision. Soviet tensions were further increased by the heavy-handed use of troops in abortive efforts to curb growing demands for national independence in Latvia, Lithuania, Estonia, Georgia, Armenia and Azerbaijan in 1989–91. From late 1990 President Gorbachev became hostage to hardliners in the security forces and in the central bureaucracy; in mid-1991, however, radical democrats regained the initiative through the landslide election of the outspoken radical, popular hero Boris Yeltsin to the new executive presidency of the Russian Federation, the reelection of free-marketeering democrats to Russia's main city soviets, and agreement between Gorbachev, Yeltsin and the rulers of eight other Union republics to work together to establish a market economy and devolve power to the constituent republics of the new confederal Soviet Union. The success of radical democrats in defeating a coup d'état by hardline Communists, the military and the KGB in August 1991 greatly accelerated the demise of the Soviet Communist Party and the disintegration of the USSR into sovereign national Republics.

▲ **Moscow's Bolshoi Ballet demonstrating its world preeminence.**

◀ **May Day in front of St. Basil's Cathedral, Moscow.**

POLAND

REPUBLIC OF POLAND

CHRONOLOGY

1918	Poland regains independence
1926	Marshal Pilsudski's coup d'état
1939	German and Soviet invasions and partition of Poland
1944	Communist-dominated Provisional Government set up
1956	Poznan riots and restoration of Gomulka
1968	Repression of student unrest
1970	Riots in Baltic ports topple Gomulka
1980	Strikes and sit-ins give birth to "Solidarity"
1981	Martial law declared
1989	End of Communist dictatorship

ESSENTIAL STATISTICS

Capital Warsaw

Population (1989) 37,875,000

Area 312,683 sq km

Population per sq km (1989) 121.1

GNP per capita (1988) US$7,200

CONSTITUTIONAL DATA

Constitution Unitary multiparty republic with two legislative houses (Senate; Diet)

Date of independence 10 November 1918

Major international organizations UN

Monetary unit 1 zloty (Z1) = 100 groszy

Official language Polish

Major religion Roman Catholic

Heads of government since independence (President) Marshal J. Pilsudski (1918–22); G. Narutowicz (1922); S. Wojciechowski (1922–26); (Military ruler) Marshal J. Pilsudski (1926–35); (President) I. Mosciki (1935–39) (in office from 1926); (Prime minister) Gen. W. Sikorski (1939–43); S. Mikolajczyk (1943–45); E. Osóbka-Morawski (1945–47); J. Cyrankiewicz (1947–48); (Party leader) W. Gomulka (1948); B. Bierut (1948–56); E. Ochab (Mar–Oct 1956); W. Gomulka (1956–70); E. Gierek (1970–80); S. Kania (1980–81); Gen. W. Jaruzelski (1981–85); (President) Gen. W. Jaruzelski (1985–89); Lech Walesa (1990–)

By 1900 the two mass movements which were to dominate Polish politics up to 1939 had emerged: the National Democrats (ND) and the Polish Socialist party (PPS). The National Democrats' leader, Roman Dmowski, dreamed of an ethnically homogeneous "Poland for the Poles". The PPS, by contrast, championed a federal Polish republic that included Jews, Lithuanians, Ukrainians and Belorussians. The leading figure in the pre-1918 PPS was Jozef Pilsudski.

On 6 August 1914 Pilsudski launched an invasion of Russian Poland, but this was rebuffed and from 1915 to 1918 Poland was occupied and plundered by the Central Powers. Pilsudski's forces supported them in the hope that Poland could thereby regain its former Lithuanian, Ukrainian and Belorussian territories. The ND and Poland's propertied classes mainly put their faith in collaboration with Russia.

In November 1918 the socialist Ignacy Daszyński proclaimed a "People's Government" in Galicia, while France recognized Dmowski's Polish National Committee as the Polish government. But the Germans hurriedly installed Pilsudski as Poland's chief of state, to whom they surrendered control of Poland and some "eastern territories". Pilsudski remained virtual dictator of Poland until November 1922. During 1919 his armies captured the Lithuanian city of Vilna, the Byelorussian capital Minsk and Ukrainian East Galicia. In May 1920 they captured the Ukrainian capital Kiev. A Soviet counteroffensive came close to capturing Warsaw, but in August 1920 Poland was saved by a Polish counteroffensive aided by France and Britain.

From late 1918 to May 1926 Poland experienced severe economic and political instability. When Peasant party leader Wincenty Witos and the National Democrats formed a center-right coalition in May 1926, amid financial panic and bloodily suppressed workers' protests, Marshal Pilsudski staged a coup. He stopped short of outright dictatorship, but preserved the *sejm* (legislature) and allowed himself to be elected president for a short time, to provide quasidemocratic legitimation of his *sanacja* ("purification" or "regeneration") regime.

Under the impact of the 1930s depression and interethnic tensions, the *sanacja* regime gravitated towards fascism. Poland signed nonaggression pacts with Stalin and Hitler in November 1932 and January 1934, respectively. Ambitious programs were launched to enhance Poland's capacity to defend itself. Unfortunately, these programs were begun too late to be of any avail against the invasion of Poland by both Nazi Germany and Soviet Russia in September 1939.

During World War II northern and western Poland became part of Germany. Here 330,000 Poles were murdered and 500,000 deported to provide "living space" for 350,000 German colonists. Poland's Belorussian and Ukrainian territories were annexed to the Soviet Union and up to two million Poles were deported eastward, mainly to Soviet labor camps.

Central Poland was ruled by the German army under Governor-General Hans Frank. Here one million Poles were deported to work in Germany and several million Poles and Jews died in Nazi concentration camps or from disease, malnutrition and injuries. The appalling conditions in occupied Poland gave rise to Europe's largest resistance movements: the Home Army and Peasant Battalions (loyal to a government-in-exile in London under General Wladyslaw Sikorski) and, from 1942, a Communist-led People's Army. The subhuman conditions in the urban ghettos gave rise to savagely suppressed

Jewish uprisings in Warsaw, Bialystock and Vilna in April to May and September 1943.

Poland's prospects were radically transformed by Hitler's reckless invasion of the Soviet Union in June 1941. Poles and Russians became uneasy allies against a common foe. After June 1941 Poles were gradually released from Stalin's labor camps and prisons and a Polish army (loyal to Poland's "London Government") was formed in the Soviet Union. In 1942 many Poles were recruited into the hard-pressed Soviet armed forces. In April 1943, however, Soviet–Polish relations were shaken by German revelations that thousands of Polish army officers had been murdered in Soviet "custody" at Katyń in 1940. When Sikorski's "London Government" demanded an investigation, Stalin broke off relations with the London Poles.

Soviet liberation of the eastern territories was accomplished in January–June 1944 and in December 1944 a Communist-dominated provisional government was set up. Meantime, the nationalist pro-Western Home Army staged its Warsaw Rising against the Germans from 1 August to 3 October 1945. Tragically, the Red Army's westward advance was halted for several months, enabling the Germans to crush the rising, kill over 200,000, punitively demolish Poland's capital city and teach the Poles an awful lesson. The crushing of the Warsaw rising largely destroyed the Polish nationalist resistance.

The Red Army and the Soviet-backed Polish Army finally entered Warsaw in January 1945, and Britain and America recognized Poland's provisional government in July 1945, in return for the inclusion of Mikolajczyk (as deputy premier) and some other "London Poles" and a promise of "free elections". By 1946 banks, public utilities and most of Polish industry had been nationalized, while land had been redistributed. The 1947–49 Three-Year Plan targeted investment on postwar rebuilding and agriculture. Moreover, as leader of the Polish Workers' party, Wladyslaw Gomulka resisted Soviet calls for rural collectivization, becoming the first postwar Communist leader openly to defy Stalin.

The parliamentary elections finally held in January 1947 were far from free. After the election, in which the "Democratic Bloc" supposedly won 80 percent of the vote, several hardline Stalinists took over key ministries. The PPS was subjected to thousands of arrests and expulsions before being dissolved. Soviet pressure forced Poland to turn down desperately needed US Marshall Aid in July 1947, and intimidation forced Mikolajczyk to flee abroad in October. In September 1948 the "Moscow Communist" Boleslaw Bierut ousted Gomulka and committed the Party to rural collectivization.

From 1949 to 1953, Bierut consolidated a one-party "proletarian dictatorship" and police state. Growing hostility to the regime erupted in March 1956. Workers' protests in Poznań in June 1956 escalated into demonstrations in which hundreds were killed or wounded by the Polish security forces. Diverse demands for cultural, economic and political relaxation, increased contact with the West and workers' control of industry coalesced into a campaign to restore Gomulka to the party leadership. The crisis was intensified by conflict within the ruling party, menacing Soviet troop movements, the unheralded

▲ **The Polish cavalry in 1914.**

attendance of Soviet leaders at a critical meeting of the party's central committee in October 1956 and open warnings that Poles would resist Soviet military intervention. The crisis was defused by Khrushchev's acceptance of Gomulka, decollectivization, relaxation of censorship and an accommodation with the Catholic Church, in return for assurances that Gomulka would maintain the "leading role" of the party and Poland's alliance with the USSR.

During the 1960s Gomulka encouraged grandiose plans for the expansion of Polish heavy industries and the economy grew respectably. By the mid 1960s, however, Gomulka was becoming increasingly authoritarian. Heavy-handed suppression of student unrest in 1968, and of workers' strikes and mass protests in Poland's Baltic ports and shipyards in December 1970, led to his removal from the party leadership.

Under the ex-miner Edward Gierek, Poland embarked on a "dash for growth" which produced promising results in 1971–74. In 1971 Gierek made a great show of visiting factories, mines and shipyards to listen to workers' grievances. He also promoted large infusions of Western capital and technology. However, Poland's indebtedness to the West rose alarmingly, and by the late 1970s, Poland was experiencing acute debt-service problems. Moreover, the growth of incomes in 1971–76 had outstripped supplies of consumer goods, especially food. A belated attempt to restore market equilibrium by raising food prices in 1976 sparked off widespread protests, inducing Gierek to rescind the price increases.

In 1976–77 the intelligentsia launched an active Committee for the Defense of Workers (KOR). Gierek's regime became increasingly repressive and mendacious. Popular alienation from the regime was further increased by a euphoric resurgence of Polish Catholic nationalism, induced by the elevation of Polish Cardinal Wojtyla to the Papacy and his triumphal return visit to Poland as Pope John Paul II in June 1979.

In July 1980 another abortive attempt to raise retail food prices triggered massive strikes and sit-ins by Silesian coal-miners and steel-workers and Baltic shipyard workers and to a rapid emergence of *Solidarność* (Solidarity), a nationwide Catholic trade union boasting 10 million members, led by Gdansk electrician Lech Walesa and closely advised by Church leaders and KOR activists. The Gdansk Agreement of 31 August 1980 established workers' rights to strike and form independent trade unions, wage rises, weekly broadcasts of Mass, relaxation of censorship, and an overhaul of the economic system. Gierek abdicated in September 1980. But his successor Stanislaw Kania was even less capable of coping with Poland's deepening crisis. Unfortunately, Solidarity encouraged workers to think that they could enjoy the potential benefits of economic and political liberalization without foregoing job security and fixed prices. It also encouraged strikes, wage demands and a ban on weekend work for miners, which exacerbated shortages and inflation and further depressed production. In successive attempts to restore "discipline", and avert potential Soviet military intervention, General Wojciech Jaruzelski became premier in February 1981 and party leader in October 1981, before imposing martial law from December 1981 to July 1983. Thousands of Solidarity and KOR activists were interned or beaten up or killed, and Solidarity was outlawed. Inflation rose sharply. Nevertheless, Church mediation and Jaruzelski's evident desire to avoid an unrestrained reign of terror allowed Solidarity to survive.

In 1983, the Pope was again allowed to visit Poland, martial law was formally lifted, most political detainees were released and economic reforms were initiated. Political stalemate continued until the *sejm* voted down the government in September 1988 and a new government was formed under the "reform Communist" Mieczyslaw Rakowski. The talks of February–April 1989 led to extensive reforms and the holding of contested multiparty elections to the *sejm* and a new upper house (senate) in June. Solidarity candidates won 99 of the 100 senate seats and all 35 of the *sejm* seats available to them. In August 1989, a Solidarity-dominated government was formed under Tadeuz Mazowiecki, a respected Catholic intellectual then close to Lech Walesa.

In 1990, Poland underwent bracing rapid moves toward a market economy. Real GDP, industrial output, real personal incomes, real personal consumption and total employment all fell. But exports rose, foreign trade and the state budget were in surplus, private sector output and private employment increased, monetary discipline was strong and inflation fell. In place of shortages, shops were full of goods which few could afford. In reward, Poland's Western creditors granted a moratorium on debt-service payments in 1990 and wrote off half Poland's debt in 1991. But many Solidarity activists and rank-and-file workers felt "betrayed" by falling real wages and mass unemployment. Lech Walesa played on this grass-roots discontent in order to win the December 1990 presidential election campaign, which brought the Mazowiecki government to a bitter and acrimonious end. President Walesa appointed a free-marketeering government under Jan Bielecki, and committed to faster privatization of Polish industry in 1991–93.

▲ **Lech Walesa, hero of the Solidarity movement.**

◀ **The Pope visits Poland in 1983.**

CZECHOSLOVAKIA

CZECHOSLOVAK SOCIALIST REPUBLIC

CHRONOLOGY

1918	Czechoslovakia is established
1935	Neo-Nazi *Sudetendeutsche Heimatfront* win 62 percent of German vote in elections
1938	British and French Premiers agree to Germany's proposed annexation of "Sudetenland"
1939	Hitler completes German annexation of the Czech Lands
1944	Slovak uprising
1945	Expulsion of over three million Germans
1948	Communist takeover under Klement Gottwald
1953	Repression of workers' protests at Plzen and Ostrava
1968	The "Prague Spring", suppressed by Soviet troops
1989	End of Communist dictatorship

ESSENTIAL STATISTICS

Capital Prague

Population (1989) 15,636,000

Area 127,900 sq km

Population per sq km (1989) 122.3

GNP per capita (1988) US$10,140

CONSTITUTIONAL DATA

Constitution Federal Republic with one legislative house (Federal Assembly)

Date of independence 28 October 1918

Major international organizations UN; COMECON; WTO

Monetary unit 1 koruna (Kčs) = 100 halers

Official languages Czech; Slovak

Major religion Roman Catholic

Heads of government since independence (President) T. Masaryk (1918–35); E. Beneš (1935–38); E. Hácha (1938–39); E. Beneš (1939–44) (Government-in-Exile); E. Beneš (1945–48); K. Gottwald (1948–53); A. Zápotocky (1953–57); A. Novotný (1957–68); Gen. L. Svoboda (1968–75); G. Husák (1975–89); V. Havel (1989–)

Before World War I Czechs and Slovaks had not aspired to independent statehood, let alone to a combined Czecho-Slovak state. The Czech lands of the once illustrious kingdom of Bohemia had been incorporated into Austria in the 17th and 18th centuries. By contrast, Slovakia remained subordinate to Hungary. During World War I, Czech and Slovak *émigrés* began to canvass the idea of a Czecho-Slovak state. To be strong enough to survive, this had to include Bohemia's "Sudeten" Germans.

Czechoslovakia was set up in October 1918. It inherited one-quarter of the population but two-thirds of the industrial capacity of the Austro-Hungarian empire. It enjoyed good educational provision and the strongest economy in interwar eastern Europe. These advantages helped to make Czechoslovakia the only state in the region in which parliamentary democracy flourished in the interwar years.

Czechoslovak stability owed much to the venerable Professor Tomás Masaryk and his close associate Dr Eduard Beneš. Even the "Sudeten" Germans became reconciled to Czechoslovakia, whose stability and prosperity contrasted with hyperinflation, mass unemployment and violent struggles between left and right in Austria and Germany. Slovaks, however, resented the large influx of Czechs into Slovakia's schools and administration.

The 1930s depression and the rise of Nazi Germany tore Czechoslovakia apart. Worst affected were "Sudeten" German industrial areas, which were heavily dependent on exports to Austrian and German markets. After 1933, "Sudeten" Germans sought salvation through incorporation into Hitler's Reich. At a September 1938 meeting in Munich, the British and French premiers agreed to Hitler's proposed annexation of the "Sudetenland". The loss left the Czechs largely defenseless when, in March 1939, Hitler completed his annexation of the Czech lands. Czech resistance, culminating in assassination of "Protector" Heydrich in May 1942, led to savage German reprisals. Slovakia was granted autonomy under a Catholic-fascist regime in October 1938 and lost territory to Hungary in November 1938. The remainder gained "independence" as a German protectorate in March 1939, but an abortive uprising precipitated full German military occupation in autumn 1944.

A government-in-exile was led by Beneš in London, while Czechoslovak forces under General Ludvík Svoboda fought alongside Soviet forces in 1942–45. In December 1943, anticipating the USSR's role in reshaping postwar Eastern Europe, Beneš expediously signed a Treaty of Friendship with the Soviet Union.

In early 1945 President Beneš appointed an antifascist National Front government to pursue a democratic "middle way" between Communist dictatorship and capitalism. In early 1946, Soviet as well as Western troops were withdrawn. This helped the Communist party to win 38 percent of the vote in the circumscribed May 1946 election and the Communist leader Klement Gottwald became the premier. In Slovakia, where Slovak Democrats won 62 percent of the vote to the Communists' 30 percent in May 1946, the Communists soon arrested or dismissed opposition leaders on trumped-up charges and established centralized Communist rule. In July 1947 Gottwald's government was summoned to Moscow and made to reverse its initial acceptance of US Marshall Aid.

In February 1948 Communist ministers purged the police

and incited trade unions to set up "action committees" to purge public administration. After Jan Masaryk fell to his death from a high window, opposition leaders fled abroad or were arrested. In May 1948 the government "won" 89 percent of the vote in single-state elections and in June Gottwald took over the presidency. From then to 1953 over 130,000 citizens became political prisoners or were executed.

From 1949 to 1960 the economy grew rapidly, as the first Five-Year Plans collectivized and mechanized agriculture and turned Czechoslovakia into a "workshop of the East".

In March 1953 Gottwald died. His hardline successor, Antonín Novotný, ruled until January 1968. But in the early 1960s obsolescent heavy industries went into decline, while economic rigors impeded the development of high technology. Technocratic demands for economic liberalization dovetailed with demands for Slovak autonomy and for "socialism with a human face". This became the catchphrase of the "Prague Spring", which installed the liberal Slovak Alexander Dubček as a hugely popular party leader in January 1968 and war-hero General Svoboda in the presidency in March 1968. The liberal "action program" of April 1968 and the infectious exuberance of the Dubček regime exposed the coercion and hypocrisy on which neighboring Communist regimes rested. On 20–21 August 1968 Czechoslovakia was invaded by a million Warsaw Pact troops. Dubček, premier Cernik, speaker Josef Smrkovský and National Front chairman Frántisek Kriegel were abducted to Moscow. They won a chance to prove, through appeals for restraint, that they alone could control Czechoslovakia. Hundreds of thousands nevertheless did demonstrate against the military occupation in 1968–69, when Dubček had to cede the party leadership to Dr Gustáv Husák.

Czechoslovakia was put into a deep freeze by Dr Husák. Half a million supporters of liberalization were sacked, demoted, imprisoned or driven into exile. In 1975 dissident intellectuals such as playwright Václav Havel began to circulate "underground" publications, and in 1977 launched "Charter 77" to publicize violations of human rights, but most people withdrew into privacy, consumerism and cynicism.

Husák retired from the party leadership in December 1987; and, in November 1989, the East German regime fell. This stimulated the formation of Civic Forum, a coalition of opposition groups. Czechoslovakia's Communist regime was brought down by mass demonstrations and strikes between 17 November and 27 November 1989. Václav Havel was elected president and 200,000 prisoners were released in December 1989. Civic Forum and its Slovak counterpart ("Public Against Violence") won nearly half the vote in the June 1990 elections and formed a government under Marian Calfa. During 1990 Civic Forum divided into a Christian Democratic wing supporting rapid transition to a market economy, and a more cautious "Liberal Club". Political infighting, inflation, falling output and mounting unemployment constrained economic liberalization in 1990–91. Czechoslovakia had no foreign debt burden, but forced Slovakia's demands for independence.

▼ **Ineffectual protest at the Warsaw Pact invasion, 1968.**

HUNGARY

REPUBLIC OF HUNGARY

CHRONOLOGY

1918	Hungary gains independence
1919	Romanian military occupation
1941	Genocide of Jews and gypsies begins
1945	Radical redistribution of farmland
1947	Communist takeover
1956	Hungarian Revolution (October) and Soviet invasion of Hungary (November)
1961	Relaxation of repression
1989	End of Communist dictatorship

ESSENTIAL STATISTICS

Capital Budapest

Population (1989) 10,580,000

Area 93,033 sq km

Population per sq km (1989) 113.8

GNP per capita (1988) US$8,650

CONSTITUTIONAL DATA

Constitution Unitary multiparty republic with one legislative house (National Assembly)

Date of independence 16 November 1918

Major international organizations UN

Monetary unit 1 forint (Ft) = 100 filler

Official language Hungarian

Major religion Roman Catholic

Heads of government since independence (President) Count M. Károlyi (provisional) (1918–19); B. Kún (Mar–Aug 1919); K. Huskár (1919–20); (Regent) Adm. M. von Horthy (1920–44); German occupation (1944–45); (Prime minister) Z. Tildy (1945–46); F. Nagy (1946–47); L. Dinnyés (1947–48); I. Dobi (1948–52); M. Rákosi (1952–53); I. Nagy (1953–55); A. Hegedüs (1955–56); I. Nagy (Oct–Nov 1956); J. Kádár (1956–58); F. Muñnich (1958–61); J. Kádár (1961–65); G. Kállai (1965–67); J. Fock (1967–75); G. Lázár (1975–87); K. Grósz (1987–89); M. Németh (1989–90); J. Antall (1990–)

Hungary entered the 20th century as an autonomous kingdom within the Austrian Habsburg Empire, with its own government, administrative, legal, and financial systems. But it did not have its own armed forces, foreign policy or control of trade regulations. These imperial prerogatives became major sources of friction between Austria and Hungary, hastening the reassertion of full Hungarian independence in 1918, after Hungary had fought and suffered defeat alongside Austria and Germany in World War I.

The radical-liberal Count Mihály Károlyi ruled Hungary in coalition with the Social Democrats from late 1918 to March 1919. Governmental inertia further radicalized Hungary's workers and peasants and catapulted to power Béla Kún's new Hungarian Communist party, establishing a Hungarian Soviet Republic which rapidly alienated the peasantry, Christians and many workers and intelligentsia. Undaunted, Béla Kún's government (which was primarily Jewish) invaded Slovakia and proclaimed a "Slovak Soviet republic". Forced to withdraw under threat of Western assistance to Czechoslovakia, it then launched an abortive invasion of Transylvania in July 1919.

On 1 August 1919 Kún and his cronies fled abroad. The ensuing counter-revolution installed reactionary Admiral Miklós Horthy as "regent" of Hungary (pending a Habsburg "restoration"). In the initial "White Terror", around 5,000 Jews, Marxists and liberals were executed, 75,000 imprisoned and several hundred thousand driven into exile. Romanian troops occupied and looted Budapest and southern Hungary and, by the end of 1919, the industrial and agricultural output was down to half 1913 levels. Budget deficits, money supply and prices were rising, and in 1922–24 Hungary experienced hyperinflation.

Count István Bethlen, Horthy's ablest prime minister (1921–31), absorbed the rising Smallholder party into his "Government party", in return for "dethronement" of the Habsburgs (1921) and a token land reform. Bethlen also curbed the White Terror and established close cooperation with Jewish financiers and industrialists. However, agricultural export earnings and industrial output fell 60 percent and 27 percent respectively in 1929–33, forcing Hungary to establish exchange controls and default on its debt service. In August 1931 Bethlen resigned, his economic strategy undermined.

Premier Gyula Gömbös (1932–36) was a self-avowed National Socialist. Although Horthy clamped down to prevent the execution of his aims, Gömbös was able to rig the 1936 parliamentary elections to produce a "Right-Radical" majority and to announce his intention to proceed to a fascist state, though he died before it could be implemented. Nevertheless, through Gömbös, Hungary became allied to Germany and Italy.

Hungary regained southern Slovakia and part of Ruthenia in 1938–39, nearly half of Transylvania in 1940 and part of Yugoslavia in 1941. Gömbös's successors, though liberal-conservatives by inclination, became Hitler's "partners in crime", obliged to enact Nazi-inspired antisemitic legislation and participate in the German invasion of the USSR. In March 1944 Hungary came under German military occupation. In October 1944 Horthy was deposed in favor of Ferenc Szálasi, leader of the Hungarian-fascist Arrow Cross. Two months later, however, Hungary was "liberated" by the Soviet Red Army. In December 1944 the Soviet military administration launched a center-left "National Independence Front", which formed a provisional government with minor Communist representation. However, many Hungarians were ready for radical change and Communist agriculture minister Imre Nagy implemented the radical land reform of January 1945.

The Communist party, headed by Mátyás Rákosi, now projected a moderate image in a drive to recruit support. In the elections of November 1945, however, the Communists won only 17 percent of the vote. The Smallholders, who won the election, honored a pre-election promise to keep the Communists in government. Communist interior minister Lásló Rajk used his control of police and security matters to intimidate the Communists' main rivals in 1947–48. Nevertheless, the Communist share of the vote only increased to 22 percent in the August 1947 election. In November 1947 the Communists made parliament vote to hold no meetings for over a year, during which time leaders of the centrist parties were forced to resign or flee abroad. Hungary became a "People's Republic" under a "dictatorship of the proletariat", headed by Rákosi. In 1949–53 there were repeated "purges" and over 150,000 people were imprisoned. Rákosi established a planned economy and promoted development of heavy industries, most of which were inappropriate to a small mineral- and capital-deficient country, while launching coercive collectivization of agriculture.

In 1953, however, the USSR forced Rákosi to relinquish the premiership to Imre Nagy, the moderate protégé of Soviet premier Malenkov. Nagy presided over a liberalization of political, cultural and economic life, but in 1955 Malenkov himself fell from grace, enabling Rákosi to oust Nagy. Rákosi was ousted in July 1956 and, from October 22 to November 4, a full-blown revolution involving millions of students, workers, peasants and intelligentsia called for free elections, withdrawal of Soviet troops and a nonaligned Hungary and brought Nagy back to power. The Hungarian revolution was crushed by Warsaw Pact forces in November–December 1956. Nagy and his closest associates were executed in 1957. Over 20,000 Hungarians were killed in the ensuing repression.

An unexpected reversal of fortunes occurred under János Kádár, party leader from October 1956 to May 1988. From late 1961 onward, after five years of repression and recollectivization Kádár set out to conciliate the intelligentsia, workers, peasants and even Catholics. Most political prisoners were released. The collective farms were granted unprecedented commercial autonomy. Collective farmers responded by vigorously expanding Hungary's food output and exports.

In 1968 the Kádár regime inaugurated a carefully-prepared "New Economic Mechanism", encouraging industrial enterprises to establish direct contracts with one another and with their final customers, to become self-financing, and to use repayable bank credits in place of outright grants. To balance increased managerial prerogatives, the party-controlled trade unions were more actively to defend workers' interests. This social contract was nevertheless vulnerable. First, the economic and cultural effervescence which it required could exceed the political limits set by Kádár and by Soviet tolerance. Second, the regime depended on its ability to deliver economic prosperity. Yet this small mineral-deficient country was very dependent on the state of world trade. The Kádár regime survived thanks to Soviet readiness to cushion the Hungarian economy against the full impact of rising world energy prices and Western willingness to underwrite the Kadarist strategy.

In May 1988 a special party conference finally ousted 76-year-old Kádár. Liberal "reform Communists", led by Imre Pozsgay and Rezsö Nyers, hoped that this would lead to a more liberal market economy and political pluralism, but they were blocked by the Party "old guard". In early 1989 the "reform Communists" acceded to opposition demands for multiparty elections. They also rehabilitated Imre Nagy and the ideals of October 1956. In October 1989 parliament "disestablished" the ruling party. The free-marketeering Democratic Forum decisively won the elections of March 1990 and formed a government headed by Jozef Antall. The Antall government continually proclaimed its commitment to plans for rapid economic liberalization, though the outdated infrastructure and economic uncertainties of 1990 delayed the actual execution of these plans.

▲ **Protestors in Budapest in 1956 burn posters of Rákosi.**

ROMANIA

REPUBLIC OF ROMANIA

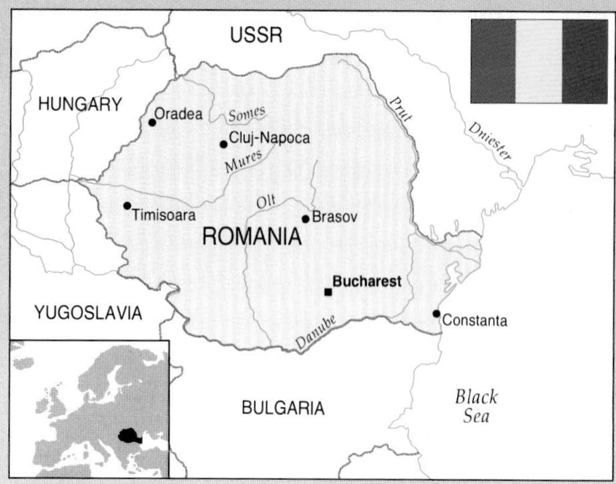

CHRONOLOGY

1907	Repression of major peasant revolt, killing over 10,000
1912	Balkan Wars against Turkey and Bulgaria
1920	Radical redistribution of farmland
1944	Romania switches to Allied side in World War II, as Soviet forces invade
1958	Withdrawal of Soviet forces from Romania
1977	Major earthquake and repression of miners' strikes
1987	Repression of workers' protests in Brasov
1989	End of Communist dictatorship and execution of Nicolae and Elena Ceauşescu
1990	Pogrom against Hungarian minority in Transylvania at Tirgu Mures
1990	National Salvation Front "win" elections

ESSENTIAL STATISTICS

Capital Bucharest

Population (1989) 23,168,000

Area 237,500 sq km

Population per sq km (1989) 97.5

GNP per capita (1988) US$6,400

CONSTITUTIONAL DATA

Constitution Unitary republic with one legislative house (Grand National Assembly)

Major international organizations UN

Monetary unit 1 Romanian leu = 100 bani

Official language Romanian

Major religion Romanian Orthodox Church

Heads of government (King) Carol I (1880–1914); Ferdinand II (1914–27); Michael (under Regency) (1927–30); Carol II (1930–40); (Prime Minister) Gen. I. Antonescu (1940–44); Michael (1944–47); (President) C. Parhon (1948–52); P. Groza (1952–58); I. Maurer (1958–61); G. Gheorghiu-Dej (1961–65); C. Stoica (1965–67); N. Ceauşescu (1967–89); I. Iliescu (1989–)

The "core" of modern Romania, the Danubian principalities of Wallachia and Moldavia, only achieved full unity in 1862 and independent statehood in 1878. Moreover, until 1918, Transylvania remained part of the Habsburg Kingdom of Hungary, while "Bessarabia" (Eastern Moldavia) was annexed to the Russian Empire from 1812 to 1918. From 1866 to 1914 the new state was consolidated under Prince Carol of Hohenzollern-Sigmaringen. Romania became a kingdom in 1881 and, as a result of the Balkan Wars of 1912–13, was able to annex South Dobruja. Industrialization was assisted by the meteoric growth of a largely foreign-owned oil industry.

In August 1916 Romania went to war on the Allied side, but capitulated in December 1917 to the Central Powers, who occupied its oilfields, and then in November 1918 it briefly reentered the war on the winning side. The acquisition of Transylvania, Bessarabia and Bukovina and reacquisition of South Dobruja doubled Romania's territory and population, including large ethnic minorities: Hungarians, Russo-Ukrainians, Jews, Germans, Bulgarians and Gypsies.

Demands for radical change were greatly increased by the upheavals of 1912–1919. King Ferdinand conceded universal suffrage in 1918, while land reform was undertaken in 1920–23. The Conservative party, the party of big landowners, never recovered from these setbacks, with the result that the Liberal party, the party of the rising urban bourgeoisie, came to power.

In 1927 King Ferdinand died. The "clean" reformist National Peasant party, led by Iuliu Maniu, was able to sweep to power in 1928 on a wave of revulsion against Liberal malpractices. The new government was strongly committed to social reform, freer trade and the attraction of foreign capital. But the 1929–38 world depression ended its plans. Crown Prince Carol, exiled in 1926, took advantage of the crisis to reclaim his throne in June 1930, to accept premier Maniu's resignation and to play Romania's deeply divided parties off against each other in the 1930s. He set up a royal-fascist dictatorship in 1938–40, during which Romania came to depend on Nazi Germany. In 1940, however, Carol was forced to cede Bessarabia and Bukovina to the Soviet Union, northern Transylvania to Hungary and South Dobruja to Bulgaria. He fled abroad in disgrace, abandoning his teenage son Michael to the tender mercies of the dictatorship of General Ion Antonescu. Romania became Nazi Germany's most important foreign supplier of vital raw materials. In 1943 Romanian antifascists formed an underground Patriotic Front. Then in August 1944, as Soviet forces started to invade, young King Michael formed an antifascist coalition government and sided with the Allies.

Although the Romanian Communist party had only 900 members in 1944, the Soviet Union imposed a Communist-dominated government in March 1945. In 1947 leaders of others parties were mostly imprisoned; the National Liberal party and the National Peasant party were outlawed, King Michael was exiled, and a "People's Republic" was proclaimed. In 1948 the purged Social Democratic party was forcibly absorbed by the Communists; and a centrally planned economy was established. However, many large enterprises became Soviet-Romanian joint-ventures known as "Sovroms", which exploited Romania's natural resources for the USSR's benefit. Agriculture was forcibly collectivized between 1949 and 1962.

In 1952 party leader Gheorghiu-Dej supplanted "Moscow Communists" and proceeded to dissolve the unpopular

"Sovroms". Romania's participation in Soviet suppression of the Hungarian revolution in November 1956 was rewarded by full withdrawal of Soviet troops from its soil in 1958. Thus freed from direct Soviet subjugation, Romania in 1961–64 resisted Soviet proposals that it should specialize in food exports and import industrial products from Comecon's more advanced member-states. Defiant independence and autarkic emphasis on the development of heavy industries continued under the dictatorship of Nicolae Ceauşescu from 1965 to 1989.

Ceauşescu's foreign policy initiatives were mistaken for liberalism and, through the 1970s, Western governments saw him as a valuable thorn in Russia's side, to be courted with "soft" loans, investments and trade preferences. But the borrowed capital was simply misappropriated by the Ceauşescu "kleptocracy" (including his extended family "mafia") or expanded inefficient energy-intensive industries far beyond Romania's capacity to feed the fast-growing nonagricultural workforce. In 1976–81 this former net exporter of oil and foodstuffs became dependent on imported oil and foodstuffs. But Romania ran into serious difficulties in servicing its US$10 billion debt to the West in 1981–82. These difficulties induced the regime to tighten its grip and repay the debt during the 1980s. Ceauşescu put 1980s Romania through a draconian austerity program. The social costs were enormous. The long decline in living standards and intensified exploitation of abysmally paid workers, evoked bloodily suppressed miners' strikes in July 1977 and labor unrest in Brasov in November 1987. In the late 1980s Ceauşescu initiated a "systemization" program to bulldoze Romanian and Hungarian villagers into soulless state-controlled apartment-blocks, subject to police surveillance. Eventually persecution of Hungarian pastor Lazlo Tokes in Timisoara evoked mass protests from Romanians as well as Hungarians and this became the catalyst of bloody confrontations between the Securitate (secret police), the army and citizens, which brought down the regime and led to the execution of Ceauşescu and his hated wife in December 1989.

A "National Salvation Front" (NSF) quickly seized power in late 1989 and early 1990. In 1990 and early 1991 it was widely felt that the same people as before were running the country using much the same methods under a new label. In May 1990 the (ex-Communist) interim NSF president Ion Iliescu "won" 85 percent of the votes in the presidential election after campaigns in which the opposition were subjected to harassment from NSF supporters. The Romanian nationalist association, Vatra Romaneasca, incited ethnic hatred. In June 1990 miners, transported in from the provinces and armed with shovels and clubs, viciously attacked opponents of the NSF encamped in Bucharest's University Square.

On the economic front, output continued its decline. The NSF repeatedly promised economic liberalization and limited privatization but reform was bedeviled by unbridgeable mistrust between rulers and ruled.

▼ Orphans infected with AIDS as a result of Ceauşescu's family policies.

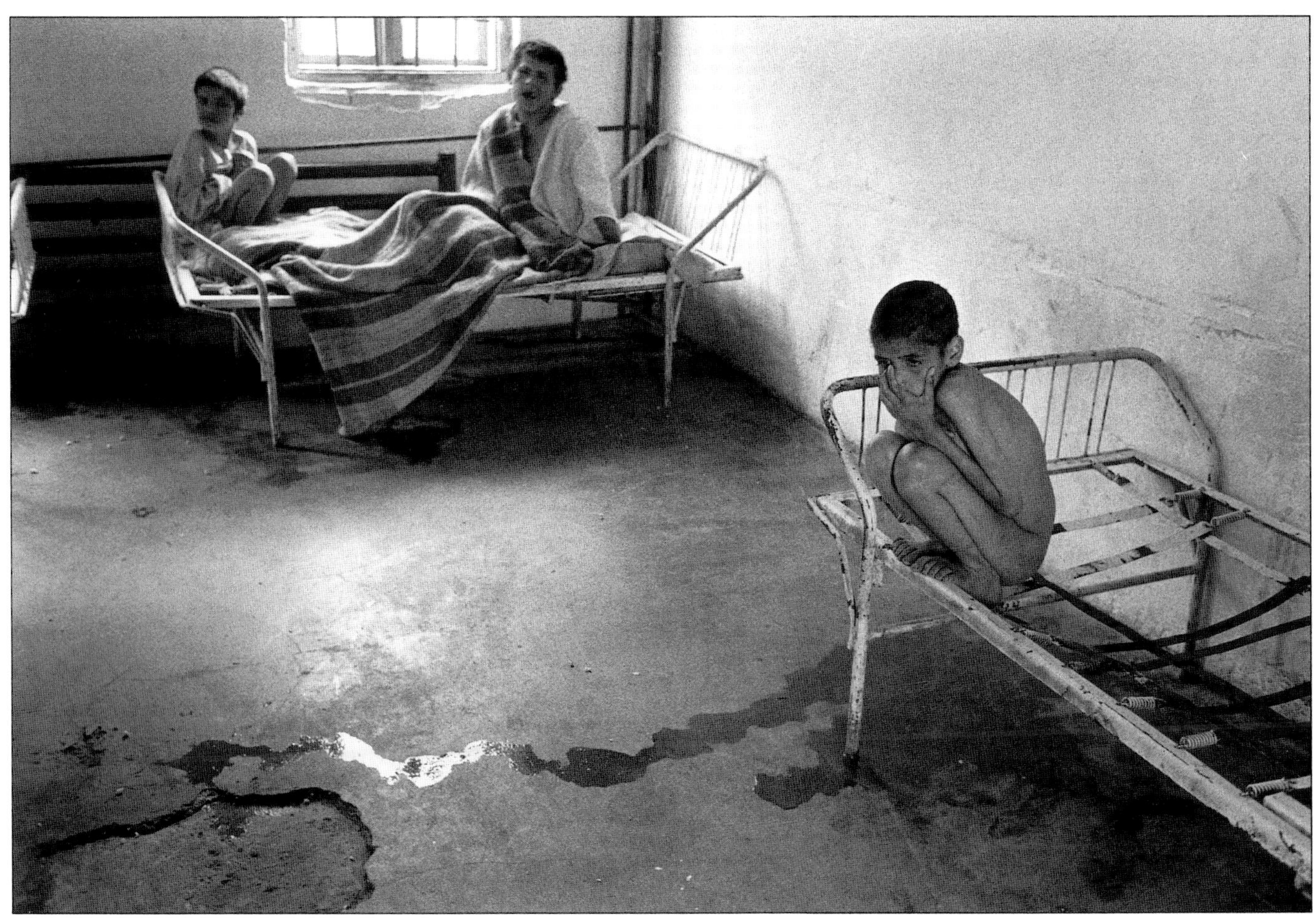

YUGOSLAVIA

SOCIALIST FEDERAL REPUBLIC OF YUGOSLAVIA

CHRONOLOGY

1903	Serbian military coup establishes constitutional monarchy under King Peter Karageorgevich
1918	Formation of Triune Kingdom of Serbs, Croats and Slovenes
1929	Royal dictatorship proclaimed
1945	Communists "win" uncontested elections – Federal People's Republic is established
1968	Albanian autonomist unrest in Kosovo
1971	Autonomist unrest in Croatia
1989	End of Communist monopoly of power in Yugoslavia
1991	Violent confrontation between Slovenia, Croatia and federal forces as former proclaim independence

ESSENTIAL STATISTICS

Capital Belgrade

Population (1989) 23,710,000

Area 255,804 sq km

Population per sq km (1989) 92.7

GNP per capita (1988) US$6,540

CONSTITUTIONAL DATA

Constitution Single party socialist federal republic with two legislative houses (Chamber of Republics and Provinces; Federal Chamber)

Date of independence 1 December 1918

Major international organizations UN; COMECON; I-ADB

Monetary unit 1 Yugoslav dinar (Din) = 100 paras

Official languages Macedonian; Serbo-Croatian; Slovene

Major religion Serbian Orthodox

Heads of government since independence (King) Peter I (1903–21); Alexander I (1921–34); Peter II (1934–45); (President) Marshal J. Broz Tito (1945–80); (Head of Collective Presidency) L. Kolisevski (May 1980); C. Mijatović (1980–81); S. Krajger (1981–82); P. Stambolić (1982–83); M. Spiljac (1983–84); V. Djuranović (1984–85); R. Vlajković (1985–86); S. Hasani (1986–87); L. Mojson (1987–88); R. Dizdarević (1988–89); M. Rožicé (1989–90); B. Jovic (1990–91); Stipe Mesić (1991–)

A multinational "South Slav" state first came into being in 1918, in the form of the "Triune Kingdom of Serbs, Croats and Slovenes", under Serbia's Karageorgevich dynasty. This brought together Serbia, Croatia, Dalmatia, Slovenia, Montenegro, Northern Macedonia, Bosnia-Herzegovina, the Vojvodina, Kosovo-Metohija and Novi Pazar. These disparate territories had never before been part of a single state. For four centuries, the Catholic Austro-Hungarian Empire had controlled the more developed Slovene and Croatian lands, while the more mountainous regions had been controlled by the Ottoman Empire, by Bosnian (Serbian), Albanian and Turkish landlords, and by the Eastern Orthodox Church. Antagonism between Serbs and Croats has bedeviled 20th-century Yugoslavia.

In 1918 Serbia became the nucleus around which Yugoslavia coalesced. Yet war-ravaged Serbia was less developed than either Croatia or Slovenia and contributed under half Yugoslavia's territory and population. Non-Serbs resented Serbian chauvinism and use of the Serbian army, monarchy and capital city Belgrade to govern Yugoslavia as if it was "Greater Serbia".

During the later 1930s Yugoslavia drifted into deep dependence on Nazi Germany until 1939 when alarm mounted at Germany's summary annexation of Austria and dismemberment of Czechoslovakia. Prince Paul belatedly tried to pull back from the encircling Axis embrace. Finally, Prince Paul and premier Tsvetkovich acceded to the Anti-Comintern Pact on 25 March 1941. Two days later, they were overthrown by a Serbian nationalist coup and Hitler ordered and effected a punitive invasion of Yugoslavia.

Spring 1941 brought German occupation of northern Slovenia, most of Serbia and the Banat; Italian occupation of southern Slovenia, Dalmatia and Montenegro; Hungarian annexation of the Vojvodina; Bulgarian annexation of Macedonia; and a puppet "independent state of Croatia", administered by the murderous Catholic-fascist *Ustaša*. The Yugoslav government and King Peter took refuge in London.

Yugoslavia produced two major resistance movements: royalist Serbian-nationalist "*chetnicks*" (bands) led by Colonel Mihailovich; and Communist-led Partisans under Josip Broz (alias "Tito"), the only movement ever to draw mass support from all sections of Yugoslav society. Both movements were almost annihilated in 1942–43. But, helped by Italy's collapse in September 1943, Tito's forces grew to 800,000 by May 1945. In November 1943 the Communist-dominated Anti-Fascist Council for National Liberation of Yugoslavia declared itself "the supreme legislative and executive body". The Partisans liberated south and central Yugoslavia in 1944 and the rest in 1945, with Red Army assistance. In August 1945 the Communists launched a "National Front" which, by intimidating opponents, won 90 percent of the vote in uncontested elections to a constituent assembly. This proclaimed a Federal People's Republic of Yugoslavia.

From 1946 to 1990 Yugoslavia was governed by a Communist police state, headed by Marshal Tito until his death in 1980. In defiance of Stalin, they actively supported the Communists in the Greek civil war and championed a Communist "Balkan Federation". In 1947, against Soviet advice, they launched a Five-Year Plan for rapid state-directed development of heavy industries and collective agriculture, expecting the Soviet Union to provide much of the resources and technical aid.

But rural collectivization and centrally planned industrialization were inappropriate for several reasons. The terrain did not readily lend itself to large-scale mechanized cultivation; and the peasants, who had been the backbone of Partisan resistance to fascism, would not meekly submit to collectivization. Meanwhile the country's meager resource endowments were not conducive to industrialization, while centralization rekindled resentments against Serbian domination.

Yugoslavia was "excommunicated" from the Soviet bloc in 1948, for trying to be more Stalinist than Stalin. This cut off crucial imports and aid from the Soviet Union and Czechoslovakia, plunging the economy into crisis. An attempt to persevere with ultra-Stalinism unaided in 1948–50 only made matters worse, and Tito's regime only survived because it was bailed out by massive Western aid in the 1950s.

The need to transcend Yugoslavia's ethnic and religious divisions made it desirable to devolve decision-making to regional or local institutions. In 1950 an embryonic new system of "workers' self-management" was launched, when elected workers' councils with key roles in enterprise management were established in almost all medium and large industrial enterprises. The 1958 program of the League of Communists of Yugoslavia (SKJ) (as the Communist party was renamed in 1953) advocated direct democracy and producers' control of the means of production. The 1963 constitution established rotation of all elective offices (other than Tito's) and prohibited individuals from simultaneously holding office in both federal and republican bodies. Until 1965, however, Yugoslavia was in practice still managed from the center, and enterprise autonomy was very restricted.

In 1965–66, however, the champions of full decentralization and enterprise autonomy triumphed and breathed new life into the self-management system. During the 1970s and 1980s, however, enterprises were circumscribed by new investment and price controls, and other strictures in a vain attempt to control the economy.

From the 1950s to the 1970s, nevertheless, Tito's increasingly decentralized federation achieved rapid economic growth (buoyed up by booming tourism and emigrants' remittances after 1960) and managed to contain regional demands for greater autonomy. But the Communist monopoly of power crumbled in 1989–90, as Yugoslavia succumbed to regionalism, resurgent ethnic conflict and economic contraction. In 1989, when inflation exceeded 2,000 percent, Yugoslavs completely lost confidence in the SKJ's capacity to deal with the crisis. Elections were held in each republic in 1990 and the SKJ disintegrated into separate national parties. Catholic nationalist governments were elected in Croatia and Slovenia, but the Communists won decisively in Serbia under Slobodan Milosevich. In 1991 Slovenia and Croatia declared independence. The army intervened unsuccessfully in Slovenia and fullscale civil war broke out in Croatia.

▲ **An Albanian peasant family in Southern Yugoslavia.**

BULGARIA

PEOPLE'S REPUBLIC OF BULGARIA

CHRONOLOGY

1908	Declaration of a fully independent kingdom
1918	Eruption of peasant anger at futile warfare
1923	Bloody overthrow of Stamboliski
1941	Czar Boris declares war on the Western Allies
1942	Jews and gypsies sent to concentration camps
1944	Soviet troops invade
1947	Communists decimate the Agrarians and establish a "proletarian" dictatorship
1949	"Moscow" Communists decimate "national" Communists
1984	Official persecution of Turks and gypsies
1989	Communist dictatorship ended amid economic collapse

ESSENTIAL STATISTICS

Capital Sofia

Population (1989) 8,987,000

Area 110,994 sq km

Population per sq km (1989) 81.0

GNP per capita (1988) US$7,510

CONSTITUTIONAL DATA

Constitution Unitary single-party socialist republic with one legislative house (National Assembly)

Date of independence 5 October 1908

Major international organizations UN

Monetary unit 1 lev (leva) = 100 stotinkí

Official language Bulgarian

Major religion Eastern Orthodox

Heads of government since independence Prince Ferdinand (1887–1918); Czar Boris (1918–43); (Prime minister) D. Bozhilov (1943–44); I. Baynanov (Jun–Sep 1944); Col. K. Georgiev (1944–46); G. Dimitrov (1946–49); V. Chervenkov (1949–56); (Party boss) Todor Zhivkov (1954–60) (Prime minister) Todor Zhirkov (1960–70); (President) Todor Zhirkov (1971–89); (Prime minister) Andrei Lukanov (1989–90); Dimitar Popov (1991–)

From 1396 to 1908 Bulgaria was formally under Ottoman overlordship. Its society was decisively shaped in the 1870s when Bulgaria achieved national autonomy, a liberal constitution and radical redistribution of almost all large landed properties to peasant proprietors. From October 1918, Bulgaria was ruled by Alexander Stamboliski's violent Peasant Union (alias Agrarian Union). In June 1923 Stamboliski and thousands of his activists were butchered in a military coup. In September 1923, however, the Communist party staged an abortive insurrection: this intensified Tsankov's campaign against the agrarian and proletarian left. In May 1934 another military coup established a government of "national regeneration". In 1935, however, Czar Boris established a quasi-fascist royal dictatorship and drew close to Germany.

In 1941 Czar Boris signed the Axis Pact, recreated "Greater Bulgaria" (by annexing Macedonia, Thrace and South Dobruja) and declared war on the Western Allies. He introduced Nazi-style antisemitic legislation and deported Jews from Macedonia and Thrace to Germany for extermination. He made similar arrangements for Bulgaria's Jews, although most of these were saved by the deteriorating Axis military situation in 1942–44. In 1943 agrarians and communists came together in an antifascist "Fatherland Front", which subsequently gained tentative support from the Soviet Union, whereas the Western nations made unsuccessful attempts to make deals with Bulgaria's royalist regime.

The USSR declared war on Bulgaria on 5 September 1944, freeing the "Fatherland Front" to overthrow Bulgaria's royalist regime. In October 1944 Churchill and Stalin secretly agreed that Bulgaria should be in the Soviet "sphere of influence". In the October 1946 election the Communists won 53 percent of the votes cast and in June 1947 Communist control was consolidated by arresting agrarian leader Nikola Petkov on charges of espionage.

In 1947–53, while experiencing the full rigors of Stalinism, Bulgaria was ruled by so-called "Moscow Communists". After Stalin's death in 1953, workers' unrest in Plovdiv in May 1953 and Soviet demands for a "New Course" helped to restore "local" Communists to favor. Todor Zhivkov gradually ousted or demoted doctrinaire Moscow Communists. Bulgaria pursued highly-centralized and comprehensive "command planning", emphasizing large-scale capital-, energy- and mineral-intensive heavy industries.

In 1984–89 Zhivkov tried to deflect mounting discontent and criticism by stirring up Bulgarian nationalism and hostility to Turks and Gypsies. In May 1989 he rashly defied Turkey to open its border with Bulgaria; it did and, by September, over 300,000 Bulgarian Turks had fled to Turkey. Finally, a police crackdown on demonstrators on 26 October 1989 precipitated a coup from within Zhivkov's ruling party on 10 November. Mladenov, the new Communist leader and President, and his new premier, Andrei Lukanov, repudiated the policies of the Zhivkov regime, legalized opposition parties, amended the constitution to "disestablish" the "leading role" of the Communist party, changed the name and image of the Communist party to "Socialist party", outlawed discrimination on grounds of race and religion, restored the Turks' right to use Turkish names and organized free elections. The socialists, however, soon split into mutually incompatible liberals and hardliners and Lukanov's government fell.

ALBANIA

PEOPLE'S SOCIALIST REPUBLIC OF ALBANIA

CHRONOLOGY

1912	Independence is achieved after the Balkan Wars
1915	Frontiers are revised leading to Treaty of London
1927	Pact of Tiranë
1939	Italian invasion and rule
1941	Albanian Labor party formed
1946	Republican constitution adopted
1961	Relations with USSR are broken
1968	Albania formally leaves Warsaw Pact
1977	Severing of relations with China
1991	Election results confirm Communist rule

ESSENTIAL STATISTICS

Capital Tiranë

Population (1989) 3,197,000

Area 28,748 sq km

Population per sq km (1989) 111.7

GNP per capita (1986) US$930

CONSTITUTIONAL DATA

Constitution Unitary single-party socialist republic with one legislative house (People's Assembly)

Date of independence 18 November 1912

Major international organizations UN

Monetary unit 1 lek = 100 qindars

Official language Albanian

Major religion Muslim

Heads of government since independence Prince William of Wied (1913–14); Essad Pasha Topdani (1914); Italian rule (1915–20); Council of Regents (1920–24); (Prime minister) Ahmed Zog (1924–28), as King Zog (1928–39); Italian rule (1939–44); (President) Enver Hoxha (1946–82); Ramiz Alia (1982–)

Albanian independence came with the liberation of peoples and territories during the Balkan Wars in the first years of the 20th century which pushed back the rule of the Turks. The small state, its frontiers contested by Greece and Serbia, preserved its sovereignty despite the rivalry of the Austro-Hungarian Empire and Italy in the region. But World War I brought stronger Italian influence to bear and Albania was under a pro-Italian provisional government in 1918. Although Italy confirmed Albanian frontiers and independence in 1920, the Fascist regime of Benito Mussolini opposed Greek claims to part of the country and landed troops in 1923 in support of Ahmed Zog, the ruler of Albania, and on Greek Corfu. This heralded a long period of Italian domination of Albania.

Strategic considerations and imperial aims encouraged Mussolini to attack Albania in April 1939. The country quickly fell into Italian hands and was annexed. But internal opposition soon developed and was organized into a liberation committee in 1942. A socialist coup led to a new independent government in 1944, itself replaced by the communist regime of General Enver Hoxha in 1946. This became a daunting and severely repressive single-party regime, destroying any connection with previous regimes, with neighboring states and with the West. Close links with the Soviet Union were developed, as they were with the People's Republic of China after 1949. When these two Marxist states came into conflict with each other, the Hoxha regime sided with China, broke off relations with the USSR in 1961 and finally ended in isolation.

A revised constitution in 1976, forming a Socialist People's Republic, and a new head of state in 1982 as Hoxha fell ill, altered little in this austere country. However the changes elsewhere in the Communist world and demands for democracy finally brought some movement. An opposition party, the Democratic party, was created in 1990, and attacks against the Communist dictatorship began. Many Albanians fled the country once frontiers were opened in 1991. Elections were held in 1991, in which the towns voted solidly for the Democratic party while the countryside retained its support for the Communists, and the government retained power.

Economic and social trends

The people of Albania, isolated in their rugged hills and secluded valleys, are a distinct Balkan people, the Skqypetars. Some were Muslim but in the North there were Roman Catholics and in the South Greek Orthodox Christians. All areas were extremely poor, with subsistence farming and little industry outside the three main cities, Tirane, Durazzo and Scutari. The two coastal cities developed manufacturing capacity with Italian subsidy between 1926 and 1939.

The Hoxha government applied a strict centrally planned control of the economy. This was not successful although minimal standards were achieved for food production and the feeble manufacturing sector. Foreign aid was minimal; Albania cut ties with Soviet international organizations and membership of the United Nations (from 1955) did not bring economic prosperity. The regimentation of society was unrelenting – churches became state museums, contact with the outside world was limited, party discipline was rigid. The changes of 1990 were not accompanied by constructive plans for economic improvement, and many hundreds of people chose to leave for Italy, hoping to find freedom and prosperity there.

GREECE

HELLENIC REPUBLIC

CHRONOLOGY

1923 Treaty of Lausanne with Turkey

1967 Military coup changes political order

1974 Restoration of democracy

1981 Greece enters the European Community

ESSENTIAL STATISTICS

Capital Athens

Population (1989) 10,096,000

Area 131,957 sq km

Population per sq km (1989) 76.5

GNP per capita (1987) US$4,350

CONSTITUTIONAL DATA

Constitution Unitary multiparty republic with one legislative house (Greek Chamber of Deputies)

Major international organizations UN; EC; EEC; NATO

Monetary unit 1 drachma (Dr) = 100 lepta

Official language Greek

Major religion Christian

Heads of government since 1900 (King) George I (1864–1913); Constantine I (1913–16); Alexander (1917–20); Constantine I (1920–22); George II (1923–24); (Prime minister) E. Venizélos (1924); G. Kaphandaris (1924); M. Papanastasiou (1924); T. Sophoulis (1924); A. Mikhalakopoulos (1924); T. Rangalos (1925–26); A. Zaimis (1926–28); E. Venizélos (1928–32); M. Papanastasiou (1932); E. Venizélos (1932); P. Tsaldaris (1932–33); E. Venizélos (1933); P. Tsaldaris (1934–35); (King) George II (1935–41); Axis occupation (1941–46); (King) George II (1946–47); Paul (1947–64); Constantine II (1964–67); Military junta (1967–73); (President) P. Ghizikhis (1973–74); (Prime minister) K. Karamanlis (1974–80); G. Rallis (1980–81); A. Papandreou (1981–89); T. Tzannetakis (Jul–Oct 1989); Y. Grivas (Oct–Nov 1989); X. Zolotas (1989–90); C. Mitsotakis (1990–)

Greek political life in the 20th century has been characterized by the clash of extremes, fluctuations between republicanism and monarchy and often violent internal conflict. European opposition to a proclamation of unification of Greece and Crete in October 1908, provoked a coup bringing Eleutherios Venizélos, a powerful ruler from Crete, to power as premier in 1909. Venizélos represented anti-German and expansionist objectives and his authority was sustained into World War I. The pro-German sympathies of King George I and his son King Constantine brought a change in succession in 1916 when Constantine abdicated, to be replaced by his brother Alexander. Venizélos kept Greece on the side of the Anglo-French Entente until 1918.

A calmer period followed the war, with a territorial agreement with the new Republic of Turkey signed at the Treaty of Lausanne in 1923, and an exchange of populations with Bulgaria and with Turkey in Smyrna. However, Italian ambitions led to the occupation of Corfu in 1923, and the ensuing crisis saw a return to domestic strife over the question of whether the Greek state should be a monarchy or republic. Coup followed coup until Venizélos returned to power in 1928, to steer the government through a series of crises, including a revolt on the island of Cyprus in 1930. But further disagreements and an attempted military coup forced Venizélos out of office. He finally retired from politics and died in France in 1936 after failing to lead a successful revolution against the government in Athens. Some stability came with a restoration of the monarchy in 1935 and strong executive government in the hands of General Metaxas from 1936. Labor troubles led to the imposition of martial law and a rapprochement with both Turkey and Germany.

War again destroyed political harmony when the country was invaded by Italian armies in 1940. Although the invasion was initially repulsed, in April 1941 mechanized German troops attacked and were in Athens within three weeks, despite British support for the Greeks. The country was occupied for three and a half years, including the outlying islands as far as Crete, and subjected to severe repression. The defeat and withdrawal of the Germans did not bring peace to Greece. Immediately after liberation, in 1944, the leftwing guerrilla movements clashed with the center-right and monarchists. British troops intervened in the north, under General Scobie, in support of the monarchy and against the Greek communists and other Marxist groups. Elections were held in 1946, with an overall victory of the Populist party led by Tsaldaris, and the monarchy was restored, endorsed by a popular vote. King George II returned but died shortly afterward, succeeded by his brother King Paul. The civil war continued until 1950, causing lasting divisions throughout Greek society.

Greece was among the poorest and most devastated countries of Europe in 1945, and the work of rebuilding the economy and the shattered society after the war was interrupted by continuing political conflict. The divisions were the old rivalry between the supporters of monarchy and those of republicanism, and now the sharper conflict between the right and the left, many of whom were exiled and looked to the newly established communist states of the Balkans for help. Many Greek communists lived in France and elsewhere, organizing for their return to Greece and for what they saw as a proper democracy. These tensions were aggravated by further hostility between Greece and Turkey, despite their joint membership of the defensive alliance, NATO. These tensions increasingly involved the island of Cyprus, where communities from both countries lived side by side and where demands for the withdrawal of the British (who had been in possession since 1878) were accompanied by some cries for union with Greece.

Political order was changed by a military coup in 1967 which suspended the 1952 constitution and forced the king into exile together with many moderate political figures. Military rule did not last and the joint demand for a democratic republic and pressure of events in Cyprus brought further changes in 1974 and 1975. There was strong support for the republic – 77 percent in a referendum – and the government had to respond to a Turkish invasion of Cyprus (which had a large Greek population) in 1974. The Greeks could not intervene militarily against their much more powerful, larger and better-equipped neighbor. The island was divided, with the line controlled by a United Nations' peace-keeping force, and Greece itself returned to democracy in 1974. This change brought more stable government, the disappearance of many of the old issues that had caused divisions in the past, and the entry of Greece into the European Community (EC). Much of the success can be attributed to Konstantinos Karamanlis, first as a minister, then prime minister and finally as president of Greece in 1980, reelected in 1990.

Economic and social trends

Political divisions in Greece have been sharpened by the serious poverty of the mainland and the islands. Many of the latter are unable to sustain their populations and yet in 1947 Greece was granted sovereignty over the many islands of the Dodecanese which it had claimed before the war. There is little manufacturing, some production of primary minerals - iron pyrites, lignite, ores of various kinds - and limited agricultural production. Wheat and maize are grown in insufficient quantities to feed the population, but olives and oil are overproduced, as is wine. Some tobacco exports have brought foreign exchange. These basic deficiencies in the economy are matched by the poor infrastructure, notably railroads and communications, in this very dispersed country.

For much of the 20th century Greek businessmen have developed one important source of wealth, shipping. Greek-registered ships and lines have been among the world's largest, but they proved exceptionally vulnerable during the wars. In 1945 some three-quarters of the Greek fleet of merchant shipping had been destroyed or damaged. It required time and funds to rebuild, but this was achieved with British, American and finally European Community help.

One further change in the economic fortunes of Greece came in the 1960s with the development of cheap flight holidays. At last even the islands were able to generate some wealth, as tourism became the most important currency earner for Greece. The air transport which brought tourists also permitted another local industry to expand, clothing, which could now be exported quickly and cheaply. This, with tobacco, forms the largest category of Greek exports to its two strongest trading partners, Germany and Italy.

Greek society and culture has to coexist with the strong memories and reminders of the great classical past. Much recent creativity and energy has therefore stressed the idea of Greek identity and shared experience with literature and music coming from the islands and the often quite substantial Greek "colonies" in cities such as Beirut and Alexandria or on Cyprus. This new "Hellenism", the awareness of Greek cultural elements spread across the Eastern Mediterranean, binds contemporary Greeks. They share a language, despite some marked dialects and variants, as from Crete, a religious form (Greek Orthodoxy, formed after independence) and a powerful and preserved folk culture. In recent years there have been loudly voiced demands that treasures from ancient Greece – such as the sculptures from the Parthenon in Athens which were taken to London in the 19th century – should be returned to the modern Greek state.

▲ **Greek Orthodoxy dominates traditional Greek culture.**

ITALY

ITALIAN REPUBLIC

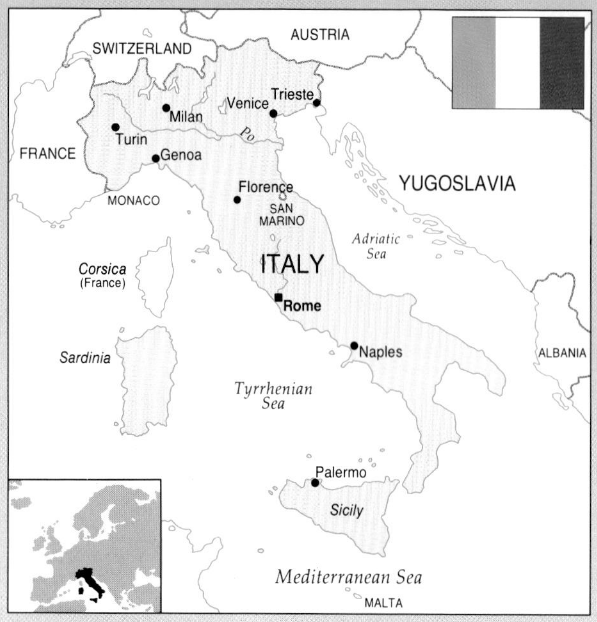

CHRONOLOGY

1922	Fascist March on Rome led by Benito Mussolini
1923	Corfu invaded
1928	Former prime minister Giolitti murdered by Fascists
1929	Fascists achieve concordat with the Papacy
1936	Formation of the Berlin–Rome Axis agreement
1945	Mussolini executed
1946	Abdication of Royal family. Republic established
1948	Christian Democrats win a majority in legislative elections
1978	Assassination of Aldo Moro, leader of Christian Democratic Party

ESSENTIAL STATISTICS

Capital Rome

Population (1989) 57,436,000

Area 301,277sq km

Population per sq km (1989) 190.6

GNP per capita (1987) US$10,420

CONSTITUTIONAL DATA

Constitution Republic with two legislative houses (Senate; Chamber of Deputies)

Major international organizations UN; EC; EEC; I-ADB; NATO

Monetary unit 1 lira (Lit) = 100 centesimi

Official language Italian

Major religion Roman Catholic

Heads of government since 1900 King Victor Emmanuel III (1900–46); (President) E. de Nicola (1946–48); L. Einaudi (1948–55); G. Gronchi (1955–62); A. Segni (1962–64); G. Saragat (1964–71); G. Leone (1971–78); A. Pertini (1978–85); F. Cossiga (1985–)

The young Italian democracy – which had been formed only in 1861 – faced great pressures at the end of the 19th century, both domestically and in foreign relations. The industrialization of the cities of northern Italy and the poor conditions of the overpopulated south created a structural imbalance which gave rise to strongly voiced demands for change. The parties of the left and trade-union based syndicalist movements, which were more anarchist than Marxist, all gained adherents. In foreign affairs the Italians were not isolated but in alliance with both Germany and the Austro-Hungarian Empire, yet felt constrained to search for a colonial empire. The Italian government sought this empire in the lands of north and east Africa, hoping also to ease overcrowding and poverty which had already led many thousands to emigrate to the United States. Libya was colonized and some attempts were made to control and annex Somalia and Eritrea, but in 1896 Italian troops were heavily defeated by Ethiopian tribesmen at the battle of Adoua.

The European war of 1914 isolated Italy, which did not join the Central Powers with which the government was in alliance. However, domestic pressures forced a decision to enter the war and in 1915 Italy joined the British and French, and attacked Austria-Hungary. But the war was not a success for Italy since the northern cities suffered and in 1917 Italian armies were defeated in one of the few decisive engagements of the war, at Caporetto. The Italian minister Sonnino managed to acquire territory from Austria in the Trentino and also Trieste, but the sense of humiliation did not vanish. This was clearly seen in 1919 when Italian soldiers, led by the flamboyant poet Gabriele d'Annunzio, marched into Fiume on the Yugoslav border and held the city for Italy. But the most serious effect of the war was to encourage violence and divisions in society. Communists clashed with nationalist and rightwing groups in the streets, particularly in central Italy where the rightwing Fascists were strongly led by a former socialist journalist, Benito Mussolini.

In 1922, Mussolini, a powerful orator, felt strong enough to challenge the constitutional regime and the Fascists marched on Rome. Mussolini was accepted by the king as prime minister and the imposition of authoritarian rule began, although the king remained head of the Italian state. Political opponents were coerced or eliminated. The Fascist regime rested on a mixture of repression and modernization, investing in new enterprise, building projects and technology. In some regions the former administration and its officials refused to bend before the Fascist state and industrialists did not permit political aims to determine the economic development of the new state corporations. Moreover the Roman Catholic Church, which came to an agreement with the Fascist government in 1929, managed to preserve its grip over education while keeping out of political debate. Among Mussolini's most determined opponents were the local interests and secret societies of the South, Mafia and Camorra. Fascist squads were ready to match local violence with that of the state they controlled.

In foreign policy Mussolini immediately tried to assert Italian rule and extend its territory. Thus in 1923 the Greek island of Corfu was invaded. The League of Nations, of which Italy was a senior member, did nothing to protest or obstruct this, and the pattern of behavior for Europe's dictators was established. But domestic policy preoccupied Mussolini and it was not until 1929 that he again turned to overseas adventures. In 1934 Italy launched a fierce attack on Ethiopia, using aircraft and gas on poorly-armed troops. An appeal to the League of Nations by the Emperor of Ethiopia resulted in the imposition of economic sanctions against Italy, but these were slow to have effect. In 1936 Mussolini authorized support for General Franco's rebellion in Spain and formed the Berlin–Rome Axis in alliance with Hitler. The Axis agreement tied Italy to Hitler's desperate course and, despite delaying until 1940, Mussolini entered the war. Italian troops attacked Greece and were engaged in North Africa and Somalia, but these actions were unsuccessful, and German assistance was necessary to prevent a major defeat in the Balkans. Italians occupied southern France and Corsica in 1942 but the Anglo-American invasion of Sicily turned Italy into a battlefield which ended when a ceasefire was negotiated by Marshal Badoglio with the king's approval in 1943. A formal change of sides occurrred, but Mussolini remained a puppet ruler supported by the Germans. The result was misery for the people and continuing war as the front moved northward. Mussolini was finally caught in April 1945 after a remarkable rescue by German paratroops. He was shot and then hanged, with his mistress, by an Italian crowd, ending as an object of public derision.

Italian postwar reconstruction was successful. A democratic system was restored and approved by referendum and, once the royal family had abdicated, a weak executive was established. The loosely organized Christian Democratic party won consistent electoral support and formed the major party of government from 1946.

Italy participated in integrated European Community institutions, became a member of the Western European Union (WEU) and NATO and avoided any rash foreign adventures. Administration was effective and government achieved a rhythm with the emergence of the Communist party as the major party of opposition.

The Italian republic became a delicate mixture of centralization and regional self-government. Special regions were created, notably in the frontier areas and where distinct minorities lived in Trentino and Aosta. The Mafiosi and similar groups welcomed the end of Fascism and continued to usurp the functions of the state in parts of Sicily and southern Italy, but they were soon engaged in a struggle with the police and military authorities of the state. In general the institutions of the state were strong, and this was demonstrated during the 1970s and early 1980s when public order was disrupted by violent groups from the far right and by the revolutionaries of the Red Brigades. The government refused to capitulate when a unit of the Red Brigades took the president of the Christian

▼ **Mussolini addresses crowds in Florence, 1930.**

Democratic party and former prime minister, Aldo Moro, hostage and finally assassinated him in 1978. They pursued the terrorist group effectively and brought large numbers to trial.

Economic and social trends

The Italian economy has had a constant difficulty: the divide between the industrial, urbanized north and the impoverished area south of Rome and encompassing the islands of Sicily and Sardinia. Industrialization and urban development in the north and center during the late 19th century brought an increase in living standards which spread to the agricultural communities of the fertile valleys in this area. Rivers, well-established ports such as Genoa, Viareggio, Livorno and La Spezia, and good road and rail communications all provided the resources for growth. Two world wars inhibited this growth, but northern Italy acquired a modern and prosperous economy through the century. The contrast with the south is sharp. Barren land, overpopulated villages, often containing many thousands of inhabitants, lack of employment and manufacturing all created dreadful poverty. Many people moved north in search of work, or went to the United States. Thus the dilemma for Italian governments in the 20th century has been to resolve this imbalance.

Early industrial development brought other distractions. Workers demanded economic and political change through a powerful syndicalist (revolutionary trade union) movement which opposed the state in the north both before and after 1914. Governments in Italy therefore had no confirmed tradition of strong and efficient economic and social policy. The Fascists tried to change this, urging state intervention and strong leadership. At first their interest was supportive of small business against the big companies and opposed to all communist, syndicalist and leftwing movements, but once in office

the regime turned to support the large and successful commercial and manufacturing sector. The state was to lead but there had to be competition in the marketplace; thus the Fascists moved in the direction of corporatism, basing enterprise on large corporations, competing in the heavy industrial sector. In other respects the Fascists did little to improve conditions for many people. Economic policy became increasingly dependent on war production, which dominated after 1934.

At the end of World War II Italy needed to rebuild the national economy, its railroads, factories and work structures, and to tackle problems of the south which had suffered additionally during the war. American finance came through the Marshall Aid program and the many family links across the Atlantic. By 1960 Italian production rivaled that of France or Germany in some areas: motor and electrical goods, aircraft and some specialized arms manufacture were of high quality, and Italy became internationally respected for industrial and fashion design.

Governments also began to tackle the problems of the south. State companies started production of steel and shipyards at

Taranto, cars in Naples, oil refineries in Sicily, bringing work and some redistribution of wealth, although sometimes it was workers from the north who came with the necessary skills. Special funding was administered through the *Casa per il Mezzogiorno*, to encourage and direct investment and to assist financing that was not so readily snatched by a local Mafia.

Cultural and scientific trends

For many people Italy is virtually synonymous with culture. The tremendous riches of the past – from pre-Roman times through the explosion of the Renaissance to the music, especially opera, of the 19th century – form the basis of contemporary culture, everpresent and a constant point of reference. Italians of the 20th century remained operagoers, enthusiastic fans of singers like Caruso and Gigli, and aware of the buildings and museums that surrounded them in every town. The Catholic Church retained a voice in education even through the years of Fascism.

However, the 20th century struck a new chord in the Italian genius. Industrial developments found Italians skilled in engineering and technology. They were among the great innovators in designing internal combustion engines, ships and aircraft. This quality was immediately stressed by the Futurist movement in the arts in their manifesto of 1913, in which they applauded creativity inspired by technology, speed and movement. Much Italian genius was repressed by Fascism, only to emerge again after 1945. Novelists such as Alberto Moravia, historians and philosophers like Gaetano Salvemini and Benedetto Croce created a new, realist perception in fiction and in analysis, relating the experience of Fascism to the Italian past and to the republican present.

Italian film directors in the postwar period developed an innovative range of interests, from social realism to the exploration of experience and sensation in the work of Federico Fellini and Michelangelo Antonioni. The cinema provided many of the most internationally famous names from Italy after 1945.

▲ **Italian football is highly successful and hugely popular.**

◀ **Mourning for Aldo Moro, murdered in 1978.**

SPAIN

KINGDOM OF SPAIN

CHRONOLOGY

1902	Alfonso XIII takes over with full regal authority
1923	Military coup by Primo de Rivera
1931	Republic proclaimed – King leaves
1936	Civil war begins, led by General Francisco Franco
1970	Burgos trial of Basque ETA militants
1977	Communist Party legalized. Elections held
1982	Spain admitted as member of NATO
1986	Spain admitted to European Community

ESSENTIAL STATISTICS

Capital Madrid

Population (1989) 39,159,000

Area 504,783 sq km

Population per sq km (1989) 77.6

GNP per capita (1987) US$6,010

CONSTITUTIONAL DATA

Constitution Constitutional monarchy with two legislative houses (Senate; Congress of Deputies)

Major international organizations UN; EC; EEC; I-ADB; NATO

Monetary unit 1 peseta (Pta) = 100 céntimos

Official language Spanish

Major religion Roman Catholic

Heads of government (Prime minister) A. Maura (1917–23); Gen. M. Primo de Rivera (1923–30); Gen. Dánaso Berenguer (1930–31); Adm. J. Bautista Aznar (Feb–Apr 1931); N. Alcalá-Zamora (Apr–Oct 1931); M. Azaña (1931–33); A. Lerroux (1933–34); R. Samper (Apr–Oct 1934); A. Lerroux (1934–39); M. Portela Valladares (1935–36); M. Azaña (Feb–May 1936); S. Casares Quiroga (May–Jul 1936); J. Giral (Jul–Sep 1936); F. Largo Caballero (1936–37); J. Negrín (1937–39); (Head of state) Gen. F. Franco Bahamonde (1939–75); (Prime minister) A. Súarez Gonzalez (1976–81); L. Calvo Sotelo (1981–82); F. González Márquez (1982–)

Spain, once one of the major powers of Europe, was unable to regain that place after its former colonies in South America had become independent during the 19th century and Cuba and the Philippines had been lost in a short war with the United States in 1898. Political instability affected the old monarchy. The form of the state itself was in question: was Spain a unitary state or a federation of its 50 provinces? Serious internal disputes isolated Spain from the rest of Europe and a regency exercised by the young king's mother during the late 1880s and 1890s weakened the country. Even when Alfonso XIII took over personal rule in 1902 stability and progress did not come; the conservative factions that wished to see a strong authoritarian government then attempted to achieve this in 1923 when Primo de Rivera staged a coup. He sought to cut through the confusions of the old politics, in favor of a Spanish nationalism. Further disputes and the growth of the socialist, Communist and anarchist movements limited de Rivera's achievements. He resigned in 1930, but political chaos persuaded the king to leave the following year without formal abdication. However it was the radicals, leftists and socialists who won the elections in 1931, establishing a republic with a new constitution and introducing a few years of enthusiasm and change. In 1936 the socialists and Communists made significant electoral gains, confirming the drift of the republic to the left.

The victory of the Popular Front of socialists and left Republicans in the 1936 elections brought a swift reaction from the army, and from traditionalists who feared a socialist state. General Francisco Franco gathered troops from the colonial base in Morocco and, with Italian help, crossed to southern Spain, thus beginning the civil war. All center, left and liberal elements combined to defend the republican government in what became a prolonged and vicious war. Appeals for help were not widely answered. Individuals came to Spain from many countries to fight for the government in the International Brigades but only the Soviet Union provided military assistance. Franco was strongly supported, however, as Fascist Italy and Nazi Germany supplied materiel and manpower. Nevertheless it took three years for Franco's armies to claw their way through Spain. The civil war cost thousands of lives, divided families and wrecked towns and infrastructure. The bombing of the Basque town of Guernica became a symbol of this new element in warfare. The volunteers returned home when Franco's armies entered Catalonia, the last major republican stronghold, in January 1939. The war ended when Madrid fell in March 1939.

Despite fears that he might do so, Franco resisted the appeals from Hitler and did not become involved in World War II, preferring to consolidate his regime. He had declared himself Head of State in August 1937 and thereafter imposed an authoritarian rule. Opposition was not tolerated and many people left the country. Leading republicans were executed, imprisoned or, in the case of many minor officials, removed to remote parts of the country. Spain became a police state with information strictly controlled and the press censored. The state became a unitary state, with all elements of regional authority removed, and traditional values emphasized. The Falange, the political party that had supported the Francoist rebellion, hoped to control policy and government, but Franco prevented even this. He distanced himself from the Falange, imprisoned their leader Manuel Hedilla and tried to reduce their role in the administration. The authoritarian regime changed in a number of respects. In 1942 the parliament, the Cortes, was restored, although this was largely a showpiece. Franco sought a reconciliation with the Roman Catholic Church and the progressive political movement of that church, *Opus Dei*, became important in government and policy-making. Several of Franco's ministers were from *Opus Dei*. Increasing concern with the succession led Franco to make an agreement with the royal family; the king's son, Juan Carlos de Borbòn, was accepted as successor in 1969.

However Franco's real concern was for international recognition. By 1953 Spain was a member of OECD and had a treaty with the United States which permitted American bases on Spanish airfields. This also opened the way for foreign investment. In 1973 the vice-president and deputy head of government, Admiral Carrero Blanco, was assassinated and Franco never managed to quell Basque demand for change. In 1975 his health failed and on his death Juan Carlos became king.

Change occurred immediately. Repression disappeared and political exiles returned, including Communists from the Soviet Union. Censorship and the trappings of authoritarian rule were removed. Spain joined NATO and the European Community and an agreement was reached with Britain over the frontier at Gibraltar, which had been closed by General Franco. Democracy returned to Spain with a new constitution and national elections, confirmed by a referendum in which 94.2 percent voted in favor. Adolfo Súarez became the new prime minister. However the supporters of the old order and of Franco did not vanish and an attempted coup, carried out in the national parliament building in 1981, was foiled by the determination of the king and the support of political leaders.

Economic and social trends

The loss of empire seriously diminished the wealth and available resources of Spain. Industry was concentrated in the north, in the Basque provinces and around Barcelona. Much of the country was devoted to inefficient agriculture, either in large estates or on small farms barely able to grub a living from the high, dry lands that fill the Spanish plateau. Little change occurred in the early decades of the 20th century, and civil war further destroyed the economy by damaging factories and

▼ **Civilians prepare for an air raid, Madrid, 1930s.**

communications and by driving enterprise abroad. Industrial strength was affected principally in the north, since this was a stronghold of the republican government. For almost forty years Spain was a divided country unable to call upon all its human resources. Moreover, Francoist Spain was initially deprived of outside help since the rest of the world was engulfed in war, and after 1945 there was little sympathy for an authoritarian regime which had been associated with the prewar governments of Germany and Italy.

However, some progress occurred. State-led projects began to modernize the infrastructure and membership of OECD in 1953 opened the way for foreign investment and trade. Cities were rebuilt and the tourist boom that started around 1955 brought foreign currency and work. A stabilization plan encouraged further investment from European states and the United States and this also froze wages to prevent inflation. The first national development plan was proposed in 1964, but this, like its successors, concentrated on heavy industry and neglected social aspects of industrialization, and such plans were anyway restricted by the near-total dependence upon imported fuels. The result was a curious mixture of outmoded methods of farming and old equipment and ultramodern projects. The state took over steelworks and companies such as Hispano-Suiza, making railroad engines. It invested in prestige ventures such as Pegaso, which were intended to create competitive luxury motor cars but ended by manufacturing commercial vehicles. But Spain had a market of 35 million people and a workforce with no history of recent labor troubles since the old unions and worker syndicates had been destroyed by Franco's dictatorship. Even before 1975 foreign companies were attracted by cheap labor, the lack of union power and image of order and progress.

The real change in the Spanish economy and society came after Franco's death. Foreign business made Spain an important manufacturing base with Ford, Chrysler, Renault and Citroën all establishing plants there. The new democracy applied for membership of the European Community and this was achieved in 1982 despite difficulties in negotiating the agreement because of Spain's large production of cereals, wine and olive oil, all of which were in surplus in the Community already. Spain then achieved some of Europe's highest growth rates and offered a rapidly expanding consumer market.

The changes that came after Franco's death affected society immediately. Traditional values were challenged. Censorship ended and a liberalization followed which initially affected the major cities and the coastal areas where foreign tourists – as many as thirty million a year – and foreign villa owners and their customs were familiar. But soon Spain was awash with all manner of consumer goods that had previously been banned, including pornographic magazines and films. The place of women in society was transformed and brought into line with that of women elsewhere in the European Community. The modernization of Spain became possible but this evolution went with a respect for established conventions, above all for the Roman Catholic Church and for the family.

It was the economic drive that characterized post-Franco Spain. The rural exodus continued and many of those who had sought refuge in the countryside or had been required to exile themselves in the depths of Andalusia or the Estramaduras returned to the cities. This new social mobility, however, exacerbated the regional disparities that had long existed. Wealth was concentrated in a few areas, or was dependent on narrow and limited sources such as tourists and foreign property speculators. Industrial wealth was confined to the north; Catalonia in particular sought investment and commercial contacts with other parts of Europe and had become one of the

▲ **Bullfighting remains a big tourist attraction throughout Spain.**

fastest expanding regional economies in the European Community by the late 1980s. But Spain remained Europe's fastest growing market, of 39 million people of whom 35 percent were between the ages of 18 and 40, working in an economy which was expanding at the rate of five percent per year.

The few years of the republic, from 1931 to 1936, were a period of literary excitement when it seemed that things might change. In particular poets such as García Lorca, Guillén and Salinas gained an international reputation and the writer Ortega y Gasset brought Spanish intellectual life into the mainstream of European philosophy. But this did not last, and one of the most serious effects of civil war and dictatorship was artistic and cultural repression. Several leading personalities left the country, others were executed or imprisoned. Pablo Casals, the cellist, Pablo Picasso and Joán Miró among a number of artists, and Luis Buñuel, the film director, all departed, refusing to compromise with the Francoist regime. The death of Spanish intellectual and cultural activity was expressed for many by Picasso's painting *Guernica* (1937), a largely monochrome portrayal of the horrors of war and the gutting of a nation, inspired by the bombing of this Basque town. Lorca's assassination in 1936 turned him into a hero and martyr to Spanish culture.

However, some artists and creative writers survived the dictatorship, even achieving fame with their symbolic and fictional critique of the violence and stagnation, such as Camilo José Cela in his novel *La familia de Pascual Duarte*. Others returned after a period of exile, such as the painter Salvador Dalí in 1948. Franco's rule offered a kind of cultural and national unity, stressing the traditions of the church and history and, above all, the common element of language. This was disseminated in journals and reviews and heavy monuments such as the great tombs of the Valley of the Fallen, which took 20 years to build. But censorship in all forms was the real hallmark of the regime. It was therefore in other cultural traditions that real vitality was shown, in flamenco dancing, in bullfighting and football. Franco himself was a fan of the football club Real Madrid and the greatest matador of all, Manolete, became a national legend after his death in the ring in 1947.

As with the economy, the liberalization that came after 1975 brought tremendous vigor in literature and art and the return of more exiles. The jumbled themes that had existed and partly existed before Franco's death could now be fully and fittingly expressed, but much the greatest energy went into political pamphlets and writings, and writing and film of an erotic or pornographic nature.

The Basques

The Basques have lived in the western Pyrenees since before Roman times, and they retain a distinct language and culture. Their long history of struggle with the Spanish unitary state concerns demands for self-government and has been both political and violent. More than two million Basques live in Spain, and 200,000 in France, in a mixture of rural communities and large industrial cities. Spanish Basques are sharply divided between richer, professional and middle classes and the increasingly impoverished workers. In 1895 a Basque Nationalist Party was formed. Hopes of political recognition during the second republic in 1931 were dashed by General Franco's victory in the civil war. The Basque lands were conquered in 1937, their language was prohibited and all autonomy removed.

However the Basques were among the wealthiest people in Spain's 50 provinces and this meant heavy taxation from Madrid. Basque nationalists began to pressure the government. Political groups were formed but troubled by divisions, often along class lines. In 1952 ETA (*Euzkadi ta Askatasuna* – Basque Homeland and Freedom) was formed and made ever more radical demands. The death at a roadblock of a young Basque in 1968 opened a period of direct action which included assassinations of police, officials and businessmen, and of a police chief, Manzanas. The trial of 11 Basque nationalists at Burgos in 1970 marked the start of a more violent campaign, met by the declaration of states of emergency in the Basque provinces, with more than 8,500 arrests between 1970 and 1979.

Spain's 1978 constitution introduced "pre-autonomy" for Basques and Catalans, and further transfer of powers came with the return of democracy which created a large measure of regional self-government. However, the nationalists were divided into a number of groups, some more socialist and others less radical, and did not obtain a majority of Basque votes; in 1979 there were 40 percent abstentions in the Basque provinces for the vote on autonomy. But disputes persisted over how the region should be governed and demands for more autonomy, especially in policing matters, continued, accompanied by violence. Serious divisions among nationalists thus followed the major achievement of Basque self-government in the new democracy.

▶ **Basque extremists in the 1980s.**

PORTUGAL

REPUBLIC OF PORTUGAL

CHRONOLOGY

1910	Monarchy overthrown
1932	Salazar governs under civil dictatorship
1966	Murder of General Delgado, opposition leader
1974	Democracy restored
1976	Constitutional government resumed
1982	Membership of the European Community

ESSENTIAL STATISTICS

Capital Lisbon

Population (1989) 10,372,000

Area 92,389 sq km

Population per sq km (1989) 112.3

GNP per capita (1987) US$2,890

CONSTITUTIONAL DATA

Constitution Parliamentary state with one legislative house (Assembly of the Republic)

Major international organizations UN; EC; EEC; I-ADB; NATO

Monetary unit 1 Escudo (Esc) = 100 centavos

Official language Portuguese

Major religion Roman Catholic

Overseas territories Macao

Heads of government (Prime minister) A. Baptista (19920–21); B. Machada (1921–22); A. da Silva (1922–23); A. de Castro (1923–25); V. Guimarães (Feb–Dec 1925); A. da Silva (1925–26);A. Carmona (1926–28); J. de Freitas (1928–30); D. de Oliveira (1930–32); A. Salazar (1932–68); M. Caetano (1968–74); A. Carlos (May–Jul 1974); V. Goncalves (1974–75); Adm. J. Pinheiro de Azevedo (1975–76); M. Soares (1976–78); N, da Costa (Aug–Oct 1978); C. Pinto (1978–79); M. Pintasiglia (Jul–Dec 1979); F. Carneiro (1979–80); D. do Amaral (Dec 1980); P. Balsemão (1980–82); M. Soares (1983–85); A. Sílva (1985–)

Portugal entered the 20th century with part of its former overseas empire intact in Africa, India and the Far East. However, change was threatened and the monarchy was under attack from those who wished to see more progressive policies and modernization of the economy and of political institutions. The monarchy was overthrown in 1910 but little change followed. Portugal did not participate in World War I, and when Spain adopted authoritarian government in 1923, demands for strong rule by an executive council took over the administration in Portugal too. By 1932 this council was controlled by Dr Antonio Salazar.

Salazar remained in power until ill health forced his retirement in 1968. His main concerns were to distance himself and Portugal from events in Spain, adhering to a strict neutrality during the Spanish Civil War and during World War II, and to try to retain the Portuguese colonies. These were both a source of revenue and somewhere to send frustrated, critical and unemployed young men and political opponents. It became increasingly difficult to maintain the empire, and when he retired from office even Angola and Mozambique, the least developed colonies, were experiencing organized insurgencies. Salazar's regime was not without its domestic opponents, but the most serious challenge, by General Delgado in 1966, was thwarted and Portugal kept good relations with Britain and the United States as a full member of NATO.

Salazar's successor as effective head of government from 1968, Professor Caetano, continued the same basic policies, but national wealth and resources were drained by the counterinsurgency operations in Southern Africa. Conscripts suffered long and difficult military service alongside regular troops in these lands far from their homes. The discontent and the demands for change, democracy and political accountability finally erupted into revolution in 1974. This spring revolt produced a flurry of demands and political groups but it was the army which led the revolution and became its guardian under the authority of General Spinola, a veteran of the colonial wars. The dictatorship was overthrown and, although there was little bloodshed, there were attacks on informers and the old secret police, the PIDES. Political extremists from a variety of leftwing groups, from Leninists to Maoists, started a critical debate about the development of the country and competed with a range of center parties. General Spinola attempted to retain a grip over events but demonstrations and uncertainty prevailed. Some lands and larger estates were appropriated by tenant farmers and landless workers, and Spinola felt obliged to resign after a few months.

A further military coup occurred in spring 1976, but this was followed by a revised constitution and more stable democracy. In subsequent elections parties of the center, socialists and Christian Democrats, gained control of the Portuguese government and the transition to constitutional democracy was successfully consolidated. The former colonial possessions became independent states.

The ending of dictatorship also allowed Portugal to apply for membership of the European Community and this was achieved in 1982. This change, which coincided with Spanish entry into the Community, meant Portugal no longer had anything to fear from its larger and more powerful neighbor and the future lay with the institutions and partners of the European Community.

Economic and social trends

The Portuguese economy for much of the century was based on weak industry, poorly organized landownership, vast estates often in the hands of absentee owners and stagnating production. Industry was found largely in the north of the country and the infrastructure was poorly maintained and often inadequate. The state survived on wealth from a few traditional investors, notably from Britain, and from its rich African colonies which produced important primary commodities. There was little incentive to change until these territories began to rebel and demand independence.

Economic and social change came with the 1974 revolution. The weakness of the economic base became apparent once the colonies were lost. Farming had to become more efficient and some lands were quickly bought by foreigners who introduced modern methods, but this meant a reduced demand for labor and increasing numbers of Portuguese emigrated to West European countries to seek work. However there was a slow expansion of industry and improvement in roads and other communications. More significantly, the growth of tourism brought immediate revenue and investment to the Algarve. A more sound and more far-sighted policy was membership of the European Community which brought development aid, and encouragement to foreign investment and manufacture.

Until the revolution of 1974 Portuguese society and its culture were static, based on traditional centers in the old universities and cities, such as Lisbon and Coimbra. The change in 1974 brought a sudden liberalization in many forms. There was a flowering of literature, much of it leftwing in tone and aimed at yet more dramatic change. This extended to areas unimaginable before 1974, such as the liberalization of the role of women. This was highlighted by the popularity of female writers, and even by Europe's first woman prime minister, Signora Pintasiglia.

▼Revolutionary mural in Lisbon, 1974.

FRANCE

FRENCH REPUBLIC

CHRONOLOGY

1947 First Cannes Film Festival

1954 French withdrawal from Indochina

1958 Military *coup* in Algiers

1968 Riots against the regime lead to general strike

1981 Socialist François Mitterrand elected president

ESSENTIAL STATISTICS

Capital Paris

Population (1989) 56,107,000

Area 543,965 sq km

Population per sq km (1989) 103.1

GNP per capita (1987) US$12,860

CONSTITUTIONAL DATA

Constitution Republic with two legislative houses (Senate; National Assembly)

Major international organizations UN; EC; SPC; EEC; I-ADB; NATO

Monetary unit 1 Franc (F) = 100 centimes

Official language French

Major religion Roman Catholic

Overseas territories French Guiana; French Polynesia; Guadeloupe; Martinique; Mayotte; New Caledonia; St. Pierre and Miquelon; Wallis and Futuna

Heads of government since 1900 (President) E. Loubet (1899–1906); A. Fallières (1906–13); R. Poincaré (1913–20); P. Deschanel (Jan–Sep 1920); A. Millerand (1920–24); G. Doumergue (1924–31); P. Doumer (1931–32); A. Lebrun (1932–40); (Prime minister) P. Pétain (1940–42); (Chief of State) P. Laval (1942–45); C. de Gaulle (1945–46); (President) V. Auriol (1947–54); R. Coty (1954–59); C. de Gaulle (1959–69); A. Poher (Apr–Jun 1969); G. Pompidou (1969–74); A. Poher (Apr–May 1974); V. Giscard d'Estaing (1974–81); F. Mitterrand (1981–)

After France was crushingly defeated at the hands of Prussia in 1870, the search for security and stability in domestic and international politics became the vital concern for French governments. Despite its colonial possessions, France was diplomatically almost totally isolated, apart from an alliance with the remote and cumbersome Russian Empire. The Third Republic, which was set up out of a civil war in Paris in 1871, was rocked by a series of crises and scandals affecting political institutions, the army and relations with the Roman Catholic Church. Revolutionary movements and anarchists threatened strikes and public disorder, and war with Britain loomed over colonial disputes. Yet shortly after the turn of the century the regime remained in place, strengthened by its very survival through the troubles; and a new understanding, the *Entente cordiale*, was established with the British in 1904. When war broke out in 1914 France faced the German Empire with Russian and British allies. Nevertheless, World War I was especially devastating for France, with much of the eastern region occupied for the duration, and such strains were imposed on men and materiel that serious mutinies broke out in 1917. British and American support prevented German victory and the appalling experience of the war encouraged prime minister Georges Clemenceau to insist on the humbling of Germany at the Peace of Paris, even at the cost of European stability. France obtained the return of the provinces Lorraine and Alsace, which had been lost in 1870, and the demilitarization of the German Rhineland. Clemenceau also insisted that Germany pay reparations for the damage done by the war.

A council member of the League of Nations with mandates in the Middle East, France seemed set for a major international role after 1919. However, this was not to be. Uncertain policies came with unstable governments. An electoral system based on proportional representation created coalition governments, shortlived and often weak in the face of constant motions of censure. The country was pulled increasingly toward the extremes, both to fervent nationalism and to the left, with the Communist party gaining strength. Civil disorder and a wave of assassinations in 1934 prevented consensus. In foreign policy a proposed treaty with the Soviet Union was abandoned, in favor of a proposed link with Hitler's Germany, by Pierre Laval in 1935. In April 1936 the French elected a coalition government of the left including Communists, known as the Popular Front. This created another change of direction but disagreements over social and economic policy and dislike of a foreign policy subservient to that of Britain brought the restoration of a center national government. This government cooperated with Britain, joining British prime minister, Chamberlain in acceding to the September 1938 Munich agreement with Hitler regarding Czechoslovakia, in a concerted attempt to avoid European war. A year later, though, France declared war on Germany after the invasion of Poland.

World War II proved to be another painful disaster for France. When the German attack came in May 1940, the French armies were defeated within six weeks. A provisional government signed an armistice which left a large part of the country occupied by the Germans and a puppet regime was installed at Vichy under the leadership of the veteran Marshal Pétain. Paris was occupied. When the Americans landed in North Africa in 1942 all of France was occupied by Germans and Italians.

However, not all French people accepted the armistice. In

June 1940 a former officer, Brigadier-General Charles de Gaulle, a longstanding opponent of the weak policies and lack of political foresight of the Republic since before the war, broadcast an appeal from London for as many French as possible to join him and the creation of the Free French. De Gaulle, tireless in the struggle against Germany and determined to ensure that France would be recognized as a Great Power and retain a voice in the future ordering of the world, was joined by French men and women from all parts of the Empire, notably Francophone Africa. His movement linked communists and priests, workers and intellectuals, soldiers and resistance fighters. He argued for France to such effect that he was accepted by Churchill and finally by the Americans (who had favored General Giraud, then in North Africa) as the head of a provisional government able to speak for France. Free French troops landed with the Allies in Normandy in June 1944 and fought for the liberation of their country.

France had again been destroyed by war. The infrastructure was in ruins after occupation, requisitions, bombings and finally by the fight for liberation itself. Although reconstruction through the United States-financed Marshall Aid scheme began in 1947, the new Fourth Republic suffered from similar difficulties to those of the prewar regime. De Gaulle gradually withdrew from political activity just as renewed war in France's colonies in Indo-China began to drain French resources. This war lasted until 1954 when the French were defeated at Dien Bien Phu by the revolutionary forces of General Giap and Ho Chi Minh, after which the prime minister, Pierre Mendès-France, negotiated a total withdrawal from the region. But the respite was brief. On 1 November the Algerian war of independence began; it lasted until 1962 and involved 500,000 troops. The fighting was bitter, vicious and brought tension with Tunisia and Morocco, only recently independent from France. Allegations abounded about the use of torture by the French, lack of successful and resolute policies and suspicions about furtive discussions between the antagonists as well as muddled direction; the controversy was such that France was again divided. The result was a military coup in Algiers in May 1958, demanding the recall of de Gaulle as president. He returned to head a provisional government and in October became the president of a Fifth Republic under a new constitution. By 1962 he had ended the Algerian war but not in the manner anticipated by many of his original supporters. Independence was conceded largely on Algerian terms, endorsed by 91 percent in a national referendum.

However, de Gaulle was unabashed and subordinated everything to his determination to restore strong government, success and dignity to France. The election of the president became direct in 1962 and elections to the national legislature by a constituency system gave more stable majorities, but presidential power was the truly new element. The prime minister was a presidential nominee, and presidential responsibilities included foreign policy and the integrity of the constitution. De Gaulle restored confidence in the economy,

▼ **A Gestapo informer is accused after liberation, 1945.**

supported a currency devaluation, took a prominent role in the early development of the European Community, and in 1963 cemented a close relationship with Federal Germany. Criticism of his dominance and lack of social policies grew after his reelection for a second presidential term and exploded in violence in the streets of Paris in May 1968, when students and workers threw up barricades against the police. Order was restored but de Gaulle recognized both the need for change and the decline in his own popularity. He placed a reform of local government before the people in a referendum in 1969, failed to gain an absolute vote of approval and resigned. His successors, regardless of political allegiance, followed many of his policies in international affairs. They upheld French independence, which de Gaulle had stressed, by withdrawing in 1966 from the military component of the North Atlantic Treaty Organization (of which France was a founder member), and retained a powerful military force and an independent nuclear capability.

Even the first socialist president, François Mitterrand, elected in 1981, changed little of this. However, domestic politics remained volatile and while the Communist party lost favor with the electorate, a far right National Front won electoral support during the 1980s. In 1986 the French tested their constitution further by electing a center-right majority and obliging the president to appoint a prime minister, Jacques Chirac, who had the majority's support. This interlude was brief and in 1988, with Mitterrand reelected as president, a socialist majority was returned to the National Assembly. Some constitutional changes were introduced to encourage decentralization and respond to regional demands, but by 1990 France had regained a strong international position and with its partners in the European Community had a healthy economy dedicated to gradual integration in Europe.

Economic and social trends

France remained essentially an agricultural country before 1914 and the ravages of war, failure to obtain adequate reparations from Germany and the general economic decline of the 1930s brought little change before World War II and its military occupation. France still had more than thirty percent of its 40 million inhabitants dependent on agriculture, living in small towns and villages dominated by peasant farming. But elements of change were at work. The French proved skilled engineers and were quick to move into and develop new industries such as aeronautics and motor vehicle manufacture. They were technical innovators, such as Peugeot, and adopted methods of mass-production very early in the factories of Renault and Citroën. France soon became internationally known for the production of high-quality and luxury goods.

Many structural changes happened after 1945. The infrastructure, almost totally destroyed during the war, was rebuilt and modernized with the help of American Marshall Aid. The industrial economy began to expand, stimulated by the needs of colonial wars in North Africa and the Far East, by the Korean War in 1950 and by French participation in the European Coal and Steel Community and the European Community (EC).

▲ The 1980s glass pyramid outside the Louvre, Paris.

▼ The Tour de France finishes in the *Champs Elysées*, Paris.

Despite the lure of agricultural farm support a drift into the towns and growing suburbs of Paris occurred. The dependence on agriculture diminished: by 1968 only 15 percent of the population lived off the land and estimates in the late 1980s suggested that this would be reduced to six percent by the end of the century. France became a major producer of nuclear power and was in the forefront of technology in this field for energy supplies and military use. French companies were innovators in aircraft and space industries and in road and rail communications. By the 1980s France possessed the most advanced, efficient and fast rail system of any country, in the *Train de Grande Vitesse* (TGV) and led Europe in missile and satellite communication, with the Ariane launcher.

Government led this advance into new production, mainly through the Planning Commissariat introduced in 1946. Most French banks had been nationalized in 1946 (the remaining few in 1981) and with this state-led initiative the French economy became competitive and sustained steady growth. The greater political stability that came with the Fifth Republic in 1958 contributed to this transformation of the economy and society. The population began to grow, assisted by prosperity and tax incentives, and became more mobile, accelerating the move into the cities, with the principal centers of population and economic growth around Paris, Lille, Toulouse, Bordeaux, Lyon and the Channel ports. By the late 20th century France had left behind the static rural world and was one of the world's largest industrial producers. However, food production remained substantial, providing all France's needs and dominating agricultural output in the European Community.

The rise in oil and commodity prices in the 1970s slowed economic growth and brought rising unemployment – some two million in the 1980s – but inflation was not high and manufacturing output remained steady, helped by consumer loyalty to domestic products and by the close links between state and industry. Government planning and investment encouraged industrial production and the state not only owned the banks but was a major holder in many key sectors, from aircraft to motor industries. Foreign relations, in particular with former colonies, were also developed with French exports in mind and Francophone Africa shared a currency that was tied to the French franc, the international value of which held steady after the devaluation of 1963.

Cultural trends

Throughout much of the century France remained in the forefront of European ideas and culture despite the political and economic difficulties, and Paris was always a center of the arts for French and foreign artists. France had long been associated with the visual arts and the post-Impressionist, Cubist and Modern movements in painting all flourished in Paris and in the south of France, seeking the light and imaginative world of the *Midi*, following in the footsteps of Cézanne and Van Gogh. French visual arts were given added importance by the arrival and later permanent residence of the Spaniard Pablo Picasso.

However, social and political stresses produced a special flowering of literature and philosophy. At the turn of the century this covered themes from the celebration of rural life in the works of Pagnol, to the exquisite depiction of Parisian society by Marcel Proust, to nationalist novels and reflective literature from Barrès, Charles Maurras and *Action française* to the socialist national pride found in the work of Charles Péguy. But much intellectual energy in the period 1930–60 was highly critical of the government and established interests, and this found inspiration in the work of a few leaders in the literary and intellectual world, Jean Genet, André Gide, André Malraux and Jean-Paul Sartre. Another group member, Simone de Beauvoir, led a new wave of feminist literature.

The social themes were also explored with particular genius by what the French called "the seventh art", cinema. Urban and rural life and the emotional drama that surrounded them became the focus of generations of French film directors. Many demonstrated fine comic quality. This outpouring of French imaginative and technical ability spanned the 1930s and 1940s, in the work of Jean Renoir and René Clair, but found its greatest exponents in the "new wave" of films and the tradition this created with directors such as François Truffaut, Jean-Luc Godard, Claude Chabrol and Jacques Tati.

Despite the social critique coming from the world of the arts and intellect, the French state encouraged and invested in the preservation and development of culture, the buildings, museums, sites of special interest, and the creation of prestigious schemes like the Pompidou Center for the arts in Paris, opened in 1977. Regional festivals also received public funding.

Corsica

The island of Corsica became the first part of metropolitan France to be given a limited form of self-government and special status as one of France's regions. This was approved in 1982 as a derogation from the 1958 constitution of the Fifth Republic. The island, consisting of two administrative departments, has a Corsican Assembly and regional executive responsible for matters such as tourism, culture, economic development, energy and communications. This change came partly as a result of pressure from Corsican nationalists.

The Corsicans, a Mediterranean people with their own language and tradition, were absorbed into the French state after the defeat of the nationalists led by Pascal Paoli in 1769. The Corsicans had briefly been independent, overthrowing their Genoese masters in a revolt that began in 1729. Although the British established a short rule on the island during the wars with France in 1794, the Corsicans became integrated with the rest of France and, like their most famous countryman Napoléon Bonaparte, made their living and fought and died for France. But the nationalist aspirations did not disappear even in the 20th century; rather they remained hidden during the struggle with fascism. Corsica, part of Vichy France, was occupied by the Italians in 1942 but was the first part of France to be liberated in 1943 – the word *maquis*, which came to symbolize the entire French resistance movement, is the word for the dense Corsican vegetation.

After 1945 demands for autonomy became stronger from Corsicans, as they did from Bretons and other nationalist minorities in France. Corsicans complained of lack of investment, colonialism applied by French governments, and the refusal to establish a university on the island or to recognize the language as a living minority tongue. The arrival of 18,000 *pieds noirs*, white French from North Africa, after Algerian independence in 1962 sharpened the tensions: nationalists felt that this group received preferential economic assistance, and that the government deliberately refused to take action when some *pieds noirs* became involved in fraudulent transactions in the local wine trade. In 1975 this issue led to riots and violence. Despite government concessions on education, language, financing, violence and bombings continued, many by the *Front de la Libération nationale corse* (FLNC). In 1982 Corsica became an autonomous region but the working of this administration was not easy. The island is poor, with only 247,000 inhabitants, and established political groups did not change while the nationalists suffered from divisions and continuing violence.

UNITED KINGDOM

UNITED KINGDOM OF GREAT BRITAIN AND NORTHERN IRELAND

CHRONOLOGY

1939	War declared over invasion of Poland
1956	Suez operation
1972	Northern Ireland's parliament suspended
1982	Falklands war with Argentina

ESSENTIAL STATISTICS

Capital London

Population (1989) 57,218,000

Area 244,110 sq km

Population per sq km (1989) 234.4

GNP per capita (1987) US$10,430

CONSTITUTIONAL DATA

Constitution Constitutional monarchy with two legislative houses (House of Lords; House of Commons)

Major international organizations UN; Commonwealth of Nations; EC; SPC; EEC; I-ADB; NATO

Monetary unit 1 pound sterling (£) = 100 new pence

Official language English

Major religion Christian

Overseas territories Anguilla; Ascension; Bermuda; Virgin Is.; Cayman Is.; Falkland Is.; Gibraltar; Hong Kong; Montserrat; Pitcairn; St. Helena; South Georgia; South Sandwich Is.; Tristan da Cunha; Turks and Caicos Is.

Heads of government (Prime minister) H. Asquith (1908–16); D. Lloyd-George (1916–22); A. Bonar Law (1922–23); J. R. MacDonald (Jan–Nov 1924); S. Baldwin (1924–29); J. R. MacDonald (1929–35); S. Baldwin (1935–37); N. Chamberlain (1937–40); W. S. Churchill (1940–45); C. Attlee (1945–51); W. S. Churchill (1951–55); A. Eden (1955–57); H. MacMillan (1957–63); A. Douglas-Home (1963–64); H. Wilson (1964–70); E. Heath (1970–74); H. Wilson (1974–76); J. Callaghan (1976–79); M. Thatcher (1979–90); J. Major (1990–)

The influence and apparent strength of the British Empire in 1900 extended to every continent and across all oceans. However, the long reign of Queen Victoria, from 1837–1901, and the vast colonies and dominions that provided support and materials, sheltered the British from realities that were to become ever more serious. The Liberal party came into office in 1905 with a sweeping electoral victory and faced the need to modernize and reform. They began this by intervening in social welfare, education and labor relations, with the cooperation of the recently formed Labour party. But international events raised fears of a clash with the German Empire, especially in Morocco in 1905 and again in 1911, and when colonial disputes with France were resolved Britain and France drew closer together, forming an *Entente cordiale* in 1904, followed by a similar agreement with the Russians in 1907. Therefore when general European war finally came in 1914 these three states found themselves in alliance against the Central Powers, Germany and Austria-Hungary.

The Liberal government had begun to improve the quality of British armaments but fighting was fierce and its scale immense. The war became concentrated in eastern France where extensive trench systems and the machine gun dominated battles such as that of the Somme in 1916, which cost 60,000 British casualties on the first day. Four years of war in Europe, in the Middle East and Africa and at sea, although ultimately victorious with American help, seriously impoverished Britain and exposed the structural weaknesses. After the war there seemed only political uncertainty, industrial and labor unrest. The longstanding issue of Irish unrest at British rule was apparently resolved, by an agreement in 1920 to grant independence to the south but not the north, but no party managed to find a secure majority until Stanley Baldwin's Conservative government in 1924. Britain had joined the new League of Nations and became a leading member, and British governments from that of the wartime prime minister David Lloyd George helped to bring Germany back from the humiliation of defeat to play a full part in European relations. Britain was a guarantor of the 1925 Locarno treaties between Germany and her neighbors, but relations with the Soviet Union remained remote and strained. However, continuing troubles in the old industries such as shipbuilding, coalmining and textiles defied easy solutions and even Labour governments in 1924 and 1929 failed to solve the economic and industrial disputes. A national coalition government was formed in 1931 after export markets and production had been further damaged by the Wall Street collapse of 1929. Policies were also weak and vacillating in the face of the European dictators; the government trusted to the League of Nations and to diplomatic agreements such as that made with Hitler over Czechoslovakia at Munich in 1938. The invasion of that state in March 1939 brought a change of policy and in September, when Germany invaded Poland, Britain declared war on Germany, despite having commenced rearmament only in 1937.

The war was more testing than that of 1914. After the defeat of France in 1940 Britain was threatened by a German invasion, which was averted by the air defense in the Battle of Britain, by naval protection of vital convoys from North America, and by the change of leadership when Winston Churchill became prime minister and inspired national resistance. He brought together dynamic and efficient politicians from all parties and backgrounds to extract the maximum from British resources. The turning point came in 1942, with Britain free from the threat of invasion, German forces defeated in North Africa and preparations underway for the Allied invasion of the European mainland. An Anglo-American-French force landed in the south of France and in Normandy in 1944. Almost a year later Germany surrendered and in August 1945 British troops fighting the Japanese also completed their successful defense of India, Burma and Southeast Asia.

Churchill promised only further efforts after the economic and human strain of war, and concern for the future was reflected in the defeat of his Conservative party in the general election of May 1945 and the victory of the Labour party. Many of the new government were experienced as members of the wartime coalition, notably the leader Clement Attlee who had been deputy prime minister. The Labour government tried to adjust to peace and rebuilding and to introduce socialist policies to Britain. The welfare state was created with a national health service, public involvement in education, nationalization of transport and some heavy industries between 1945 and 1951. Recovery was slow, food and other key commodities were still rationed for much of this period, but a positive effort was made to create stable institutions abroad. The United Kingdom gave independence to India, the heart of the old Empire, in 1947; became a permanent member of the United Nations Security Council; forged defensive agreements with France and the governments of Holland, Belgium and Luxembourg, an agreement that was to include Germany and Italy as the West European Union (WEU); and took the initiative in the foundation of NATO (North Atlantic Treaty Organization) which bound the United States and Canada to a united defence of democracy. Thus the Labour government established the principles for postwar British foreign and defence policy by associating European and North American interests and by investment in an independent nuclear deterrent. By 1960 many of the former colonial possessions had achieved their independence while remaining members of the Commonwealth, and only a few territories still looked to British rule and protection. An attempt to prevent loss of control of the Suez Canal involving France and Israel in 1956 was unsuccessful and

▼ **Upper-class relaxation in Edwardian England.**

Northern Ireland

The Government of Ireland Act 1920 created Northern Ireland with a separate executive and parliament at Stormont, a civil service and administration subsidized by Westminster. This part of Ireland, with about one million Protestants and Unionists, and half a million Catholics, was governed by an unchallenged Unionist majority. The south became the Irish Free State (later Eire). A military campaign by the Irish Republican Army (IRA) in the 1950s to overthrow the government in the North and create a united Ireland was foiled by Belfast and the Dublin government. In 1968, with Loyalist violence against Catholics, the "troubles" and rioting began again. The British government sent more troops, then imposed curfew and internment. By 1970 the communities were sharply divided, subject to coercion from extremists. Casualties mounted. In March 1972 self-government in Northern Ireland was suspended and direct rule from Westminster followed. The restructuring of the Royal Ulster Constabulary and a military security operation drastically reduced casualties after 1973 but nationalists and Loyalists continued to clash. By 1990 all propositions foundered on opposition from nationalists with some support from the Dublin government and from Unionists who rejected any proposal for Irish unity, seeing little advantage in being a minority in a predominantly Roman Catholic republic of united Ireland and preferring the union with Britain. The Anglo-Irish Agreement of 1985 associated the Dublin government with the administration of Northern Ireland, but this too was denounced by the Unionists who were not consulted. Meanwhile assassinations and violence continued, both in Northern Ireland and, in the case of the IRA, elsewhere, even on the European continent.

divided the country to such an extent that a reappraisal of British foreign interests followed. Many commitments "east of Suez" were eliminated or reduced and gradually a closer link was sought with European states. Initially this was through the European Free Trade Association (EFTA) but by 1963 British membership of the European Economic Community (EC) was a priority although it was not achieved until 1974.

During the 1960s greater prosperity and growth was apparent, but neither Labour nor Conservative governments managed to control inflation, curb trade union unrest, prevent growing unemployment and restore a solid economy. Strong executive authority was again possible in 1979 when the Conservatives broke the series of close election results. Their subtantial majority enabled a new leader, Margaret Thatcher, to attempt fresh policies in the search for economic recovery and defense of British interests overseas. Her government renegotiated the financial terms of British membership of the European Community, adhered to the basic principles of defense policy by maintaining a close association with the United States, wholehearted commitment to NATO and to Britain's nuclear capability. An attempt by Argentina to annex the Falkand Islands in 1982 was firmly resisted by war, but in other disputes diplomatic settlements ended difficult and ambiguous situations, notably with Spain over access to Gibraltar and with China over the return of Hong Kong when Britain's lease on the territory expires in 1997. The decade of the 1980s found the British relatively more prosperous, increasingly European in outlook with the building of a Channel Tunnel underway, and with a new emphasis on private initiative. The United Kingdom had completed the transition from imperial to European power without losing all influence with the Commonwealth or all of the "special relationship" with the USA.

Economic and social trends

Britain, the first industrial nation, faced great competition as the 19th century closed and newly emergent states such as the United States, Japan and Germany challenged the declining British Empire. However, this was not fully apparent until the war of 1914–18 exposed the weaknesses in the economy, although labor unrest, demands for improved conditions, greater welfare and pay had brought widespread strikes and litigation even before 1914.

Women had seen their place in society change dramatically during World War I when they worked in factories and other jobs. They obtained the vote in 1918, a recognition of their part in the war effort and of the strength of feeling shown in the women's suffrage movement before 1914; and more relaxed styles of behavior in the 1920s brought large numbers of women into employment. The effects of these changes and the structural defects of much industry were not removed by the first Labour government in 1924, and a strike in the coal mines in 1926 turned into a brief but damaging general strike.

World War II again imposed tremendous strains on the economy, although resources were efficiently mobilized for war and production reached record levels with full employment. The transition to peace, though, was difficult. Manufacturing relied on outmoded equipment and techniques. While management lacked the imagination and incentive to spend on research, development and marketing as many rival economies were doing, the workforce looked to the Labour government for better conditions and more pay. American aid helped the immediate recovery but the government invested heavily in an expanding public sector, in transport and welfare. As a slow improvement came through the 1950s and 1960s, British consumers turned to foreign goods, which were often cheaper and regarded as of better quality. Domestic farming had become more efficient during and after World War II, bringing a significant social transformation. Modern techniques and equipment meant a drop in the demand for farm labor, thus accelerating the drift into the cities.

These shifts in population, together with a steady increase in the total population from a postwar baby-boom and immigration from ex-colonies, and the economic improvement which came with a prosperous construction industry as towns expanded, preceded a consumer boom in the 1960s.

This change in society and economy helped to bring a structural change which was not destroyed by the oil and energy crisis of 1973, nor by the rise in commodity prices and resulting inflation fired additionally by constant demands for wage increases. Britain sustained the economy by the development of new resources, natural gas and oil in coastal waters which brought investment and tax revenue as well as support for the currency. London revived as an international finance center. By 1990 some sixty percent of the people owned their own homes and many had bought shares in the previously state-owned companies, which the government opened to private investors. However, this growth and wealth had penalties. It was concentrated in the southeast and around London, and was most prominent in energy resources and the financial sectors rather than in manufacturing.

Membership of the European Community from 1974 encouraged the redirecting of British trade toward Europe and away from former colonies, and it made British society more European in outlook. Sterling remained outside the European Exchange Rate Mechanism (ERM) until 1990 and misgivings remained about the speed of political and economic integration. Moreover, manufacturing output was still lower than that of many rivals and inflation was higher.

Cultural and scientific trends

Much of British culture was focused in London. However, one powerful British contribution to the creative arts had a wider basis: its deeply rooted literary tradition, still alive in the 20th century. British poets particularly retained their loyalty to the countryside. World War I produced its poets (Rupert Brooke and Wilfred Owen among the best known) and writers of war-based fiction, but after 1918 there was a renewed energy exploring modernity in writing and art. Much of this was linked to innovative personalities in London like Edith Sitwell and her "new poetry"; but the realism of declining Empire was found in the work of Joseph Conrad (Polish by birth but writing in English), its moral dilemmas were explored by G. K. Chesterton and H. G. Wells and a truly agonizing description and feeling of a nation uncertain of its role and culture imbued the poetry and novels of W. H. Auden, T. S. Eliot, Graham Greene and Evelyn Waugh. Thus the dominance of the literary form and the nation's decline characterized British culture at this level.

▲ The victory parade after the Falklands War of 1982.

◀ London burns during the German bombing (Blitz) of 1940.

IRELAND

REPUBLIC OF IRELAND

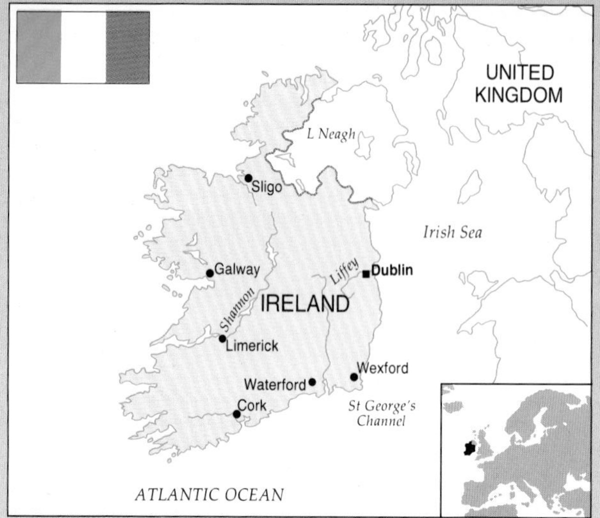

CHRONOLOGY

1905	Formation of Sinn Féin party
1919	Armed struggle for independence
1921	Partition of Ireland
1922	Irish Civil War
1925	Agreement on boundary with UK
1937	Establishment of Eire as an independent republic
1948	Republic of Ireland is created
1985	Anglo-Irish Agreement
1990	First socialist party member, Mary Robinson, elected to Presidency

ESSENTIAL STATISTICS

Capital Dublin

Population (1989) 3,515,000

Area 70,285 sq km

Population per sq km (1989) 51.0

GNP per capita (1987) US$6,030

CONSTITUTIONAL DATA

Constitution Unitary multiparty republic with two legislative houses (Senate; House of Representatives)

Date of independence 6 December 1921

Major international organizations UN; EC; EEC

Monetary unit 1 Irish pound (I£) = 100 new pence

Official languages Irish; English

Major religion Roman Catholic

Heads of government since independence (Prime minister) W. T. Cosgrave (1922–32); Éamon de Valéra (1932–48); John A. Costello (1948–51); Éamon de Valéra (1951–54); John A. Costello (1954–57); Éamon de Valéra (1957–59); Sean F. Lemass (1959–66); John M. Lynch (1966–73); Liam Cosgrave (1973–77); John M. Lynch (1977–79); Charles Haughey (1979–81); Garret Fitzgerald (1981–82); Charles Haughey (Mar–Dec 1982); Garret Fitzgerald (1982–87); Charles Haughey (1987–)

At the start of the 20th century Irish politics were in a state of turmoil. The largely Protestant, northern-based unionists wanted to remain within the United Kingdom. The mainly Catholic, rural-based republicans and nationalists who formed the party of Sinn Féin demanded independence.

The outbreak of World War I precipitated the crisis as republicans first opposed conscription into the British armed forces then, on Easter Monday 1916, led an abortive armed uprising in Dublin. The executions and imprisonments which followed deepened divisions and three years later Sinn Féin constituted a separate elected assembly, the Dáil Éirann. Two years of armed conflict persuaded Britain to partition the island. The six northern counties became the province of Northern Ireland within the United Kingdom, separated from the other 26 counties. In 1922 this arrangement was ratified by the Dáil, leading to the formation of the Irish Free State. Under Éamon de Valéra the Republicans opposed partition, however, and the new oath of allegiance to the British Crown, leading to a further year of civil war. Not until 1927 did de Valéra enter the assembly, but five years later his Fianna Fáil party gained power. He succeeded in breaking away further from Britain by introducing a new constitution in 1937, which replaced the Crown-appointed governor-general by a popularly elected president. The new state of Eire, which was neutral during World War II, lasted only ten years. In 1948 another constitutional change took Eire out of the Commonwealth, gave full control over external affairs to Dublin, and created the Republic of Ireland.

Political power in the new state was shared by the two main parties, Fianna Fáil and Fine Gael, the successor of the 1922

government. In 1973 it joined the European Community (EC) at the same time as Britain. Both major parties were obliged to turn to smaller parties such as Labour and the Progressive Democrats to form coalitions.

Relations with Britain were often strained over the issue of partition, though the Anglo-Irish Intergovernmental Council (1981) and the New Ireland Forum (1983–84) led to the signing of the Anglo-Irish Agreement in 1985. This established a permanent Cabinet-level conference between the two countries in which the Republic could express views and make proposals on Northern Ireland affairs. It also involved measures to coordinate cross-border security.

Social and economic trends

In 1900 the island's main industrial base was centered on Belfast, while the south was largely a supplier of cheap labor and food. In 1926, with Belfast now the capital of Northern Ireland, less than ten percent of the new state's workforce were in industry, while the 1903 Wyndham Act, which enabled landlords to sell to their tenants on mutually favorable terms, had created a large class of land-owning farmers. Denied free access to British markets until 1938, the government first protected selected industries to substitute for imports then, under de Valéra, introduced widespread quotas and tariffs in a policy of economic nationalism. The failure to provide enough manufacturing jobs to absorb surplus rural labor led to economic crises in the 1950s; in 1959 Ireland switched to the strategy of attracting foreign manufacturing companies. Membership of the EC stimulated the country's agricultural sector, which still in 1990 employed a quarter of the workforce, 80 percent of which was based on cattle and dairy products. Membership of the European Monetary System (EMS) in 1979 and the creation of EC-wide milk and beef surpluses soon cut farm revenues.

These policies rapidly changed Ireland from a poor rural country into an export-oriented manufacturing economy based on electronics, pharmaceuticals, mechanical engineering and food and drink. Nonetheless, Ireland remained the poorest country in northwestern Europe, its GDP only two-thirds the EC average.

Cultural trends

The new country formed in 1922 was a socially homogeneous, largely rural one; almost 70 percent of the population lived in rural areas. Over 90 percent of the population were Catholic, and religion formed a key element in the new national identity. In the first 20 years laws restricting contraception, abortion and divorce were added to the censorship of film and publications. The state desired a Gaelic cultural revival. Its Anglo-Irish heritage included internationally-known figures such as James Joyce, W. B. Yeats and G. B. Shaw.

The 1960s and 1970s were decades of liberalization and social diversity, involving reductions in censorship, laws protecting women employees and a shift towards bilingualism as official policy. By the 1980s only one percent of the population were native Irish speakers, though one-third could speak it owing to official encouragement.

▼ **Dublin, ravaged by fighting, Easter, 1916.**

BELGIUM

KINGDOM OF BELGIUM

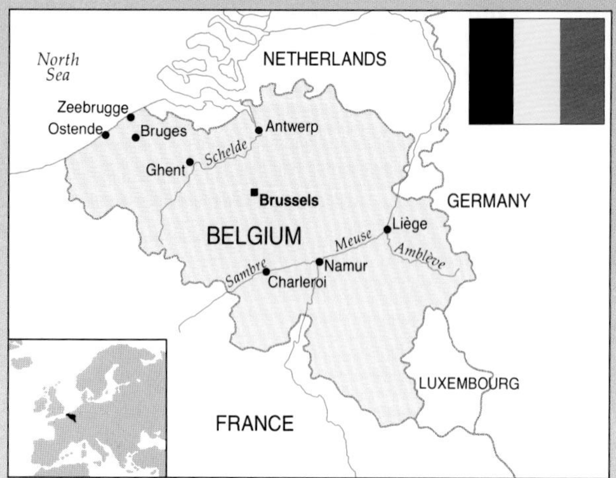

CHRONOLOGY

1908	Government takes over Congo region from King Leopold II
1914	Belgium is occupied by German troops
1940	King Leopold III orders surrender to German troops
1950	Referendum is held on "royal question"
1957	Belgium becomes a founder member of European Community
1962	Legislation officially divides Belgium into four language areas

ESSENTIAL STATISTICS

Capital Brussels

Population (1989) 9,878,000

Area 30,518 sq km

Population per sq km (1989) 323.7

GNP per capita (1987) US$11,360

CONSTITUTIONAL DATA

Constitution Constitutional monarchy with two legislative houses (Senate; House of Representatives)

Major international organizations UN; EC; EEC; I-ADB; IDB; NATO

Monetary unit 1 Belgian franc (BF) = 100 centimes

Official languages Dutch; French; German

Major religion Roman Catholic

Heads of government (Prime minister) C. de Broqueville (1911–18); G. Cooreman (Jun–Nov 1918); L. Delacroix (1918–20); H. de Wiart (1920–21); G. Theunis (1921–25); A. Vyvere (May 1925); P. Poulett (1925–26); H. Jaspar (1926–31); J. Renkin (1931–32); C. de Broqueville (1932–34); G. Theunis (1934–35); P. van Zeeland (1935–37); P. Janson (1937–38); P. Spaak (1938–39); H. Pierlot (1939–45); A. Van Acker (1945–46); P. H. Spaak (Mar 1946); A. Van Acker (Mar–Jul 1946); C. Huysmans (1946–47); P. H. Spaak (1947–49); G. Eyskens (1949–50); J. Duvieusart (Jun–Aug 1950); J. Pholien (1950–52); J. Van Houtte (1952–54); A. Van Acker (1954–58); G. Eyskens (1958–61); T. Lefevre (1961–65); P. Harmel (1965–66); P. Vanden Boeynants (1966–68); G. Eyskens (1968–72); E. Leburton (1973–74); L. Tindemans (1974–78); P. Vanden Boeynants (Nov–Dec 1978); W. Martens (1979–81); M. Eyskens (Apr–Sep 1981); W. Martens (1981–)

Belgium is a divided society. The region has been culturally divided since the Roman Empire, which produced a division between those areas where Romance languages still flourish and those where Germanic languages are spoken. In Belgium, the language boundary runs east to west roughly through the middle of the country, dividing it into a French-speaking area to the south known as Wallonia and a Dutch- (often called Flemish-) speaking area to the north known as Flanders.

What had been an informal boundary for centuries became official through legislation in 1962–63, which created four language areas. In addition to legal recognition of these two areas, Brussels was declared bilingual and a small German-speaking region was created. One can hardly speak of a Belgian identity. Even in 1990, most inhabitants thought of themselves as Flemings or Walloons first and Belgians second.

Belgium was created in 1831 as a neutral nation. Most of the country was nevertheless occupied by Germany in 1914 and the trench warfare on Belgian soil brought the population to the brink of starvation. The war also fueled the Flemish movement. The inability of French-speaking officers to communicate orders to Flemish-speaking troops caused needless suffering. Hostility between the language groups increased as some Flemish groups saw the German occupation as an opportunity for their movement.

The war brought several lasting effects to politics. The socialist Belgian Workers' party was no longer excluded from governmental participation. Following the war, universal male suffrage was introduced, helping the socialists to become the second largest party and the largest in Wallonia.

The period between the wars brought increased demands from the Flemish movement. The principle of territoriality, notably the exclusive use of language in an area, was first introduced. A Flemish university, but with a French-speaking section, was created in Ghent.

At the beginning of World War II, Belgium was again invaded by Germany, and conquered in only 18 days, in May 1940. King Leopold III personally ordered the surrender and ignored the request of the government to go into exile. His actions and his discussions with Hitler concerning the future of Belgium in a Nazi-dominated Europe disqualified him to reign in the eyes of many. Allied troops liberated Belgium in September 1944, and during the campaign the Germans moved the king to Austria.

After the war, the "royal question" became a major political issue. In 1950 a referendum was held in which 58 percent of the population approved of the king's return. However, only 48 percent of Brussels and 42 percent of Wallonia approved, and he was forced to abdicate in favor of his son Baudouin.

In the 1950s the "school question" returned to the forefront of politics. The question of state support to Catholic schools was an issue that had been important in the 19th century. After years of bickering, the "schools pact" (1958) provided for subsidies for Catholic schools, but parents were guaranteed the opportunity to choose between state and Catholic schools. Although the pact was intended to be permanent, it could be reviewed after 12 years, and the school question on occasion continued to be a problem.

By far the greatest political problem in recent decades has been the relationship between the language groups. Increased

Flemish demands produced a backlash and creation of a political movement in Wallonia. By the end of the 1960s not only had the older political parties split, but new parties representing the nationalist movements had emerged.

In 1970 the constitution was amended to provide new political institutions to help reduce conflict between the language groups. The linguistic boundaries were given a constitutional basis. Recognition was given to the "cultural communities" (Francophone, Flemish, and German-speaking), and political institutions were created to provide these communities greater opportunities for self-determination. The "cultural councils" were given legislative powers in matters related to the cultural life of the community. Because Brussels is a bilingual area within Flanders, three regions – Flanders, Wallonia, and Brussels-Capital – were created.

The 1970s and 1980s saw the development of Belgium toward a federal structure. Constitutional revisions in 1980 and 1988, together with subsequent legislation, expanded the powers for the communities and regions. Most educational matters were placed in the hands of the communities, which also had powers in cultural affairs. The regions had responsibility for ten major areas: town and country planning, environmental protection, nature conservation, housing, water policy, energy policy, employment policy, and certain aspects of public works, transportation, and the economy. Budgets for these bodies were increased accordingly. Additional reforms were planned for the 1990s: the senate was to be reformed along federal lines and members of the cultural and regional councils elected directly.

Economic and social trends

Belgium was affected early by the industrial revolution. Major industrial centers for coal and steel production, textiles, zinc, and glass developed, mainly in Wallonia, as Flanders remained more predominantly agricultural. However, Walloon industries were hit hard by the two wars and the depression. After World War II, Flemish industry began to catch up. New prosperity in Flanders helped to increase the friction between the two regions.

Although not a traditional colonial power, Belgium had acquired the Congo through the personal initiative of King Leopold II. The king became the sovereign ruler of the Congo in 1884, but increasing financial problems brought a government takeover of the region in 1908. Raw materials and wealth from the Congo helped the country recover from the devastation of World War I. After World War II, anticolonial movements forced independence in 1960. However, subsequent violence and civil war led to military intervention by Belgium and the United Nations.

After World War II, Belgium was at the forefront of movements toward European integration. It linked with Luxembourg and The Netherlands, and was a founding member of the European Coal and Steel Community (ECSC) and of the European Economic Community. Brussels is now the headquarters of the executive of the latter, and so attracts a large population of civil servants and officers from all the countries of the Community.

▼ **Linguistic issues arouse passion in the 1970s.**

NETHERLANDS

KINGDOM OF THE NETHERLANDS

CHRONOLOGY

1917	"Great Compromise" provides state support for Catholic and Calvinist schools, and male suffrage for over-25s
1920	Women over 25 are granted the vote
1944	"Hunger Winter", 15,000 die in food shortage
1947	First Dutch "police action" carried out in Indonesia
1949	Sovereignty handed over to Indonesia after Round Table Conference
1973	Dutch economy suffers over oil boycott
1979	Unemployment reaches 17 percent

ESSENTIAL STATISTICS

Capital Amsterdam

Population (1989) 14,846,000

Area 41,863 sq km

Population per sq km (1989) 434.4

GNP per capita (1987) US$11,860

CONSTITUTIONAL DATA

Constitution Constitutional monarchy with one parliament (States General) comprising two legislative houses (First Chamber; Second Chamber)

Major international organizations UN; EC; EEC; I-ADB; NATO

Monetary unit 1 Netherlands guilder (f.) = 100 cents

Official language Dutch

Major religion Roman Catholic

Overseas territories Aruba; Netherlands Antilles

Heads of government (Prime minister) N. G. Pierson (1897–1901); A. Kuyper (1901–05); T. H. de Meester (1905–08); T. Heemskerk (1908–13); P. van der Linden (1913–18); R. de Beerenbrouck (1918–25); H. Colijn (1925–26); D. de Geer (1926–29); R. de Beerenbrouck (1929–33); H. Colijn (1933–39); D. de Geer (1939–40); P. Gerbrandy (1941–45); W. Schermerhorn (1945–46); L. Beel (1946–48); W. Drees (1948–58); L. Beel (1958–59); J. de Quay (1959–63); V. Marijnen (1963–65); J. Cals (1965–66); J. Zijlstra (1966–67); P. de Jong (1967–71);B. Biesheuvel (1971–73); J. den Uyl (1973–77); A. van Agt (1977–82); R. Lubbers (1982–)

As the 20th century began, two major problems dominated the Dutch political agenda. Both were closely associated with the development of the party political system. One of these was the "social question", associated with the rise of Social Democratic Workers' parties in virtually all of Europe. This struggle focused not only upon the working and living conditions of the rising working class, but with expansion of the franchise.

The second major problem was the "school question". During the previous century, fundamentalist Calvinist groups had, together with Catholics, pushed for state support of religious schools. Liberals and Social Democrats favored state schools.

In 1916-17 a historic compromise on both issues was reached in a parliamentary committee composed of all political parties. This Great Compromise provided that Catholic and Calvinist schools would be funded at the same level as it cost the state to run its own schools.

Completing the agreement was the decision to extend suffrage to all male citizens above the age of 25. Women were extended this right in 1920. The electoral system was changed from a district system to proportional representation, and to ensure that proportionality correctly reflected the desires of the electorate, compulsory attendance at the polls was introduced.

The arrival of universal suffrage did not have the impact that the social democrats, in particular, had expected. Rather than achieving the hoped-for majority, the Social Democratic Workers' party gained no more than about one-fourth of the vote. The Catholic party was particularly successful in

appealing to all social classes and received the vote of more than eighty percent of all Catholics. The Anti-Revolutionary party was equally successful in obtaining the vote of those orthodox Calvinists organized in the "Gereformeerde" churches. The Christian Historical Union received most of its voters from the more devout members of the Dutch Reformed Church. After having been dominant during periods of the previous century, the liberals were now reduced to a much smaller role.

For 50 years after the Great Compromise, these five parties dominated Dutch politics. The three religious parties generally held a majority, but preferred to form a coalition with either the socialists on the left or the liberals on the right. The particular Dutch brand of democracy, with emphasis on autonomy within the societal segments and compromise between them has been called "consociational democracy".

In the 1960s, however, the Catholics and Christian Historicals began to lose votes dramatically and eventually the three religious parties merged into a single Christian Democratic Appeal.

In the 1990s, the following parties (ordered roughly from left to right) were represented in parliament: Green Left, Labor party, Democrats 66, Christian Democratic Appeal, People's Party for Freedom and Democracy (generally referred to as the Liberal party), three small orthodox Calvinist religious parties, and the Center Democratic party. Although they held fewer seats than in the past, the Christian Democrats still controlled the process of government formation and in 1991 rule in coalition with the Labor party.

One of the major political problems of the 20th century was the colonial problem. By far the most important colonial possession was Indonesia, but the Dutch also held the Caribbean islands of the Netherlands Antilles and the small colony of Surinam on the coast of South America.

After World War II, Indonesian nationalist attempts to set up an independent state led to violent conflicts. Two Dutch "police actions" were carried out, in 1947 and 1948. These produced protests in the United Nations Security Council, which led to a Round Table Conference and the handing over of sovereignty to the United States of Indonesia in 1949.

Surinam was granted independence in 1975. The Dutch continue to show interest and concern for the developments in this new nation, in part because of some 200,000 Surinamers who opted for Dutch citizenship at the time of independence, or emigrated to the Netherlands. The Netherlands Antilles continued to be a part of the kingdom of the Netherlands.

In its relations with other countries it has been said that the Netherlands' foreign policy was guided by concern for "peace, profits, and principles". At the beginning of the century, the country had a policy of neutrality which it managed to sustain throughout World War I. After the war, the former German Kaiser was even allowed to settle in the country. Since the 17th century, the Dutch have had great interest in international law, and it was therefore appropriate that the International Court of Justice was founded in The Hague, in 1946.

Dutch hopes to remain neutral in World War II were crushed by the German invasion of May 1940. The country suffered greatly during the occupation despite an active resistance movement. After the war, the Netherlands altered its position of neutrality and became a member of NATO. Through the early 1960s it was an extremely loyal member, but in part because of the influence of Vietnam, it began to take a more critical stand and by the early 1980s was being accused of "hollanditis" – a leaning again toward neutralism.

The Dutch were in the forefront in European integration, for example as founding members of the European Community. Public and governmental support for integration has been among the strongest in Europe. The Dutch also played an important role in United Nations' activities in the area of human rights and are strong supporters of organizations such as Amnesty International.

Economic trends

The Industrial Revolution was rather late in arriving in the Netherlands. However, during the 20th century the transition from a predominantly agricultural society to a major industrial power was completed. Major multinational companies such as Phillips, Akso, Unilever, and Royal Dutch Shell are Dutch firms or Dutch–British joint ventures.

At the beginning of the century there was still concern that the country should be agriculturally independent. To help achieve this the most ambitious reclamation project in Dutch history was planned – the draining of the Zuyder Zee. This project took most of the century, but may never be completed according to the original plans because of changes in priorities. Membership of the European Community made possible the importation of food products from other member countries.

◀ **Queen Beatrix, like all of the Dutch royal family, is very accessible to her people.**

The economic crisis of the 1930s and the German occupation of 1940–45 left Dutch industry in a sorry state. Many machines and transportation facilities had been either destroyed or dismantled and shipped to Germany. To aid economic recovery the government took stern measures to restore the currency and to control wages and prices. Marshall Plan aid from the United States provided the necessary funds for reconstruction. In order to prevent social unrest a Labor Council was set up in which workers, employers and the government were represented. In this body, and in the Social-Economic Council, these "social partners" still cooperate to produce a healthy economic climate.

By the 1960s the recovery had been so successful that stern measures were no longer possible. Wages rose and the Netherlands entered a period of great prosperity. Rotterdam developed into the greatest port in the world and because of its location at the mouth of the Rhine the Netherlands again played a major role in transportation and trade. Prosperity was aided by the discovery of natural gas in the north.

During the 1950s and 1960s the Netherlands developed social welfare programs among the most extensive in the world with a collective sector second only to that of Sweden.

Rising wages and the extension of the welfare state finally produced problems in the 1970s. The economy received a setback from the oil boycott of 1973, but suffered even more after 1979. Government payments for social welfare programs had become almost unbearable. Moreover, unemployment reached intolerable heights (up to 17 percent). The government budgetary financing deficit became far too high.

The decade of the 1980s became therefore one in which attempts were made to reduce unemployment and government spending. Some of the measures taken were quite drastic, but as the last decade of the century began, they seemed to have had sufficient effect to make the future again appear brighter.

Social and cultural trends

Since the 17th century the Netherlands has been divided, socially and culturally, by religion. Nationalism and Protestantism combined to produce the revolution that provided independence and the Union of Utrecht for the northern provinces in 1579. By 1648, the provinces of Limburg and Brabant had been added to the Republic, but here the Counter-Reformation had been successful. The Netherlands emerged divided by religion – almost exclusively Catholic to the south, but heavily Protestant to the north. In the census of 1899, 35 percent listed themselves as Catholic, 48 percent as Dutch Reformed, and 8 percent as "Gereformeerd" (orthodox Calvinist).

Although Catholicism was never forbidden, restrictions were placed upon the degree to which it could be practiced and Catholics were subjected to various forms of social and political discrimination. In the 19th century a movement began to emancipate Catholics from their position as second-class citizens. With the accession to power of liberals, the church hierarchy was reestablished in 1853.

Within the Dutch Reformed Church, certain members also felt themselves to be the object of forms of discrimination. The "Kleine Luyden" (literally "small people", referring roughly to the class of shopkeepers and clerks), who were more orthodox Calvinist in their beliefs also began a movement for social and political emancipation. Their great leader, Abraham Kuyper, organized a new church, the "Gereformeerde Kerk", a newspaper, a university, and the Anti-Revolutionary party, which opposed the idea, which had expressed itself bloodily in the events of the French Revolution, that sovereignty came from the people rather than from God.

This pattern of establishing organizations specifically oriented toward the members of the group became the pattern for Dutch social life. The Calvinist, Catholic, and socialist workers' movements all provided organizations for every possible aspect of social and cultural life, including their own political party, trade union, mass media, healthcare organizations, youth groups and sports clubs. Calvinist and Catholic schools and universities were set up. The various organizations within the group were themselves generally tied together by "interlocking directorates" in which members of the elite served on the boards of more than one organization. Contact between members of groups was thus often quite limited.

This pattern prevailed throughout the first half of the 20th century. Then, for reasons that remain somewhat obscure, in the 1960s the system began to break down. Many organizations became more independent and in some cases even merged with sister organizations from other groups.

In order for the segments to live together peacefully, tolerance has always been necessary. Today this can be seen not only in the attitudes of the Dutch toward others, but in policies toward drugs, prostitution, and pornography and society's openness on issues such as abortion and euthanasia. The Dutch tend to adopt quite pragmatic solutions to problems and avoid the complications of rigid ideologies.

This freedom and tolerance also produced a culture in which the arts have flourished. Dutch painters have been famous for centuries; in the 20th century this tradition has been continued by such world-famous artists as Piet Mondrian and Karel Appel. Architects such as Berlage have had considerable influence. In the performing arts, the Amsterdam Concertgebouw orchestra, Rotterdam Philharmonic, Nationale Ballet and Nederlands Dans Theater have gained international recognition. The arts have received large subsidies from the government. One of the most controversial programs was one in which visual artists were provided a level of fixed income in return for a contribution of their work.

Because of the inaccessibility of the language to most foreigners, Dutch literary and theatrical figures have received less international recognition than exponents of the other arts. Nonetheless, these areas are full of vitality and of great importance within the country.

Among the most popular sports in the Netherlands are soccer, cycling and speed skating. In addition to famous sports heroes such as footballers Johan Cruyff, Marco van Basten, Ruud Gullit, Joop Zoetemelk and Ard Schenk, the Dutch have also produced champions such as Fanny Blankers-Koen in athletics, Anton Geesink in judo, and Max Euwe in chess. Dutch field hockey teams have won Olympic and world championship gold medals.

▲ The Dutch "way with water-courses" shows in this delta plan, in Zeeland.

GERMANY

FEDERAL REPUBLIC OF GERMANY

CHRONOLOGY

1918	Revolution in Kiel and Berlin. Republic declared
1926	Treaty of friendship with USSR
1933	Hitler is appointed Chancellor
1935	Nuremberg Laws deprived Jewish community of rights
1939	Germany invades Czechoslovakia and Poland
1949	Formation of Federal Republic on Basic Law, and of German Democratic Republic
1953	East Berlin rising
1990	Unification of East and West Germany

ESSENTIAL STATISTICS

Capital Berlin

Population (1990) 77,744,000

Area 357,042 sq km

Population per sq km (1990) 227

GNP per capita (1988) E. Germany US$12,430; (1987) W. Germany US$14,460

CONSTITUTIONAL DATA

Constitution Federal multiparty republic with two legislative houses (Federal Council; Federal Diet)

Major international organizations UN; EC; EEC; I-ADB; NATO

Monetary unit 1 Deutsche Mark (DM) = 100 Pfennige

Official language German

Major religion Christian

Heads of government (Emperor) Wilhelm II (1888–1918); (President) F. Ebert (1919–25); P. Von Hindenburg (1925–34); (Chancellor and Führer) A. Hitler (1934–45); Allied occupation (1945–49); (East) (President) W. Pieck (1949–60); (Chairman of the Council of State) W. Ulbricht (1960–73); W. Stoph (1973–76); E. Honecker (1976–89); M. Gerlach (1989–90); (West) (Chancellor) K. Adenauer (1949–63); L. Erhard (1963–66); K. Kiesinger (1966–69); W. Brandt (1969–74); H. Schmidt (1974–82); H. Kohl (1982–90); (United) H. Kohl (1990–)

The German Empire, which had been created after the defeat of France by Prussia in 1870 and the annexation of the provinces of Alsace and Lorraine, became an influential state with a fast-growing population, healthy industrial economy and well-trained and formidable army. Kaiser Wilhelm II was keen to find and develop colonial territory, especially in Africa, and to create a navy capable of matching that of Britain with which he had a strong sense of rivalry. This sense of competition between Germany and both Britain and France prevented an improvement in relations between these states, and tensions between them rose with every international crisis especially those which involved the German presence in Morocco and Turkey, between 1905 and 1913. These tensions exploded into war in 1914 when the German Empire was drawn into the European crisis, initiated by a local conflict between Serbia and the Austro-Hungarian Empire, Germany's ally.

The strategic position of Germany exposed the Empire to attack from the east and the west simultaneously. The long-prepared response – the Schlieffen plan – depended on a rapid defeat of France to the west before facing Russian armies in the east, thus preventing German "war on two fronts". But in

September 1914 a defense on the river Marne by French and British troops prevented a quick victory in the west and ended in a static war in trenches along France's northeastern frontier, while the Russian forces were already engaged by Germany in the east. The Germans were unable to break a naval blockade by the Allies despite a naval engagement with the British at Jutland (1916) and submarine warfare in the Atlantic. When the Russian Bolsheviks negotiated a ceasefire in the east at Brest-Litovsk on terms favorable to Germany in March 1918 it seemed that Germany might yet snatch victory. Their armies were undefeated in Italy and in East Africa, and a fine military command under generals Ludendorff and Hindenburg suggested further success; but diminishing food and oil supplies and the arrival of American troops to help the British and French halted the 1918 spring offensive in the west. By the beginning of November 1918 the German war effort was all but exhausted.

The failure to win the war brought anger and disillusion. Excited by events in Russia, Marxist revolutionaries threatened revolt in Germany, first in a naval mutiny at Kiel and then in Berlin in November 1918 when socialist rebels attempted a coup. The Kaiser abdicated two days before the armistice (11 November), which was widely regarded in Germany as "a stab in the back". The revolutionaries of the Spartakist movement in Berlin were crushed, their leaders murdered. Out of this confusion a new republic emerged with frontiers much reduced after the Treaty of Versailles, which also imposed severe military restrictions on German armed forces. Former German lands were given to Poland, Lithuania and Czechoslovakia and returned to France, Belgium and Denmark. Social Democrats were initially the most successful political group in the new Weimar Republic, so called after the town where the Constituent Assembly met. The regime survived despite plotting from Communists and an emerging right wing between 1920 and 1924. Massive reparations were imposed by the Versailles Treaty which could not be paid; they gave rise to hyperinflation in 1922, and a French occupation of the Ruhr industrial region in 1923. A new currency, flourishing political parties and proportional representation created a lively political debate, and international frontier agreements made in the Locarno Treaties of 1925 contributed to growing stability. But in 1929 Gustav Stresemann, an able and respected politician, died suddenly and the economy was shortly afterwards ruined by the Wall Street collapse. Production and employment suffered immediately and this accentuated the political polarization between the Communists and the new Nazi party, led by Adolf Hitler. The Nazis, officially the NSDAP, won a landslide victory in 1930 and became the largest party in the parliament, the Reichstag, in 1932. Failure to control violence between Communists and Nazis and to tackle the economic difficulties destroyed government after government until, in March 1933, president Hindenburg invited Hitler to become Chancellor or head of government. More violence accompanied this move, including the burning of the Reichstag building. By the end of the year other political parties had been abolished, opponents eliminated and the totalitarian state was in place, with full

▲ Germany's heavy industry before World War I.

◀ Hitler laid emphasis on his appeal to youth.

"enabling" powers given to Hitler, the nation's *Führer* (leader). The first anti-Jewish race laws were passed in 1934, a policy that led to concentration camps, and from 1941, to the death camps and gas chambers.

Not all Nazi policies were repressive, and employment and production increased as a result of road building programs for the first *autobahnen*, and investment in housing, new forms of transport such as the airship, and motor vehicles including the *Volkswagen* or "people's car". A rearmament program also assisted economic recovery. But in 1936 Hitler moved to reclaim German lands lost since 1919. The Rhineland was reoccupied without resistance and in 1938 Austria was annexed to Germany. Then, despite the Munich agreement made with Britain and France in 1938, Hitler's troops moved into Czechoslovakia in 1939 as they did in Memel. A nonaggression pact between Germany and the Soviet Union in August 1939 allowed the invasion of Poland but Britain and France declared war on Germany on 3 September 1939. Initial German military success in Europe and North Africa, including the defeat of France in six weeks in May and June 1940, made Hitler master of Europe from the Atlantic to the Soviet Union, although British air defense prevented invasion across the Channel. Further advances in Greece followed, and in June 1941 the Germans attacked their former ally Russia without warning, reaching the suburbs of Moscow. Fortunes changed in 1942 with both Russians and Americans now in the war against Germany. German troops in the east were defeated at Stalingrad in 1942–43 and in North Africa, while Hitler faced continental war in Italy from 1943 and in northern France from 1944. Late attempts to snatch victory by novel means, using rockets and a bold counter-attack in the Ardennes, all failed to avert the total and ignominious defeat of Germany. Hitler committed suicide in his bunker in April 1945, and Germany's unconditional surrender was signed on 8 May 1945.

Institutions were destroyed, the economy and society were in tatters and a completely new period of German politics began with military occupation by the victorious powers, the Soviet Union, the United States, Britain and France. Total defeat and disruption required a new political system. Moreover the frontiers were again reduced and all traces of Hitler's Third Reich were to be erased by denazification. But the costs of occupation were heavy: the Americans, British and French amalgamated their zones in 1947, including their sectors of the capital city, Berlin, which had been divided among the victors. The decision to introduce a new currency, the Deutsche mark, ensured that the Western zones and West Berlin would be part of the free market system, while the Soviet zone was tied to the Communist closed economic system. An attempt to blockade West Berlin by the Soviets failed in 1948–49 and the division of Germany was complete. In 1949 two new states were created. The Federal Republic, or West Germany, was based on the Basic Law, a new constitution; and the Soviet zone became the German Democratic Republic or East Germany. Berlin remained under military administration. By 1955 the two Germanies were militarized and members respectively of NATO and the Warsaw Pact.

The Federal Republic, with strong regional or *Land* institutions and a weak executive, became politically stable, developing a party system based on a mixture of proportional representation and first-past-the-post elections which prevented a proliferation of parties and yet was responsible to changes of opinion. The shaping of these institutions and new practices was the achievement of the German political leaders, notably the first president, Theodore Heuss, and the Christian Democrat Konrad Adenauer. Many decisions were made at regional level by the governments of the eleven *Länder*, who also had a voice in the federal system through a second chamber, the *Bundesrat*, created as a check on the federal government by the regions. The president held few political powers and was indirectly elected; and the whole structure was subject to the jurisdiction of a Constitutional Court. Further stability was ensured by membership of the European Community (EC) and the West European Union (WEU).

The Democratic Republic also acquired new institutions based on a centralized system dominated by the communist SED, Socialist Unity Party. The links with the Soviet Union were close but substantial support came from West Germany both through government and commercial aid and from individual family support. Prosperity in the Federal Republic came to be so great that many thousands fled across the borders from the East. Many came through Berlin, where access was easier, until 1961 when the Eastern authorities surrounded the Western part of the city by a defended wall. Relations between the two Germanies and between Federal Germany and the Soviet Union improved by 1970, encouraged by the Social Democratic chancellor and former mayor of West Berlin, Willi Brandt. Soviet president Mikhail Gorbachev's program of *perestroika* (economic liberalization) brought rapid developments in the Democratic Republic. Demonstrations for change in politics and economics were the first expressions of general public demands for German unification. These demands came in the major cities of Dresden, Leipzig and Berlin in 1988 and 1989, and ended in the breaching of the Berlin Wall in November 1989. German unification occurred within a year, unopposed by the East German authorities. Monetary union came in July 1990 and gradually all the services and running of the German Democratic Republic passed to the Federal Republic's administration based in Bonn.

◀ **The Berlin Reichstag, captured by Soviet troops in 1945.**

Economic and social trends

Germany entered World War I as one of the most powerful states in the world. However the war proved very damaging. Economic blockade caused shortages of many supplies, but the worst effects of war, the bitterness and suffering, came after 1918 when many Germans were cut off from their compatriots as a result of the redefinition of the frontiers. Germany was burdened with heavy reparations and yet was discouraged from expanding industry in order to raise the revenue necessary for payments. The outcome was inflation of monumental proportions in 1922–23, an experience not forgotten among Germans of later generations. The French occupied the industrial area of the Ruhr in 1923 in an attempt to enforce payment of reparations, but were met by noncooperation and a strike. The French withdrawal was followed by American financial

Divided Berlin

Berlin, largest industrial city in Germany, former capital of Prussia, the German Empire and the Third Reich, was heavily bombed during World War II. Almost 75 percent of the buildings were destroyed by bombs and shelling and many of the rest were damaged, but the symbolic value of the city was such that the victorious powers agreed to divide Berlin into four sectors, even though the city was 190 kilometers (120 miles) inside the Soviet-occupied zone of Germany. It was administered by an Allied Commandatura, making the four military commanders the highest authority.

When tension grew between the Soviet Union and the Western Allies, the Russians stopped attending meetings of the Commandatura. With the introduction of the Deutschemark into the three linked sectors of West Berlin in 1948, the Russians cut all land communications along the approved corridors between West Berlin and the Western zones of Germany. The response was an airlift supplying West Berliners. The blockade was lifted in 1949.

The commitment of the Allies to West Berlin had been demonstrated. When the Federal Republic was founded, West Berlin was given a separate regional (*Land*) government, but could not enjoy the same constitutional freedom as other *Länder* because its formal status was still that of military occupation to abandon this would invite the Russians to demand the Allies leave the city.

West Berliners enjoyed a growing prosperity which attracted thousands of East Germans to West Berlin, a flow that began in 1945 and became swifter after a rising in East Berlin was put down by Soviet tanks in June 1953. Eventually, in 1961, the East German government built a wall along the inner-city border. This stopped the flow of refugees and began a tense period of inertia until the 1971 Four-Power Agreement regularized communications within the city but reaffirmed the partition. Thus things remained until November 1989 when the wall was removed by the East German government itself. In 1990 the city was again united and the wall, once an invitation for amateur artists on the Western side to display their talents, had largely disappeared.

▼ **Berlin celebrates reunification in 1990.**

underwriting of Germany's economy in 1924. This agreement produced an easing of economic and social stress. Reissued currency, work and a freely working democracy created an excited and enthusiastic mood as the Germans rebuilt their state. This was shattered by the international monetary crisis of 1929 which resulted in a rapid rise in unemployment and falling production. By 1933 unemployment reached six million and exports slumped.

Hitler's Third Reich changed economic and social conditions in Germany in response to this collapse. New demand was created by expansion in the armament industries and by investment in prestigious projects from public building to roads.

By 1942 the strains of war were draining an economy which lacked oil and needed open sea lanes for supply. Bombing destroyed production and the Soviet counteroffensives deprived Germany of the food production of Eastern Europe after 1943. Ore continued to reach Germany from neutral states such as Sweden. Workers were found among defeated peoples to rebuild and sustain output. But this could not last and when the war ended the German economy was totally destroyed. The cities and factories were heaps of rubble, the infrastructure almost nonexistent and society disrupted, with many Germans removed from their homes by war. Germany had virtually ceased to exist.

After 1945 Germany was rebuilt – through an "economic miracle" in the Federal Republic and economic and social reconstruction in East Germany. In the West the economy was restored by rebuilding the cities out of piles of bricks. The program was helped by American aid and by the effort of other occupying powers, but mainly by a completely new fiscal and economic structure. New currency was issued in 1948 and a graduated taxation system adopted that encouraged the growth of income. A trade-union system based on 16 main craft unions was established – Hitler had destroyed the previous system – and this allowed regular and orderly negotiation of national wages. The economic reconstruction of the Federal Republic included a clear policy: the creation of a social market economy, the free market and convertible currency controlled by government to protect the weak and prevent market distortion or corruption. This successful economy was enhanced by German participation in the European Coal and Steel Community (ECSC) and the Common Market (EC). The result was great prosperity for West Germans, a steady rate of annual growth, low inflation and interest rates, and an ever-strengthening currency. Until the mild recession of the 1980s, unemployment was not serious but it grew and continued to do so with the labor surplus coming from German unification, and migration of workers from east to west, in 1990. In earlier years the demand for labor had brought thousands of "guest workers" to Germany, many of them Turks but also Portuguese and Yugoslavs. Some social tensions developed from this immigration but there was great sensitivity about racialism which tended to hide such pressures, occasionally finding an outlet amongst the far right.

The German Democratic Republic had a less successful development, struggling to create industry in what had been a largely agricultural region before 1945. There were some successes in state-controlled manufacturing, but the economy suffered from Soviet demand for foreign currency from East German exports, and from shortages of oil and other commodities. However a solid but state-dominated welfare and education system was created, and women were encouraged to work by equal pay, state-provided creches and abortions. The deficiencies could be disguised for some years but eventually demands for more freedom of choice and movement, for better and more consumer goods, undermined the regime, the economy and the society.The reunification of 1990 was the result, but there were high costs in modernizing the industry and infrastructure of the old East Germany and adapting these to free market competition. In some areas there had been little progress since the end of the war, in pollution control or standards of building. By 1991, as unemployment rocketed in the east, many people were questioning the wisdom of so speedy a unification of these disparate economies.

▲ **The Mercedes – symbol of the postwar economic miracle.**

Culture and science

The traditional strength of German culture lay in the field of music and in the quality of its technical sciences. These flourished in the years of empire and affluence before 1914. The composer Johannes Brahms died only in 1897 and the general interest in the performance of classical music remained alive. Small towns built concert halls and opera houses.

The sciences too were a vital element in German creativity, though it was the practical application of invention to production and consumption that proved German excellence. Toys, armaments, aircraft and motor engines, electronics and rocketry were consistent examples of this achievement. The foundations were laid in the technical high schools and middle level education which also survived World War II and were prominent again after 1945 in both Germanies. In view of this practical strength it is not surprising that industrial and architectural design after World War I turned Germany, and particularly Berlin, into a cultural center rivaling Paris for the writers, artists and designers of the 1920s. Artistic and scientific skills were ruined after 1933 when many leading intellectuals fled the country, such as the playwright Bertolt Brecht and the composer Kurt Weill, while those who stayed were stifled.

After 1945 German abilities reappeared in the practical application of science and research to production, and in the enthusiasm for Germany's traditional culture and music. This was true for both Germanies. Opera in East Berlin was of high quality, although academic and scientific life was restricted by the political limits set by the state. But social and political criticism characterized literature and film among the arts in West Germany from the late 1960s, where the most serious themes were a criticism of material possessions and growth and the lack of a German identity. In the East the bitterest critics of the Communist repression were often artists, musicians and writers who frequently came to the West.

AUSTRIA

REPUBLIC OF AUSTRIA

CHRONOLOGY

1916	Emperor Franz Joseph dies
1918	First Republic is established
1933	Parliamentary regime replaced by authoritarian regime under Dollfuss
1938	*Anschluss* with Germany
1945	Proclamation of Independence and establishment of the Second Republic
1955	State treaty, ending four-power occupation and establishing neutrality
1966	End of Grand Coalition and establishment of Austrian People's Party (ÖVP) government
1970	Socialist Party (SPÖ) government formed
1986	Return to Grand Coalition
1989	Application for membership of EC

ESSENTIAL STATISTICS

Capital Vienna

Population (1989) 7,603,000

Area 83,857 sq km

Population per sq km (1989) 90.7

GNP per capita (1987) US$11,970

CONSTITUTIONAL DATA

Constitution Federal multiparty republic with two legislative houses (Federal Council; National Council)

Date of independence 30 October 1918

Major international organizations UN; I-ADB

Monetary unit 1 Schilling (S) = 100 Groschen

Official language German

Major religion Roman Catholic

Heads of government since independence (President) Dr. X. Seits (1918–20); Dr. M. Hainisch (1920–28); Dr. W. Miklas (1928–38); German Reich (1938–45); (President) K. Renner (1945–50); T. Körner (1951–57); A. Scharf (1957–65); F. Jonas (1965–74); B. Kreisky (Apr–Jul 1974); R. Kirchschläger (1974–86); K. Waldheim (1986–)

The Republic of Austria that was founded as a new state in 1918 encompassed about one-sixth of the territory of the Austro-Hungarian monarchy. The republic continued the existing legal order, inherited the large-scale bureaucracy created by the Hapsburgs and further developed government intervention in economic and social life. In these respects there was important continuity.

From the outset the political life of the new republic was beset by the turmoil of democratic and nationalistic pressures that, along with World War I, had brought about the collapse of the Hapsburgs' dynastic rule. The republic lacked a stable sense of national identity. Above all, some political leaders favored regional separation; others, a Danubian confederation in which Austria could be encompassed; and others, union (*Anschluss*) with Germany.

Ultimately, the most powerful of these views was that which denied the existence of an Austrian nation separate from the German nation. The most enthusiastic supporters of the new regime were the Social Democrats whose power base was in Vienna. However, from 1920 to 1931, and with the fascist *Heimwehr* afterwards, the Christian Social party dominated government. The Social Democrats found themselves excluded from power. Street fights between party-affiliated paramilitary forces were frequent.

The rise of Nazism in Germany, and the impact of the deep economic recession after 1929, subjected the regime to new pressures. In 1933 Engelbert Dollfuss became Chancellor and established an authoritarian regime. A brief, bloody civil war in 1934 was followed by the dissolution of the Social Democratic party and a new corporatist constitution. Dollfuss himself was killed in the Nazi putsch of July 1934. The illegal Nazi party continued to be active inside Austria, while externally Mussolini forfeited his protector role in the interests of a closer relationship with Hitler. The consequence was the annexation of Austria by Germany in March 1938. In 1939 Austria ceased to exist even as a distinct administrative unit in the German state.

Postwar Austria felt a desire to return to the "spirit" of the 1920 constitution, and a firm consensus about the value of "social partnership", giving significant influence to the trade unions. The two main political parties, the Austrian People's Party (ÖVP) – the successor to the Christian Social party – and the Socialist Party of Austria (SPÖ), were united in a Grand Coalition by a fundamental aim: to work for an early end to Allied occupation, and to use a declaration of Austria's neutrality as a means to this end. This aim was achieved in the state treaty of 1955, after which Soviet forces withdrew. Since then Austria has used its neutrality to attract a number of major international organizations to base themselves in Vienna, including for instance the International Atomic Energy Agency and the United Nations Development Organization. In 1989 Austria applied for membership of the EC. Party political life in Austria has been dominated by three political parties. The ÖVP appealed to devout Catholics. The SPÖ remained attached to a proud tradition (its predecessor the Social Democratic party had been established in 1889) but was less ideological and inward-looking than the Social Democratic party had been. The Austrian Freedom Party (FPÖ) was associated with radical right and anticlerical themes but shifted gradually in time towards neoliberalism and support for the policy of entry into the EC.

Social and economic trends

A central question at the birth of the First Republic was whether the new Austrian state was economically viable. Austria lacked domestic capital as well as basic raw materials (the ore resources of Styria were an exception) and was adversely affected by the drive to economic self-sufficiency during the Great Depression and by dependence on foreign loans. The middle classes were hard hit by the huge inflation of 1922 and subsequent austerity measures; the Great Depression led to nearly one-quarter of the employable workforce being unemployed by 1933. These developments played a key role in the breakdown of the First Republic and the events leading up to the *Anschluss* of 1938. Absorption into the German economy resulted in new capital investment, an advantage that was offset by huge wartime damage and dislocation.

After 1945, trade agreements and capital aid under the Marshall Plan for European reconstruction prevented mass unemployment and hyperinflation. Domestic economic policy gave priority to the social dimension of maintaining full employment, an aim that survived the post-1973 economic recession. A guarantor of the effectiveness of this approach was the moderation of the Grand Coalition, and more generally of party politics; and the new "parity commission on prices and wages", established in 1957. The "parity commission" represented in effect a new corporatism, bringing together employer, employee and government representatives in the pursuit of wage and price restraints. Nationalization became more widespread than in any other West European economy. In the 1980s, however, the budgetary costs of sustaining employment in the public sector led to a new climate of criticism.

Austria is home to prestigious cultural institutions: notably, the Staatsoper (where Gustav Mahler conducted at the beginning of the century); the Musikverein, the home of the Vienna Philarmonic Orchestra; and the Burgtheater, a seat of classic German theater since the 18th century. High culture attracts enormous public interest. In particular, music enjoys great prestige. Standards of teaching at Vienna's conservatories are extremely high. The love of culture goes along with an attachment to the traditional formalities of dress, address, balls and cafe society. Caution, stability and an attachment to the past are hallmarks of Austrian culture, in a country that once had a central role in Europe but by 1990 was peripheral.

▼ **Hitler's troops march through Vienna in 1938.**

SWITZERLAND

SWISS CONFEDERATION

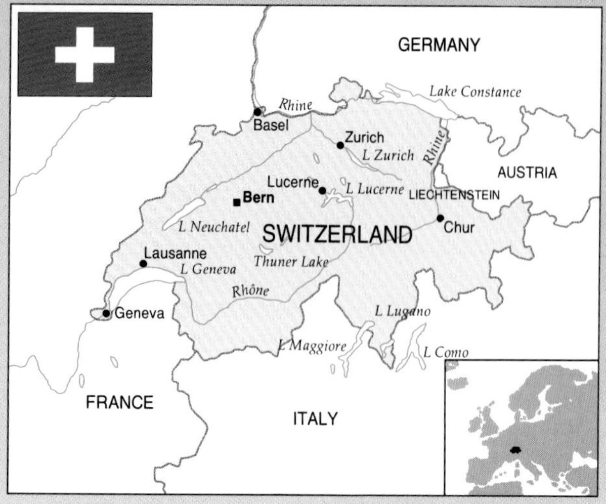

CHRONOLOGY

1919	The introduction of proportional representation
1937	A peace treaty (*Arbeitsfrieden*) signed by the trade unions and the employers' organizations in the metal industry
1944	Social Democrats first join the federal government
1959	Social Democrats enter the government on a continuing basis
1971	Women's suffrage granted

ESSENTIAL STATISTICS

Capital Bern

Population (1989) 6,689,000

Area 41,293 sq km

Population per sq km (1989) 162

GNP per capita (1987) US$21,250

CONSTITUTIONAL DATA

Constitution Federal state with two legislative houses (Council of States; National Council)

Major international organizations UN; I-ADB

Monetary unit 1 Swiss Franc (SW F) = 100 centimes

Official languages French, German, Italian

Major religion Roman Catholic

Head of government President of the Confederation elected annually by Federal Assembly

The peculiar characteristics of Switzerland stem from the combination of three enduring factors: the country's numerous social divisions, its position at the center of Europe and as an axis of communications between north and south, and its inaccessibility as an Alpine bastion. The legacy has included a well developed preference for compromise and consensus; a militia army which is highly trained and which has a central part in Swiss society; and the principle of active neutrality in international affairs. No other continental European society has escaped the convulsions of war like the Swiss.

The major historical divisions in the country have been religious, linguistic and regional. The principle of parity between Protestants and Catholics had been established in several important cantons in order to prevent involvement in Europe's religious wars of the 16th and 17th centuries. This principle was reinforced by the constitutionally guaranteed strong representation of the cantons at the national political level. Informally, a proportional representation of Switzerland's different language communities (French, German and Italian, plus Romansch) was also observed. Since the middle of the 19th century majority rule (*Majorz*) has given way to proportionality (*Proporz*). At the federal level, the constitution of 1848 contained provisions for a "collegial" rather than a majority-based executive branch. Swiss political development during the 20th century has reflected two sustained priorities: the continuing search for domestic political accommodation, and the principles of neutrality and solidarity that have informed its foreign policy. In consequence, it has not been involved in either of the two World Wars, though its domestic political development has been substantially affected by the hardships imposed by wartime conditions. War has served not to disrupt domestic arrangements but, along with the Great Depression, to strengthen them. Nevertheless, the 20th century has seen a reorientation of political development. Against the background of the Great Depression, and the shift toward self-sufficiency across Europe, the Swiss state supported cartelization. This public-sector-led approach did not, however, endure for long. As liberalization of the international economy offered new opportunities from the late 1950s, Switzerland reverted to a traditional preference for a private-sector-led approach.

A striking feature of the 20th century has been the consolidation of the power of Switzerland's relatively centralized interest groups. These groups have emerged as the decisive actors in Swiss policy making, notably the Vorort, the most important association of the Swiss business community. Trade associations have also grown in importance, particularly in vocational training. Broadly speaking, the Swiss state – in which socialism and trade unionism have been marginal – has preferred to rely on a private-sector approach to compensate people for the costs of economic adjustment. The Swiss welfare state has been built on a private insurance system and the encouragement of individual savings: per capita expenditure on insurance is the highest in the world, the level of individual savings only behind Japan.

Since 1959 Switzerland's four main political parties have formed a coalition in the federal executive. This all-party government meant that elections did not produce clear winners or losers, reflecting a historical distaste for adversarial and competitive politics.

Economic and social trends

Switzerland is the bastion of international finance capitalism and a tourist mecca, two factors that underline the extent to which the service sector is unusually well developed there. Also, it is the home to a large number of major international companies; indeed, no other advanced industrialized society is so well represented by these companies in relation to size. This international orientation is characteristic of the big banks, in particular the Union Bank of Switzerland, the Swiss Bank Corporation and the Swiss Credit Bank. The direct foreign investments of these corporations are immense, on a per capita basis larger than those of the United States and Britain. The huge assets of these corporations make them both independent of the national government and very privileged players behind the scenes of national politics. With the appreciation of the Swiss franc in the 1970s, to combat inflation, manufacturing operations were deployed abroad on a large scale and service sector employment further strengthened. This policy measure reflected the historical value placed on investors' confidence in Swiss currency. Financial confidence was further cultivated by a sustained image of political stability, military neutrality and the secrecy of banking activities. Historically, the secrecy of Swiss bank accounts was adopted to protect political refugees and their relatives from Nazi espionage in the 1830s.

The Great Depression in particular was important in focusing the attention and policies of the Swiss government on export promotion and vocational training. Together, these two policy instruments were central to the development of the Swiss economy: and in both cases reliance is placed on the private sector to organize their use. Industrial research and development has been organized by industry itself, and by the end of 1970, Switzerland was second only to the United States as an exporter of technology. The post-1973 recession saw a huge increase in export promotion activity.

Cultural trends

Switzerland is probably the most divided West European state in terms of religion, and second only to Belgium in ethnolinguistic division. In fact, linguistic divisions cut across religious, regional and class lines, so that subcultural groups remain relatively open. Swiss society is also noted for its localism and particularism, reflecting its organization into "cantons" which existed prior to the 1848 constitution. Yet, this localism – and accusations of "narrowness" – live alongside a reputation for humanitarian action, symbolized by the fact that Switzerland is home to the Red Cross.

In the 20th century Switzerland has contributed some important creative artists: for instance, Paul Klee and Alberto Giacometti in the visual arts, and Friedrich Dürrenmatt and Max Frisch in literature.

▼ The Swiss army in training.

GAZETTEER

ANDORRA

(Co-Principality of Andorra; the valleys of Andorra) **Capital** Andorra la Vella **Population** (1989) 50,000 **Area** 468 sq km **GNP per capita** (1983) US$9000.

Andorra, in the Pyrenees, has been independent since the days of Charlemagne, and a co-principality since the 13th century AD. Its joint princes (heads of state) are the bishop of Urgel in northern Spain and the president of France, who are represented by their *vegeurs* (provosts). Since World War II, the predominantly agricultural and pastoral economy has been transformed by the development of commerce (Andorra, which has been exempt from French and Spanish import duties since 1867, has attracted a flourishing trade in duty-free consumer goods) and tourism, especially winter sports. Women in Andorra did not achieve the right to vote until 1971. Throughout the 1980s the traditional political structure based on the Council General of the Valleys (CGV), which gave no voice to people born outside Andorra, was dominated by ancient families, and in any case had no legislative power, came under attack from those seeking a more democratic system. In 1981 the process of reform commenced, when the legislature and the executive arm of government separated; legislative power was now to be vested in the 28-strong CGV, which would appoint an executive council. In 1982, at a formal Consultation, an overwhelming majority demanded electoral reform, with proportional representation; this was not implemented. In 1985 the first elections of the CGV took place under the new system.

CYPRUS

(Republic of Cyprus) **Capital** Nicosia **Population** (1989) 564,000 **Area** 5896 sq km **GNP per capita** (1988) US6260 **International organizations** UN, Commonwealth, EC (associate member) **Not internationally recognized** Turkish Republic of Northern Cyprus **Capital** Lefkosia (Nicosia) **Population** (1989) 169,000 **Area** 3355 sq km **GNP** n/a

The island of Cyprus was annexed by Britain from the Ottoman empire in November 1914 and became a Crown Colony in 1925. The population was divided between a Greek majority and a Turkish minority, and calls for union with Greece in the 1920s led to riots in Nicosia in 1931. World War II left the island almost unscathed, and in the 1950s calls for union (*enosis*) with Greece were renewed, under the leadership of Archbishop Makarios III, head of the Greek Orthodox Church in Cyprus, and supported by a bombing campaign organized by the National Organization of Cypriot Struggle (EOKA). In 1959 Greece and Turkey arrived at a compromise over the government of the island, which became an independent republic in 1960 with Archbishop Makarios as president, and the Turkish Cypriot leader, Fazil Kutchuk, as vice president. Renewed and more serious fighting between the two communities after a constitutional dispute in 1963 nullified the new constitution and resulted in the arrival of a UN peacekeeping force (UNFICYP) in 1964. The number of Greek troops on the island increased alarmingly until a mutually agreed reduction in numbers of both Greek and Turkish troops in 1967. Makarios was reelected overwhelmingly in 1968 and 1973, although his abandonment of the cause of *enosis* lost him support, and in July 1974 an EOKA-led coup against him resulted in his exile, and an invasion of the island by Turkey. The coup was short-lived, and Makarios returned in December 1974; he died in 1977 without having regained the ascendancy in Cypriot affairs. For by his return the northern third of the island had been proclaimed the Turkish Federated State of Cyprus, with the aim of seeking a federated solution to the problem of the island's government. Turkish Cypriot leader Rauf Denktash abandoned the idea of federation and renamed it the Turkish Republic in 1983. Makarios's successor, Spiros Kyprianou, came under severe criticism after renewed UN negotiations on federation broke down in 1985, but the election of Georgios Vassiliou as president in 1988 brought a new impetus to resolving the differences between the flourishing Greek community and the poorer Turkish north.

ESTONIA

Capital Tallinn **Population** (1990) 1,573,000 **Area** 45,100 sq km **GNP per capita** N/A

In 1989 the Estonian populace was 61 percent indigenous and 35 percent Russian, Byelorussian and Ukrainian, and mostly Lutheran Protestant. A pawn in the game of European power politics throughout the 20th century and before, Estonia aspired to the freedom and prosperity of Finland. It struggled for freedom from the Russian empire in the revolution of 1905–06, and suffered cruel retribution. In 1918, along with the other Baltic states, Estonia seceded from the new Union. Land reforms followed, and a reorganization of society along Danish socialist lines promised prosperity, but political inexperience brought instability, and in 1934, after 21 governments in 15 years, Konstantin Päts imposed an authoritarian nationalist regime. After Soviet occupation in World War II, and rigged elections, the three Baltic states were "voluntarily" restored to the Soviet Union, and many of their people killed and deported. This mayhem continued through collectivization, until Estonia was subdued into cooperation; it became the wealthiest Union republic. In the 1980s anti-Soviet feeling grew, not only on nationalist but also on ecological grounds. Estonia led the other Baltic republics toward economic autonomy in 1987–88, but was less insistent than they in demanding Baltic independence, which was finally achieved in 1991.

ICELAND

(Republic of Iceland) **Capital** Reykjavik **Population** (1990) 254,000 **Area** 103,000 sq km **GNP per capita** (1988) US$20,160 **International organizations** UN, NATO.

In 1381 Iceland and its ruler, Norway, came under Danish rule; and Iceland remained so at the beginning of the 20th century. In 1903 it won autonomy in internal affairs. In 1918 Iceland was recognized as an independent state under the Danish crown, an arrangement that broke down in World War II when Denmark was occupied by the Germans, and Iceland by the Allies for use as an airbase. The union formally ended in 1944 and the independent republic was set up which has endured since. In 1980 Vigdis Finnbogadottir became the world's first popularly elected female president. Iceland's economy has always depended heavily on the sea. In 1975–76 a series of disputes (Cod Wars) with Britain resulted in the extension of the fishing limits around

Icelandic coasts; and in the 1980s the country came under international pressure to stop commercial whaling. In 1989, after 74 years of prohibition, the sale of beer was legalized; and economic unrest manifested in a strike by State employees. In 1990 they and the unions agreed a wage deal with government, in an effort to keep inflation under 10 percent.

LATVIA

Capital Riga **Population** (1989) 2,681,000 **Area** 63,701 sq km **GNP per capita** N/A

The Baltic Latvians are mostly Lutheran Protestants, as are the Finnic Estonians. Like the other two Baltic states, Latvia entered the 20th century in Russia's grip, and was commercially exploited by the Polish and German landowning and merchant class. A strong nationalist surge led to violent repression after the revolution of 1905–06, continuing until secession in 1918, to be confirmed by bilateral Baltic-Soviet treaties in 1920. The new peasant and Social Democrat ruling parties inaugurated wide-ranging social, educational and agricultural reforms, but political instability hampered their execution and in 1934 an authoritarian regime took power, under Karlis Ulmanis. In 1940 the Soviet Union took possession of the Baltics, and reestablished this in 1944 after Nazi occupation. Many nationalist sympathizers were deported or killed under Stalin. Industrialization and the concomitant environmental damage were suffered unwillingly; and the Green movement grew in the 1980s, as did the nationalist popular front, which, despite the 42-percent Russian component of the population (52 percent being Latvian (1989)), achieved its goal of independence finally in 1991 as the Soviet Union "went west".

LIECHTENSTEIN

(Principality of Liechtenstein) **Capital** Vaduz **Population** (1989) 28,300 **Area** 160 sq km **GNP per capita**(1985) US$16,500 **International organizations** EFTA.

The Principality of Liechtenstein, which lies between Austria and Switzerland, is made up of two counties, Schellenberg and Vaduz, and was closely allied to the Austrian empire until 1919. Its present constitution dates from 1921: a single-chamber parliament can be called and dismissed by the hereditary prince. Also in 1921 Liechtenstein adopted the Swiss currency, and in 1923 entered into a customs union with Switzerland. It has no heavy industry, and the economic importance of agriculture has declined since 1945 in favor of light industry. In 1938 Prince Franz Josef II came to the throne; in 1984, the year that women gained voting rights, he handed over executive power to his son, Prince Hans Adam, who succeeded him in 1989.

LITHUANIA

Capital Vilnius **Population** (1989) 3,690,000 **Area** 65,190 sq km **GNP per capita** N/A

Lithuania is unique among the Baltic States in having a mainly Roman Catholic population. Once a great nation, it was annexed by Russia in the 18th century, and commercially exploited by the state and by immigrant landowners and merchants. Bloody struggle against Russian rule was answered with repression, and this sharpened nationalism and interethnic and class conflict; at last, in 1918, came secession, ratified by treaties in 1920. Socialist social and agricultural reorganization followed, overshadowed by a long territorial battle over the capital, Vilnius, annexed by Poland in 1919, as well as the fear of losing Klaipeda, Lithuania's only port, to Germany, which took it at last in 1939. Political fragmentation resulted in 1926 in a coup, which brought in an authoritarian regime under Antanas Smetona. After World War II, and a period of Nazi occupation, the Soviet Union appropriated the Baltic states, and Stalin purged thousands he considered anti-Soviet, as Russians "colonized" the Baltic states en masse. Lithuanian nationalism was proof against attempts at cultural colonialism, and, sharpened by disgust with the harmful side-effects of Soviet-imposed industrialization, spearheaded the movement for Baltic independence which finally achieved its aim in 1991.

LUXEMBOURG

(Grand Duchy of Luxembourg) **Capital** Luxembourg **Population** (1989) 377,000 **Area** 2586 sq km **GNP per capita** (1988) US$22,600 **International organizations** UN, NATO, EC.

The modern Grand Duchy of Luxembourg was established in 1815, and was under German control throughout World War I. In 1921 Luxembourg entered into an economic union with Belgium, and was occupied again in May 1940 by Germany. After liberation in 1944, Luxembourg abandoned its traditional neutrality. It joined the Benelux Economic Union (1948) and the EC (1956). The Grand Duke is a monarch with executive power, wielded through a Council of Ministers led by a president, and legislative power expressed through the Chamber of Deputies. In 1984 a Socialist/Christian Social party coalition was in government. As the long-established steel industry fell into decline, Luxembourg became an international banking center, and home of several EC institutions.

MALTA

(Republic of Malta) **Capital** Valletta **Population** (1989) 349,000 **Area** 316 sq km **GNP per capita** (1988) US$4948 **International organizations** UN, Commonwealth, EC (Associate member).

The Treaty of Paris in 1814 ceded Malta to Britain. During World War II the island, an important naval base, was besieged and bombarded by the Axis powers, and in 1942 the island was awarded the George Cross, the highest British civilian decoration. Full independence was granted in 1964. The rundown of the British presence brought economic dislocation; from 1971 the Maltese Labor party adopted a policy of nonalignment and special friendship with China. In 1974 Malta broke its last remaining ties with Britain and became a republic. Also in this year the island signed a cooperation accord with Libya in respect of economic affairs and, especially, security. In 1987 the House of Representatives was dissolved, and the president resigned; the pro-Western Nationalist party won back power, with the declared aim of taking Malta into the EC and encouraging Western European investment. In 1989 Malta hosted the first summit conference between US president Bush, and Mikhail Gorbachev, leader of the Soviet Union.

MONACO

(Principality of Monaco) **Capital** Monaco-Ville **Population** (1989) 29,100 **Area** 1.9 sq km **GNP per capita** N/a

Monaco is an independent principality on the southern coast of France, ruled since 1297 by the Grimaldi dynasty; it has been under French protection since 1861. In 1911 a new constitution allowed for a National Council and a Communal Council. Prince Rainier III, who married the American film star Grace Kelly in 1956, suspended the constitution in 1959, and initiated a new, liberal one three years later. Monaco's reputation as a winter (later as a year-round) resort for the wealthy began to develop in the 19th century, and its casino at Monte Carlo is world-famous. Monaco does not levy income tax, and several international businesses have their headquarters there.

SAN MARINO

(Most Serene Republic of San Marino) **Capital** San Marino **Population** (1989) 22,900 **Area** 61 sq km **GNP per capita** (1987) US$8590

San Marino was the world's smallest republic until the independence of Nauru, in 1968. The constitution dates back to 1600, with an elected 60-member Great and General Council, two of whose members are appointed regents every six months. In 1906 a peaceful revolution restored the right to household suffrage, and in 1909 universal male suffrage was introduced. Women were given the vote in 1960. During World War II, San Marino was formally neutral, but suffered from bombing. It had a Communist government for 12 years after the end of the war. Economically, San Marino relies on light industry, agriculture and, increasingly, tourism, as well as the production of postage stamps. Income tax was introduced in 1922. In 1989 it became a permanent observer at the United Nations, and joined the Council of Europe.

VATICAN CITY

(Vatican City State) **Population** (1988) 1000 **Area** 0.4 sq km **GNP per capita** N/a.

When the Italian Papal States were incorporated into the emerging Italian state in the mid-19th century, a dispute arose as to the status of the Papacy with regard to Italy. This was resolved in February 1929 by the Lateran Treaty between the Italian government and the Vatican. The treaty was threefold: political; in this respect, the sovereignty of the Holy See in the Vatican State was recognized; a Concordat, under which the position of the Catholic Church in Italy was assured by the government; and financial; the government agreed a financial settlement with the Holy See. It was incorporated into the constitution of Italy in 1947, and was revised in 1985. The Pope is the sovereign authority in the Vatican City, which maintains diplomatic relations with foreign countries and has permanent observers at the UN. The Vatican maintains its own coinage and postal services, a broadcasting network, and a railroad system linked to the national one; it retains a police service and an army which includes the 100 Swiss Guards. It has its own judiciary, and is tax-exempt in Italy. 1989–90 saw the restoration of relations between the Vatican and Eastern European countries, and the establishment of contact with the Soviet Union.

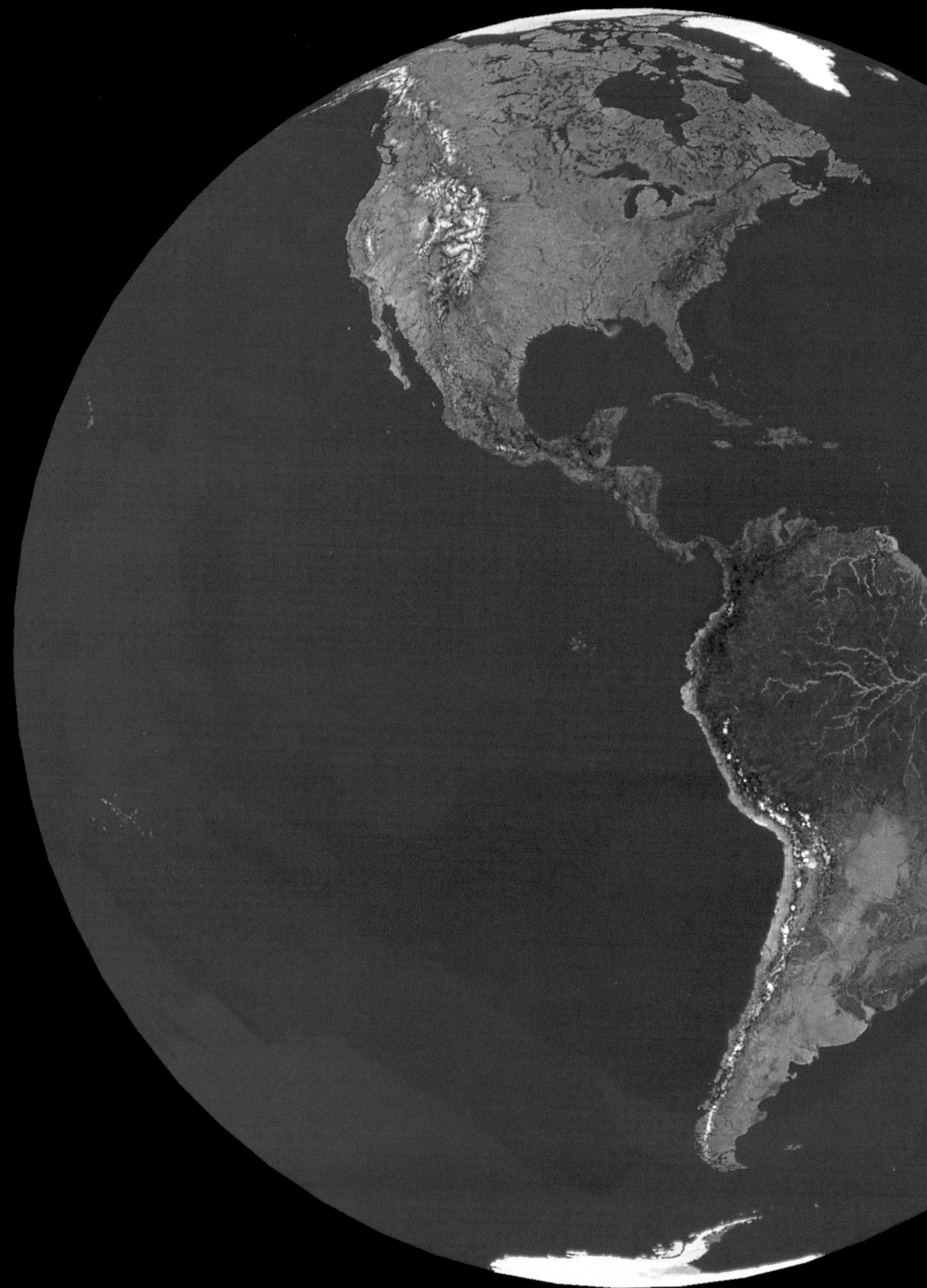

THE AMERICAS

INDEX OF COUNTRIES

CANADA

CANADA

CHRONOLOGY

1905	Alberta and Saskatchewan created out of N.W. Territories
1919	Women granted the vote
1921	Canada negotiates first international treaty with USA
1931	Statute of Westminster grants legislative autonomy to Canada
1965	Maple-leaf flag adopted as national flag
1970	FLQ terrorist campaign
1980	Referendum in Quebec; majority vote against separatism
1987	Meech Lake Accord prepared – signed 1988
1988	Free Trade Agreement with USA
1990	Failure of all provinces to ratify Meech Lake Accord

ESSENTIAL STATISTICS

Capital Ottawa

Population (1989) 26,189,000

Area 9,970,610 sq km

Population per sq km (1989) 2.8

GNP per capita (1987) US$15,080

CONSTITUTIONAL DATA

Constitution Federal multiparty parliamentary state with two legislative houses (Senate; House of Commons)

Major international organizations UN; Commonwealth of Nations; I-ADB; NATO

Monetary unit 1 Canadian dollar (Can$) = 100 cents

Official languages French; English

Major religion Roman Catholic

Heads of government since 1900 (Prime minister) Wilfred Laurier (1896–1911); Robert Borden (1911–20); Arthur Meighen (1920–21); W.L. Mackenzie King (1921–26); Arthur Meighen (1926); W.L. Mackenzie King (1926–30); R. B. Bennett (1930–35); W.L. Mackenzie King (1935–48); Louis St. Laurent (1948–58); John Diefenbaker (1957–63); Lester Pearson (1963–68); Pierre Trudeau (1968–79); Joseph Clark (1979–80); Pierre Trudeau (1980–84); John Turner (1984); Brian Mulroney (1984–)

The twelve provinces and territories of Canada were only combined into the Canadian Union at the beginning of the 20th century. Canada is a giant state with peoples of many different origins, and Canadian politics have therefore been dominated by the relationship between the federal government in Ottawa and the provincial governments. Although the constitution of Canada is officially known as a confederation, it has been largely regarded as a federal constitution.

The century opened with a strong assertion of the identity and rights of the provinces, especially when the Liberal party, under the French Canadian leader Wilfred Laurier, defeated the Conservatives in 1896. In 1905, following the expansion of the grain lands across the prairies and the development of mineral resources, new provinces – Alberta and Saskatchewan – were carved out of existing territories. However, federal authority was reasserted during World War I, when the Canadians made a great contribution to the defense of the British Empire, losing more men than the United States and distinguishing themselves in famous actions on the Western Front such as that at Vimy Ridge, near Arras, in 1917. War regulations permitted the extension of federal authority through the Union. When W. L. Mackenzie King became premier in 1921, the policy of "Canada first" was strengthened. This meant not only the stress on federal government and unity but also a growing distance between Canada and Britain, whose government could no longer expect slavish obedience. This was shown in the explicit refusal by Mackenzie King to commit Canadian troops to assist the British in the "Chanak affair" in Turkey in 1922. Although Canada participated in the trade agreements known as Imperial Preference, which were agreed at Ottawa in 1932 in response to the protectionism of the government of the United States, a degree of dominion independence from Britain came with the Statute of Westminster (1931) which allowed dominion parliaments, such as that of Ottawa, to make amendments of a constitutional kind. The staunchly "Canadian" policies of Mackenzie King were again asserted in 1938 when he refused to rubber-stamp the British policy over the German seizure of Czechoslovakia in 1938 and to commit Canadian troops unequivocally. Even in 1939 the Canadian government only declared war on Germany and the Axis Powers one week after the British declaration, and as a result of a debate and vote in the Ottawa parliament. However, as in World War I, Canadians fought without stint. They fought with special distinction in the Dieppe raid in 1942, when they suffered some 3,000 casualties out of 5,000 troops engaged, and during the Normandy landings of 1944. This commitment was confirmed after the war when Canada became a member of the NATO alliance and continued to offer extensive training facilities for the Allies.

Just as the interwar years had seen the growth of federal authority, so the period after 1945 led to further centralization. At this time also the Supreme Court became the highest court of appeal for the Canadian Union, by an act of 1949. New forces were developing in Canada around 1960 however, notably the birth of a nationalist movement in the largest of the provinces, Quebec. Here came a challenge to central authority which grew among the French Canadians, looking to their Roman Catholic heritage and to a tradition of 19th-century rebellion against the Anglo-Saxons. In 1967 the *Mouvement souveraineté-association* was created with more extreme demands of the federal government. The French president Charles de Gaulle's cry for a free French Canada, coming during a tense official visit in 1967, was greeted by tremendous applause by the French Canadians; he did not mince his works: "*Vive Montréal. Vive le Québec. Vive le Québec libre.*" The following year the various movements came together to form the *Parti québécois* (PQ), with René Levesque as president. This movement was confronted by resolute federal governments, especially those of Pierre Trudeau who had himself some French Canadian origins. Trudeau was reluctant to consider any concession to provincial separatism or nationalism; he was determined to uphold the union which had only recently adopted an official flag, the red maple-leaf emblem.

There soon emerged a violent wing of Quebec nationalism, FLQ, *Front pour la libération de Québec*, which kidnapped a British diplomat and assassinated the minister of finance from Quebec. Quick and decisive action by the federal government stopped the terrorists. The PQ however continued to flourish, becoming a truly mass movement with 150,000 members at its height. The movement began to draw support away from the Liberal regime in Quebec, led by Robert Bourassa, which sought cultural autonomy and greater provincial self-government but within the federal system. Not enough was achieved and in 1976 the people of Quebec voted for a PQ government led by René Levesque, with a mandate for political sovereignty. This proved more difficult to achieve than the leaders anticipated, especially in the face of a determined federal prime minister, Pierre Trudeau. French became the sole official language for business, administration and education, but little was gained in the way of constitutional concessions. In 1980 only 40 percent voted for separatism in a Quebec referendum. The province stood aside during the 1982 discussions on the patriation of the Canadian constitution, but in 1985 the PQ fell and the more cautious Liberal party of Robert Bourassa was returned to form the government of Quebec.

The years of Trudeau governments indicated the three major issues of Canadian politics for the future. The first issue was that of free trade with the United States, a policy that was finally achieved by the agreement of 1988. This was expected to advantage the more central provinces and not the Atlantic provinces of Canada, but it did not necessarily strengthen the federal government at the expense of the provinces, since implementation is a provincial matter.

The second issue was that of aboriginal peoples and their rights in the union, covering mainly the Indian peoples and the Eskimos (Inuit). This had not been the subject of agreed policy by the provincial governments and there were growing signs of irritation, especially on questions of land settlement and claims and the creation of separate native police agencies. Mohawk Indians staged a series of demonstrations in the provinces of Quebec and Ontario.

The third issue stemmed directly from the nationalist demands in Quebec, the constitutional form of the Canadian Union. A series of inquiries into the constitution and into provincial identities, a Royal Commission on Bilingualism and Biculturalism in 1963 and the Tremblay Commission on the Constitution (1953) culminated in the proposal for total Canadian autonomy over its constitution and amendment, dispensing with any further need to defer to the Westminster Parliament. Quebec refused to adhere to this so-called "patriation" of the Canadian constitution. Trudeau initiated the Macdonald Commission to report on the constitutional issue in 1985. An attempt to bring all provincial governments and the federal government together on this matter was made in the Meech Lake Accord (1988) which suggested that the distinct cultural and social identity of Quebec should be recognized by all provinces by June 1990. This did not occur, as Newfoundland and Manitoba refused to ratify and Quebec remained firm on the use of French as the sole official language. The question thus remained open about the conditions on which the province of Quebec would remain within the Canadian Union. However there were signs that, despite the pressures, there was a will to keep this system together, to recognize the specificity of the French Canadians, and the government of the union gained a clear mandate in the elections of 1984 and 1988 which returned the government led by Brian Mulroney.

▲ **The Canadian call to battle – World War I.**

Economic and social trends

The Canadian economy suffers from a lack of balance and consistency. Quebec and Ontario are overwhelmingly the most populous and wealthiest provinces. Some of Canada's vast mineral wealth and resources have brought consistent benefit to certain locations, but other production has suffered from variations in market price. These have notably affected the oil production of regions such as Alberta and Saskatchewan. Nevertheless Canada has the advantage of great natural riches from the prairies and grain lands, from the variety of mineral and petroleum deposits, covering commodities ranging from gold and uranium to coal and industrial minerals such as ores and zinc, copper and lead. There are also extensive resources of fish and special natural products like furs and wood. In some of the provinces, not uniformly distributed, there are successful manufacturing industries, amounting to about thirty-three percent of the GNP. But there is little specialization, except in the areas where the extraction of primary commodities dominates. These economic realities have helped to give Canada strongly marked regional loyalties, with much organized on a provincial basis. National trends do not predominate.

The economy and the society are divided on both an east-west and a north-south axis, and the latter steadily became stronger as the economic pull exerted by the United States grew. Apart from the interlude of US protectionism during the 1930s, the attraction of the US market and the provision of goods and capital from the United States boosted those Canadian provinces closer to the frontier. This trend was accentuated by the Free Trade Agreement of 1988. Trade with the United States overtook trade with the British Empire in 1945 and by 1990 accounted for 75 percent of all Canadian trade.

However, the federal government gradually asserted its authority over the east-west axis of the union. The railroad network, which grew dramatically in the late 19th century, crossed provincial frontiers and brought an association between the two main networks, the Canadian National and the Canadian Pacific in 1933. Federal investment in the major artery of the St Lawrence Seaway benefited several areas but again cemented the link with the United States. But it was through federal taxes that the Ottawa governments exercised their power. After World War II, three-quarters of taxes paid by Canadians went to the federal authorities. The spread of motor cars and the many revenues that followed, licenses, fuel taxes, and federal intervention in many matters from education and welfare to broadcasting and insurance, tightened the grip of the federal government over the economy.

The diversity of the economy reflects the great variety of peoples who live in Canada. The traditional major groups are the British and the French, who live mainly but not exclusively in Quebec, and the indigenous peoples, not large in numbers but recently regarded as especially politically sensitive. The regulation of land claims and improvement of the established system of reservations are being examined. But there are many other immigrants who came largely from the turmoil of Europe after World War I. Substantial populations of many nationalities live in Canada, have Canadian citizenship but retain their different languages and cultures. Many of these groups have helped to open the economies of remoter provinces, such as the Poles and Slovenes who live in central regions and the Germans in Vancouver. The federal government indeed has special responsibility for minorities, in particular where a religious minority is involved, and this extends to federal involvement in education. Federal grants to provincial universities started in 1952. Thus many of these different peoples look to the union for their welfare. Significant changes have affected the Canadian economy and the mixed peoples of the state. The steadily more important relationship with the United States is now recognized in the Free Trade Agreement, despite fears that this will benefit the United States disproportionately.

▲ **Disaffected French Canadians at a nationalist celebration in Montreal.**

◀ **Canada plays the Soviet Union at ice hockey – 1988.**

Moreover this Agreement has obliged the Canadian federal government to take special measures to support the economies of the more peripheral provinces, as in the Atlantic Accord for the eastern regions. A second significant change has been the growing urbanization of the Canadian population, coming with the rural decline, the mechanization of agriculture and even the drop in federal support for offshore drilling – partly in acknowledgement of the strong "Green" lobby.

Canada's diverse cultural achievements echo the mixture of peoples who live there. The dominance of the British declined during the 20th century, although the religious rigidity of some of the more extreme Protestant sects remained both in the Atlantic provinces such as Newfoundland and Nova Scotia, and in remote communities in the prairies. There is growing interest in the indigenous culture of the relatively small communities of Indians and Eskimos. Their music, sculptural and plastic arts and skilled crafts have been exhibited and marketed beyond their reserves as part of the Canadian tradition. This has been encouraged with the expansion of environmental concern in Canada, which has led to fresh pride in the natural beauties of this richly varied country.

During the 1970s other provinces became interested in their uniqueness. This developed with provincial responsibility for education which was only subsidized by federal grants at the higher levels; but the dominance of the English language prevented the continuation of this trend in the majority of provinces, and left Quebec and the French Canadians in a special position. Their origins as Canadians were as old as those of the British communities, but the development of nationalism brought a strong revival and flourishing culture based largely on the theater, film and especially literature which focused attention on French. This outpouring came before the formation of a political movement. Maurice Constantin-Weyer won the Prix Goncourt with his novel set in French Canada in 1928 (*Un Homme se penche sur son passé*) and others followed with accounts of the French Canadian past. A string of writers and poets were published – Germaine Guevremont, Claude Grignon and Roger Lemelin – and yet more distinctively French Canadian subjects drew their interest. Writers such as Charles Gill created a tradition and the contemporary French Canadian writers look to the poetry of Frechette and the novels of Le May and William Chapman.

UNITED STATES

UNITED STATES OF AMERICA

CHRONOLOGY

1929	Stock Market crash
1950	President Truman commits US troops to war in Korea
1954	Supreme Court bans segregation in public schools
1964	Escalation of war in Vietnam is authorized
1969	America puts men on the moon
1972	Watergate burglary
1986	"Irangate" scandal emerges

ESSENTIAL STATISTICS

Capital Washington DC

Population (1989) 248,777,000

Area 9,529,063 sq km

Population per sq km (1989) 26.1

GNP per capita (1988) US$19,860

CONSTITUTIONAL DATA

Constitution Federal republic with two legislative houses (Senate; House of Representatives)

Major international organizations UN: OAS; SPC; I-ADB; NATO

Monetary unit 1 US dollar (US$) = 100 cents

Official language English

Major religion Christian

Overseas territories American Samoa; Guam; Virgin Islands

Heads of government (President) W. McKinley (1897–1901); T. Roosevelt (1901–09); W.H. Taft (1909–13); W. Wilson (1913–21); W.G. Harding (1921–23); C. Coolidge (1923–29); H.C. Hoover (1929–33); F.D. Roosevelt (1933–45); H.S. Truman (1945–53); D.D. Eisenhower (1953–61); J.F. Kennedy (1961–63); L.B. Johnson (1963–69); R.M. Nixon (1969–74); G.R. Ford (1974–77); J.Carter (1977–81); R.Reagan (1981–89); G.Bush (1989–)

Victory for the North in the Civil War of 1861–65 ushered in an era of rapid industrial expansion in the United States. Politicians took a back seat while industrialists and financiers developed the economy. Secure behind its ocean boundaries, the United States proceeded rapidly to lay the foundations for its emergence as the dominant economic and political power of the 20th century. Two events symbolize the start of that century for the United States. Internationally, a war against Spain in 1898 revealed the growing political power of the new industrial giant. Not only were the remnants of Spanish rule in the Americas removed but the United States found itself with new colonies such as the Philippines. Thereafter the United States would be, despite its adherence to a policy of isolationism, drawn, by virtue of its size and strength, increasingly into world affairs, whether as peacemaker (as in the conference after the Russo–Japanese War of 1904–05) or as active participant as with its intervention in 1917 in World War I.

Domestically, the opening of the century was marked by the assassination of President McKinley in 1901. This thrust into the presidency a man with very different ideas from his immediate predecessors. Theodore Roosevelt had an activist vision of the office. He embodied the growing belief that governments had a responsibility to deal with the consequences of *laissez-faire* economics.

Roosevelt was disappointed by the performance of his successor, William Taft, and tried to win from him the Republican presidential nomination in 1912. When he failed in this he broke away to stand as a third-party candidate under the banner of the Progressive party. Taft was pushed into third place and the beneficiary of the split was Woodrow Wilson, who became only the second Democrat to hold presidential office since the Civil War.

In office Wilson enacted a number of reforms but his presidency became overshadowed by the outbreak of war in Europe. He was reelected in 1916 under the slogan "He Kept Us Out Of The War". However, German attacks on American shipping contributed to the United States' entering the war in April 1917, less than a month after Wilson's second inauguration. American intervention proved decisive: the Allies benefited from the fresh resources and by November 1918 the war was over. President Wilson took a prominent part in the subsequent peace negotiations and in the establishment of the League of Nations. His attempts, however, to persuade his fellow-countrymen to abandon their isolationism and join his brainchild, the League of Nations, were unsuccessful.

Wilson himself was a broken man for the last months of his presidency and the office returned to what seemed its natural ownership in the election of 1920. The Republicans then presided over "a return to normalcy" and the 1920s were a period of self-indulgence and prosperity. That mood was abruptly broken by the Wall Street Crash of 1929, which ushered in an economic depression and helped to break the hold of the Republican party on American politics.

President Hoover, who had been elected in 1928, seemed overwhelmed by events and in the election of 1932 the country turned instead to the Democrats in the shape of Franklin Roosevelt. This marked the beginning of a period of Democratic ascendancy. Between 1932 and 1968 there was to be only one Republican president, and from 1932 to 1992 there were only four years out of sixty in which Republicans held a majority in the House of Representatives and only 10 years in which they controlled the Senate.

Roosevelt immediately made an impact on an increasingly despairing country. Working at breakneck speed, especially during the first "Hundred Days", his administration produced a series of measures which have come to be known as the "New Deal". How successful economically these were is still debated by historians, but politically they contributed to a landslide reelection victory for Roosevelt in 1936; his Republican opponent won only two states.

In 1937 Roosevelt tried to defeat the Supreme Court which had been blocking much of his New Deal legislation. Despite his overwhelming election victory the year before, Roosevelt was defeated on the issue but almost simultaneously the Court gave way and began to confirm the constitutionality of the New Deal legislation.

As the decade ended the United States once again found itself watching war develop in Europe. Although the mood in Congress was to keep well clear of the war, Roosevelt was able to take a number of steps to aid the Allied cause and in particular to help Britain. However, an attack by Japan on the American fleet in Pearl Harbor, Hawaii, on 7 December 1941 brought the United States into the war. Once in, Americans played a major part in both the war in Europe and in Asia.

Roosevelt had broken the convention that no president would serve more than two terms, by offering himself for reelection in 1940. Again in 1944 Roosevelt ran successfully. His subsequent illness and death in April 1945 brought to office the relatively unknown Truman. The latter quickly found himself dealing with the aftermath of the war in Europe, as well as taking momentous decisions involving the dropping of two atom bombs to speed up the end of the war against Japan.

Truman's administration saw the start of the Cold War and the efforts to help Europe, economically via the massive injections of cash in the Marshall Plan, and militarily through the formation of the North Atlantic Treaty Organization (NATO). Truman responded decisively in 1950 to the invasion of South Korea by North Korea, and the United States formed the backbone of the United Nations military force there, although the Korean war was to drag on until 1953.

▼ Immigrants arrive in New York early in the century.

The Republicans won back the presidency in 1952 with a candidate of great personal popularity, General Eisenhower. In many respects Eisenhower continued the broad lines of policy set out by his immediate predecessors. He presided over a period of relative calm in domestic affairs and did his best to keep the United States out of conflict abroad. The early years of his presidency coincided with the culmination of a period of vitriolic anti-Communism which had begun in the late 1940s. The Cold War and the victory of the Communists in China had produced a mood of paranoia in which the suspicions voiced by Senator Joseph McCarthy flourished.

If with hindsight the 1950s appear as an oasis of calm, the 1960s look very different. The election of 1960 was one of the closest in American history. It brought to the presidency Senator John F. Kennedy, who embodied a youthful spirit of reform. He achieved only limited success in persuading Congress to enact his proposals, especially in the field of civil rights. Internationally his greatest success was in the Cuban missile crisis of 1962, when he forced the Soviet Union to back down from its plan to install nuclear missiles in Cuba. The promise of the Kennedy presidency was unfulfilled: his assassination in November 1963 remains one of the most traumatic events of modern history.

His successor, Lyndon Johnson, persuaded Congress to enact not only much of the Kennedy program but also his own "Great Society" legislation. Together these transformed the law in areas such as civil rights, voting rights, medical provision for the old and the poor and added a variety of other welfare programs. The domestic successes of the Johnson presidency were in marked contrast to its growing foreign policy difficulties as American involvement in the war in Vietnam became unpopular with large sections of American society. His difficulties over this issue led Johnson not to run for reelection in 1968. His successor, Richard Nixon, promised to end the war in Vietnam but in fact did a great deal to escalate it, and not until 1973 was American involvement in the war ended.

At home the Nixon presidency was marked by an attempt to reverse many social programs associated with the Democrats. His presidency revealed a mood of suspicion and insecurity which was to prove its undoing. Although reelected easily in 1972, his victory was hollow. Within a short time the president became enmeshed in what has become known as the Watergate scandal. What began as the investigation of what appeared to be a simple break-in to the Democratic National Committee headquarters during the 1972 election eventually revealed a story of cover-up and evasion on the part of the president that was to lead to his resignation in August 1974.

The combined effect of Vietnam and Watergate produced a reassessment of the presidency. Nixon's immediate successors, Ford and Carter, operated a less "imperial" style and Congress found itself being more assertive. However that style did not last: the election of 1980 brought to office a man who was determined to do what he could to reassert the prerogatives of the presidency and restore its prestige. Over the next eight years Ronald Reagan remained personally popular and did much to restore American self-respect. His reputation was damaged by revelations about the supplying of arms to the virulently anti-American regime in Iran and plans to divert the profits from the operation to aid the rightwing Contra rebels in Nicaragua. More important in his presidency, however, was a change in

▲ Hollywood filmmakers created a new art form.

▶ Anti-Vietnam demonstrators challenged the power of the state.

relations with the Soviet Union which produced the first major steps toward ending the Cold War.

At home the Reagan presidency set itself the task of undoing half a century of New Deal social democracy. In this it had some limited success but it left behind major failures, notably a mounting budget deficit. Reagan's successor George Bush showed little taste for domestic politics but a surer touch in foreign policy, notably in his firm handling of the Gulf crisis of 1990-91.

Although intervention in the Gulf had been a divisive issue within the United States (many feared being drawn into another Vietnam), the successful military conduct of the war silenced most domestic criticism. Still, as the century moved toward its close there was much discussion of whether American economic power was waning, in the face of mounting evidence of the growth of Asian and European economies.

Economic and social trends

By the beginning of the 20th century the United States economy was already the world's largest. At the end of World War II America dominated world economics. Since then it has retained its position as world leader but has seen a relative decline as other countries, notably Germany and Japan, have grown enormously.

American economic power derived from a combination of factors. The size of the country provided a large domestic market, abundant natural resources made her relatively self-sufficient, while a liberal immigration policy provided a ready pool of labor. Added to these was a spirit of openness and a political stability which provided fertile conditions for growth.

Compared with most of her rivals the United States has been characterized by a relatively low level of government involvement in the economy. However, the optimism that the economic system could be left to regulate itself suffered a severe blow in the Depression of the early 1930s.

Up to that time government had intervened little in economic affairs apart from controlling the currency and establishing tariffs. Now it seemed as if the whole system might collapse and business and industry were powerless to correct matters. Government proved to be the savior of the free-market system. In a series of "New Deal" measures the Roosevelt administration made acceptable the idea that government would take some responsibility for the system of banking and credit, agriculture, industrial production, labor relations, the care of the elderly and a host of other aspects of the economic and social life of America.

New York and the United Nations

▲ The United Nations building, Manhattan.

The idea of a world body to replace the discredited League of Nations emerged gradually during World War II.

The first regular session of the United Nations General Assembly, held in London in January 1946, accepted an invitation to establish a permanent home in the United States. Initially the UN moved to Hunter College in New York, then in August 1946 the Secretariat was installed at Lake Success on Long Island, while the General Assembly met in Flushing Meadow. A permanent home was found in Manhattan with the aid of Rockefeller money, which paid two-thirds of the purchase price of a site near the East River. Construction began there in September 1948 and by mid-1950 the 39-storey secretariat building was ready. By early 1951 the UN was firmly established on the 17-acre site. Since that time it has had its headquarters there, operating almost as a ministate inside the United States (issuing, for example, its own postage stamps). It also has its own flag, featuring a globe wreathed with olive branches. The image of UN as peacekeeper has been upheld by the assiduity of successive Secretary Generals in pursuing peace, as did Javier Perez de Cuellar in the 1990 Gulf crisis.

These changes involved also the growth of the power of the federal government relative to that of the individual states. The acquisition in 1913 of the power to levy personal income tax had given the federal government a revenue-raising potential far greater than any it had previously possessed. Not until the New Deal period was that potential realized. From then on the states would look to Washington for help with a variety of economic and social problems and Washington would use its power to equalize resources between states.

In the years after World War II the American economy grew rapidly and American affluence became legendary. The nature of that economy was, however, changing rapidly. The United States has moved from an economy based on manufacturing to one based on service industries. Today the percentage of the workforce employed in agriculture is around three percent, while a substantial majority of the nonagricultural workforce is employed in the service sector.

In 1900 the population of the United States was a little over 75 million; by 1990 this had risen to 250 million. A major part of this growth has been due to mass immigration into the country. Between 1905 and 1914 an average of one million people a year entered the country. Thereafter, especially with the tighter rules imposed in 1920 and the effects of depression and war, the numbers dropped only to rise again in the latter part of the century. The post-World War II immigration was both legal and illegal. The latter proved difficult to control, especially because of the long border with Mexico. Efforts were made from time to time to regularize the position of those who were illegal aliens. In the early part of the century the majority of immigrants were from Europe; toward the end of the century substantial numbers were arriving from Latin America and Asia. Between 1980 and 1990 the number of Americans of Asian origin more than doubled. By 1990 blacks constituted about twelve percent of the population, Hispanics about nine percent and Asians about three percent, and the original inhabitants, so-called "American Indians", and Aleuts around half a percent.

Within the United States the century saw great movements of population. In part, this has been from rural areas to the cities and then out to suburbs and to some extent back to the countryside. A feature of this has been the "flight" of the white population from the central core of many cities and their replacement there by minorities, especially blacks and Hispanics. There was also movement from the northeastern and upper midwestern states to those around the southern rim.

Of all the ethnic problems with which the United States has had to grapple, none has proved more intractable than that of the American negro. Despite the promise of equality contained in a series of constitutional amendments passed after the Civil War, American blacks suffered under a system of legal segregation in southern states during the first half of the century. Segregation of the races in the armed forces was not ended until after World War II and, although some aspects of segregation in higher education were tackled in the late 1940s, it was not until the decision in the Brown case that the provision of separate facilities in education generally was outlawed by the Supreme Court. That decision was followed by a long, slow process of trying to ensure compliance on the part of many southern communities. The civil rights reforms of the mid-1960s did much to end segregation in other walks of life. However *de facto* segregation in northern states, arising largely from patterns of residence, proved no less difficult to deal with. The process of trying to secure genuine equality of treatment involved a variety of mechanisms, from bussing children to schools in different districts to affirmative action programs.

Despite these changes and despite the growth of the black middle class, blacks remain among the poorest Americans. Those concentrated in inner-city areas are part of a problem which seems largely insoluble. Other ethnic groups who have arrived more recently, notably from Asia, have succeeded in leapfrogging over blacks. After blacks the largest minority group are Hispanics (mainly from Mexico, Puerto Rico and Cuba). Their arrival in large numbers posed special problems because they challenged the central place of English as the language which all immigrants must learn. For that reason the status of the Spanish language remains deeply controversial.

Cultural trends

In many areas of culture the United States has been concerned to assert its own identity. In literature one modern author has parodied this as a search for "the great American novel". A major claimant to be the author of such a work would be Ernest Hemingway. Other important figures include John Steinbeck, whose *The Grapes of Wrath* so vividly portrays the suffering of the Okies driven west to California by the dust bowl conditions of the 1930s; F. Scott Fitzgerald with such novels as *The Great Gatsby*, and John Dos Passos. Among playwrights the major figures are Eugene O'Neill, Tennessee Williams and Arthur Miller. America has produced a number of distinguished poets in this century, including Robert Frost and Ezra Pound.

In classical music there are major figures such as Aaron Copland and Charles Ives. Composers who reached a wider audience include Leonard Bernstein and George Gershwin. However, America's greatest musical contribution has been jazz. Whether originating in New Orleans or elsewhere, it was developed in the interwar period by both white and black musicians, notably such legendary figures as King Oliver, Bix Beiderbecke, Paul Whiteman and Fletcher Henderson. In the postwar period jazz became somewhat fragmented: the devotees of the traditional version diverging sharply from those of the more experimental or modern variety associated with such figures as Charlie Parker, John Coltrane, Stan Getz and Charlie Mingus.

Probably no aspect of modern society is more associated with the United States than the motion picture. The development of the cinema in both its silent and talkie phases owed a great deal to American energy, invention and promotion. For many "Hollywood" – the center of the film industry in Los Angeles – became synonymous with the industry generally. The great stars of the silent screen included Keaton, Pickford

and Chaplin, while its preeminent director was D. W. Griffith. Some performers, like Chaplin, made the transition to the talkies but the new form produced its own stars. Of all the many types of film emerging from the United States none has attracted more attention than the musical.

In sport, too, America went its own way. Whereas the rest of the world became dominated by soccer, the United States developed its own game of (American) football. Originally popularized as a college game, after World War II it became enormously successful at the professional level, thanks substantially to television.

Notwithstanding that success, the national game of the United States remains baseball. Developed as a professional sport in the latter part of the 19th century, it retained in the 20th its associations with the very essence of American society. Although long regarded as the national game it was not until the 1950s that at the highest level it spread beyond the northeast and midwest. Then the departure of two New York teams, the Giants and the Dodgers, to San Francisco and Los Angeles respectively, produced a major upheaval in the sport.

The 20th-century contribution of the United States to science has been immense. Inventors like Thomas Edison changed the face of America, as did industrialists such as Henry Ford. The development of great universities able to attract talent from all over the world, as well as the movement of scientists seeking refuge in the United States have combined to produce great achievements. The happy combination of resources and brain power has made the United States the world leader in many branches of science. The development of nuclear weapons (originating in the Manhattan Project in World War II) is but one example of this supremacy. In the postwar period the application of science to space exploration has continued to sustain a characteristic belief in the solubility of problems. No other country has won so many Nobel Prizes.

▲ The space shuttle, an American technological achievement, in 1985.

◀ Exuberance infuses most American activities, especially sports.

MEXICO

UNITED MEXICAN STATES

CHRONOLOGY

1901	First major oil discovery
1910	Mexican Revolution and Civil War breaks out
1917	Mexican Constitution is drawn up
1928	Ruling *Partido Nacional Revolucionário* (PNR) established by Plutarco Calles.
1938	Expropriation of 17 US and UK oil companies in Mexico
1942	Mexico declares war on Axis Powers in return for US aid
1968	Repression of student unrest during and after Olympic Games, hosted by Mexico
1976	Oil boom, with rampant corruption and inflation
1982	Debt crisis and severe economic recession triggered by falling oil prices
1985	Devastating earthquake in Mexico City

ESSENTIAL STATISTICS

Capital Mexico City

Population (1989) 84,275,000

Area 1,958,201 sq km

Population per sq km (1989) 43

GNP per capita (1987) US$1,820

CONSTITUTIONAL DATA

Constitution Federal republic with two legislative houses (Senate; Chamber of Deputies)

Major international organizations UN; OAS; I-ADB

Monetary unit 1 peso (Mex$) = 100 centavos

Official language Spanish

Major religion Roman Catholic

Heads of government (President) P. Diaz (1884–1911); F. Indalecio Madera (1911–13); V. Huerta (1913–17); V. Carranza (1917–20); Gen. A. Obrégon (1920–24); P. Calles (1924–28); E. Portes Gil (1928–30); P. Ortiz Rubio (1930–32); A. Rodríguez (1932–34); Gen. L. Cárdenas (1934–40); Gen. M. Ávila Camacho (1940–46); M. Alemán Valdés (1946–52); A. Ruiz Cortines (1952–58); A. López Mateos (1958–64); G. Díaz Ordaz (1964–70); L. Echeverría Alvárez (1970–76); J. Portillo y Pacheco (1976–82); M. de la Madrid Hurtado (1982–88); C. Salinas de Gortari (1988–)

Mexico entered the 20th century under the corrupt, repressive regime of General Porfirio Díaz, who had risen to prominence as a leading general in the Liberal armies which had defeated the old Catholic-royalist Right in the mid 19th century. The Díaz regime presided over an oil boom which, by 1920, had given rise to the world's second largest oil industry. Mexican mineral output and foreign trade increased tenfold, manufacturing output trebled, and ranching and plantation crops boomed. By 1910 23 percent of Mexico's 15 million inhabitants lived in towns and cities swollen by the large-scale expulsion of Mexican Indians from their traditional village lands by big estate owners, ranchers and foreign-owned land companies; half Mexico's "modern sector" assets were foreign owned, and over 40 percent of farmland American owned.

Amid festering resentment of repression, corruption, dispossession and foreign domination, Díaz was ousted in 1911 by Francisco Madero, a rich northern landowner who promised to observe the Liberal constitution of 1857 and to restore usurped Indian lands to their rightful owners. Madero's presidency (1911–13) was more liberal than revolutionary. But he unwittingly unleashed a bloody revolution and civil war, involving an overall population decline of nearly one million and social forces beyond his control. In early 1913 Madero and his leading associates were captured and murdered on the orders of Mexico's chief of staff, General Victoriano Huerta, who tried to rally conservative forces behind a military dictatorship. Huerta was defeated in July 1914 by a combination of forces: northern "constitutionalist" landowners, industrialists; former cowboys, miners and adventurers, led by ex-bandit "Pancho" Villa; industrial workers; and a radical agrarian movement (the *agraristas*) led by Emiliano Zapata.

Having, together with General P. Calles, led the revolutionary coalition to victory over Huerta from March 1913 to July 1914, top man Venustiano Carranza and General A. Obregón won over Mexico's labor leaders with bribes and promises of higher wages, freedom of association and protective labor legislation. This alliance freed the wealthy US-backed "constitutionalists" to inflict decisive military defeats on Villa's cavalry and Zapata's *agraristas* in 1915. Zapata and Villa retreated into their respective native strongholds of Morelos and Chihuahua, where Zapata was murdered in 1919 and Villa in 1923. By then the *agraristas'* immediate goals had been included in the constitution of 1917. Article 27 provided for land to be redistributed to peasants and villages dispossessed of their communal property before the Revolution and to settlements with insufficient land or water for their inhabitants' needs.

Once installed in the presidency (1917–20), Carranza only paid lip service to the constitution's radical commitments. His priorities were to restore order, business confidence and his supporters' financial fortunes in a Mexico devastated and bankrupted by revolution and civil war. From 1910 to 1920, Mexico's foreign debt trebled; much of its infrastructure and industrial capacity was run-down or ruined; production of staple foods almost halved, mining was in crisis, and the military appropriated the bulk of state revenue. Through the 1920s and beyond, the great mass of the population remained, materially, in conditions worse than before the Revolution.

Carranza's successors accomplished rather more. In response to the country's agrarian commitments 7.4 million hectares of farmland were redistributed to peasant households from 1920 to 1934, mostly as the inalienable communal property of *ejidos* (village communes), and a further 18 million hectares under Cárdenas (1934–40), so that *ejidos* accounted for 47 percent of all cropland by 1940.

Carranza's successors also promoted widespread (but still far from universal) expansion of secular urban and rural public education, inducting Mexican Indians to a more Hispanic and national culture and identity. But secularization of the state and education, nationalization of Church properties, the secular 1917 constitution and state regulation of the clergy brought the new regime into bloody conflict with Catholic opposition, mainly in 1926–29, causing hundreds of deaths on both sides.

Carranza and Obregón were murdered in 1920 and 1929, respectively, so it fell to Calles to organize the *Partido Nacional Revolucionário* (PNR) in 1928. This ruling party, reconstituted as the *Partido de la Revolución Mexicana* (PRM) in 1938 and as the *Partido Revolucionário Institucional* (PRI) in 1946, has selected presidents and monopolized political power without interruption since 1928, creating the most remarkably stable and continuous regime in Latin America. Since 1934, every Mexican president has served a single uninterrupted six-year term at the head of an increasingly educated, self-perpetuating technocracy and an exceedingly corrupt party and union bureaucracy.

From the 1940s, industrialization proceeded apace, aided by political stability, rapid population growth, rampant urbanization and an abundance of cheap young labor. Mexico's population quadrupled from 20 million in 1940 to over 85 million in 1990. The urban share doubled from 35 percent in 1940 to 71 percent in 1990, and Mexico City expanded from 1.8 million inhabitants in 1940 to 18 million in 1982. Nearly half Mexico's population was under 15 in the 1980s.

Illusions of social peace and harmony were shattered by violent clashes between the police and students while Mexico hosted the 1968 Olympic Games, followed by a massacre of hundreds of protesting students by security forces on 2 October 1968. The PRI has been placed on the defensive ever since, as its methods and achievements have been increasingly called into question by social critics.

In 1982 world oil prices fell and Mexico had to devalue its currency drastically and reschedule the payments due on its US $80 billion foreign debt. It underwent painful austerity programs, economic liberalization and some major privatizations of state enterprises, aggravated by a devastating earthquake in Mexico City in 1985, further falls in oil revenues in 1985–86 and US recession in 1990. Real wages halved and there was zero or negative economic growth from 1982 to 1990.

▼ Mural of the revolution by David Alfáro Siqueiros.

NICARAGUA

REPUBLIC OF NICARAGUA

CHRONOLOGY

1909	President Santos deposed in US-backed coup
1933	US marines leave Nicaragua after truce
1956	Somoza Garcia is shot dead by nationalists
1972	Earthquake in Managua kills 10,000 people
1974	Nicaragua in state of civil war
1979	Sandinista guerillas overthrow Somoza dynasty
1987	Anti-Sandinista forces begin to attack across borders
1990	Sandinistas defeated in general election

ESSENTIAL STATISTICS

Capital Managua

Population (1989) 3,745,000

Area 130,700 sq km

Population per sq km (1989) 31.1

GNP per capita (1987) US$830

CONSTITUTIONAL DATA

Constitution Unitary multiparty republic with one legislative house (National Assembly)

Major international organizations UN; OAS; I-ADB

Monetary unit 1 Nicaraguan new córdoba (C$) = 100 centavos

Official language Spanish

Major religion Roman Catholic

Heads of government (President) J. Santos Zelaya (1894–1909); J. Madriz (1909–10); J. Dolores Estrada (1910–11); A. Diaz (1911–16); E. Chamorro Vargas (1917–20); D. Manuel Chamorro (1920–23); B. Martinez (acting) (1923–24); C. Solorzano (1924–26); E. Chamorro Vargas (Jan–Nov 1926); A. Diaz (1926–29); Gen. J. Moncada (1929–32); J. Bautista Sacasa (1932–36); C. Brenes Jarquin (acting) (1936–37); Gen. A. Somoza García (1937–47); L. Argüello (May 1947); B. Lescayo-Sacasa (May–Aug 1947); V. Roman y Reyes (1947–50); A. Somoza García (1950–56); L. Somoza Debayle (1956–63); R. Shick Gutiérrez (1936–66); L. Guerrero Gutiérrez (1966–67); Gen. A. Somoza Debayle (1967–72); Three-man military junta (1972–74); Gen. A. Somoza Debayle (1974–79); Five-man junta (1979–81); Three-man junta with D. Ortega (1981–85); D. Ortega Saavedra (1985–90); Violeta Barrios de Chamorro (1990–)

Independent from Spain since 1821, torn by civil wars and frequently raided by buccaneers, Nicaragua entered the 20th century under President José Santos Zelaya, who expelled the British from the Atlantic coast, bringing the whole country under central government's control. Zelaya also antagonized the United States by offering Germany and Japan concessions to build a canal across the isthmus and was deposed in 1909 by a Conservative coup supported by US troops. Under fire from nationalist guerrillas, the US troops occupied the country until 1933, when they turned over command of the National Guard to Anastasio Somoza García. Within three years, Somoza forced president Juan Bautista Sacasa to resign and, with the National Guard and Congress under this control, got himself elected president. That inaugurated 42 years of rule by the Somoza family, tolerated by the USA because of the Somozas' loyalty to US interests. Somoza García was finally shot by a nationalist in 1956. His son and successor, Luis Somoza, tried to separate the family from the formal exercise of power in order to protect the large fortune they had amassed, by getting proxies elected to the presidency, but that did not prevent in 1959 the example of the Cuban revolution prompting the development of guerrilla groups. These became organized into the Frente Sandinista de Liberacion Nacional (FSLN).

By 1974 the country was in a state of civil war between the National Guard and the Sandinista guerrillas. The murder of Conservative leader Pedro Joaquin Chamorro in 1978 led to massive support for the FSLN guerrillas, who entered Managua in July 1979. They appointed a Junta of National Reconstruction to govern the ravaged country. But tense relations with Washington gradually helped Sandinista hardliners to take over, eliminating initial attempts at pluralistic rule. By 1981, the US was training anti-Sandinista forces, or Contras, in neighboring Honduras and the government was looking to Cuba and the Soviet Union for support. Contra insurgents began to attack across the borders with Costa Rica and Honduras, the United States suspended all aid and trade links with Nicaragua, while United States government agencies became involved in clandestine operations against the Sandinista government. Elections conducted in February 1990 gave victory to an anti-Sandinista coalition, with Violeta Chamorro – widow of Pedro Joaquin – becoming president. The Contras agreed to disarm and to be relocated throughout the country and the Sandinistas relinquished control of the armed forces.

Economic and social trends

The Sandinistas attempted a radical transformation of the economy by nationalizing the property of the Somozas, who had come to own a quarter of the arable land, plus sugar mills, industrial installations, transport and commercial firms. An agrarian reform began by creating larger state farms and cooperatives, but ended by distributing land to individual peasants. The civil war, however, caused massive damage to agriculture, as did costs and the loss of US markets and credit to the economy. By 1988 inflation had reached 33,000 percent a year; although a drastic austerity program brought inflation down in 1989 that came too late for the Sandinistas' popularity to recover. However, the failure of US aid to materialize prevented Chamorro's government from rewarding former contras, or meeting workers' demands for wage rises. In the 1990s the potential for violence in Nicaragua remained high.

PANAMA

REPUBLIC OF PANAMA

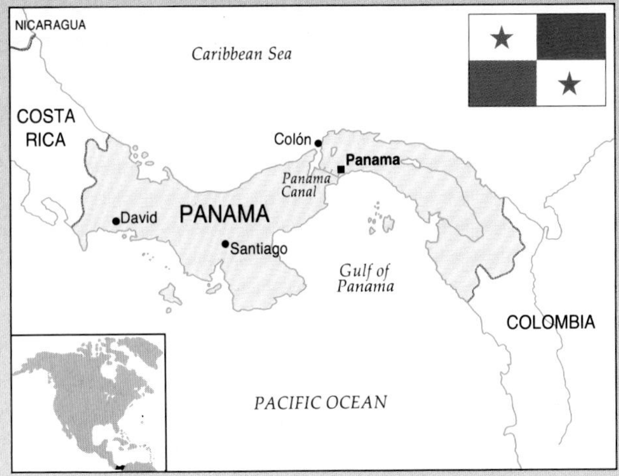

CHRONOLOGY

1902 US take over building of Panama Canal

1958 Panamanian flag is to be flown over Canal Zone

1988 General Noriega is charged with drug-trafficking

1989 Noriega declares war against United States

ESSENTIAL STATISTICS

Capital Panama City

Population (1989) 2,370,000

Area 77,082 sq km

Population per sq km (1989) 30.7

GNP per capita (1987) US$2,240

CONSTITUTIONAL DATA

Constitution Military regime

Date of independence 3 November 1903

Major international organizations UN; OAD; I-ADB

Monetary unit 1 balboa (b) = 100 cents

Official language Spanish

Major religion Roman Catholic

Heads of government since independence Revolutionary junta (1903); (President) M. Amador (1904–08); J. D. de Obaldia (1908–10); P. Arosemena (1910–12); B. Porras (1912–16); R. M. Valdés (1916–18); B. Porras (1918–24); R. Chiari (1924–28); F. H. Arosemena (1928–31); R. Alfaro (1931–32); H. Arias (1932–36); J. D. Arosemena (1936–39); A. Arias (1940–41); R. A. de la Guardia (1941–45); A. Jiménez (1945–48); D. Diaz (1948–49); J. A. Remón (1952–55); E. de la Guardia (1956–60); R. Chiari (1960–64); M. A. Robles (1964–68); (Military dictator) Lt.-Col. O. Torrijos (1968–77); (President) A. Royo (1978–82); R. de la Espriella (1982–84); N. Ardito Barletta (1984–85); E. A. Delvalle (1985–88); (Military dictator) Gen. M. A. Noriega (1988–89); G. Endara (1989–)

In 1880 a French company, the *Compagnie Universelle du Canal Interocéanique*, began to build a canal across the isthmus of Panama. However, this company went bankrupt in 1889, and was offered for sale in 1898. In 1902 US$40 million was made available to buy up the assets of the defunct French company, and to build a canal in Panama; the following year a treaty was made with Colombia on the canal. Panama declared independence on 3 November 1903 and the presence of US naval forces in the canal zone deterred the Colombians from crushing the rebellion. A new treaty was negotiated between the United States and Panama, gaining for the former the exclusive occupation and control of the Canal Zone. The constitution adopted by Panama provided for possible military intervention by the United States if necessary.

The first president of the republic was Manuel Amador Guerrero. Subsequent elections were accompanied by unrest, and there was frequent internal instability. Arnulfo Arias Madrid was elected president in June 1940, and demanded compensation for allowing US bases to be built outside the Canal Zone. He was removed in October 1941, and replaced with Ricardo Adolfo de la Guardia, who followed the United States into the war after the attack on Pearl Harbor on 7 December 1941, and allowed the transferral of the bases; of the 134 bases, 98 were returned to Panama after the war.

The postwar years saw political chaos in Panama but 1955 saw an increased flow of capital into the country, including a loan from the World Bank to finance the Inter-American Highway. In 1958, President de la Guardia requested that the Panamanian flag be flown in the Canal Zone, and the Spanish and language be given equal status there also, and this led to US president Eisenhower recognizing Panama's titular sovereignty over the zone and allowing the display of both Panamanian and US flags in certain parts of the Canal Zone.

The impeachment of president Mario A. Robles in May 1968 by Arias Madrid over the worsening economic situation led to the overthrow of the former; Arias Madrid was president for eleven days himself, before being removed in a coup. In 1974, the United States agreed that the Canal Zone would be eventually passed over to Panama, in the face of hostility from the US Congress.

Further conflict with the United States began toward the end of the 1980s. In February 1988 the leader of the Panamanian armed forces Manuel Noriega was charged by a US Federal court for drug-trafficking and dismissed from his post by president Eric Arturo Delvalle, though this move was ineffective due to Noriega's control of the government. Noriega replaced Delvalle with Manuel Solis Palma; the reaction of the United States was to freeze all Panamanian assets held in US banks. The situation became graver still with demonstrations during May 1989 against vote-rigging in the Panamanian elections. The election was condemned by the Americans, who severed ties with Panama after Noriega appointed Francisco Rodriguez as the new president in September. US president George Bush had ordered 2,000 more troops into the Canal Zone in May to reinforce the 11,000 already stationed there. This, and an attempted coup in October, unnerved Noriega, and on 15 December he declared war on the United States. This was the excuse the Americans needed and on 21 December a US invasion ousted the Panamanian dictator, who was taken to the United States to stand trial.

CUBA

REPUBLIC OF CUBA

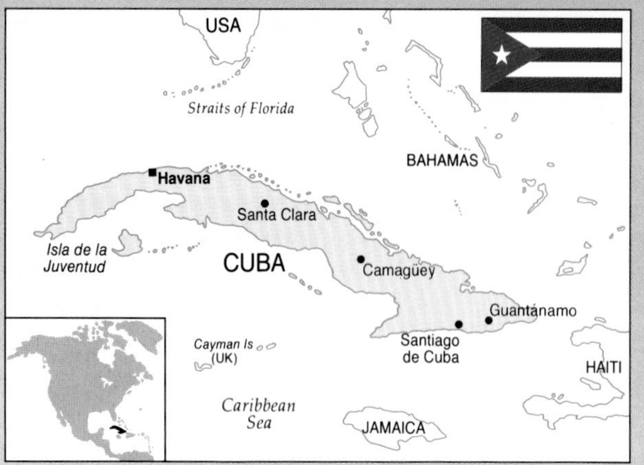

CHRONOLOGY

1902	US proconsul withdraws from Cuba
1906	US intervention following insurrectional activity
1909	US returns government to an elected administration
1929	President Gerardo Machado amends constitution
1933	Sergeants' coup led by Fulgencio Batista
1958	Fidel Castro's guerilla campaign overpowers Batista's dictatorship
1962	Cuban missile crisis
1973	Cuba and US sign five-year antihijacking pact
1980	Over 100,000 emigrés leave Cuba for the USA
1989	Castro warns Gorbachev of dangers of excessive reforms

ESSENTIAL STATISTICS

Capital Havana

Population (1989) 10,540,000

Area 110,861 sq km

Population per sq km (1989) 95.1

GNP per capita (1984) US$2,690

CONSTITUTIONAL DATA

Constitution Unitary socialist republic with one legislative house (National Assembly of the People's Power)

Date of independence 20 May 1902

Major international organizations UN; OAS; COMECON

Monetary unit 1 Cuban peso (CUP) = 100 centavos

Official language Spanish

Major religion Roman Catholic

Heads of government since independence (President) T. Estrada Palma (1902–06); U.S. administration (1906–09); J. M. Gómez (1909–13); M. García Menocal (1913–21); A. Zayas y Alfonso (1921–25); G. Machado y Morales (1925–33); A. Herrera (12 Aug 1933); C. de Céspedes y Quesada (Aug–Sep 1933); Military junta (Sep 1933); R. Grau San Martin (1933–34); C. Hevia (acting) (Jan 1934); C. Mendieta (1934–35); J. Barnet y Vinageras (1935–36); M. Gómez y Arias (May–Dec 1936); F. Laredo Bru (1936–40); F. Batista y Zaldívar (1940–44); R. Grau San Martin (1944–48); C. Prío Socarrás (1948–52); F. Batista y Zaldívar (1952–59); F. Castro (Prime minister 1959–62; Party leader 1962–76; President 1976–)

Cuba had been the center of the Spanish empire in the Americas. In 1898 it was lost by Spain as a consequence of the US-Spanish War and the invasion of the island by the Americans. In fact, a lengthy and bloody nationalist insurrection had by then confined Spanish forces to the main cities; the US invasion served to abort that struggle for independence. During three years of direct US rule, Washington imposed on Cuba the Platt amendment, a US congressional rule which asserted the right of the United States to intervene in Cuban affairs and to keep naval bases on the island. In 1902 the US proconsul withdrew, but renewed insurrection prompted another US intervention in 1906. After a wave of strikes and insurrection, the United States ended its intervention in 1909, returning government to an elected administration.

In 1929, President Machado amended the constitution so as to stay in office. A wave of violence and rebellions ended in 1933 with a military revolt overthrowing Machado, followed by a sergeants' coup led by Fulgencio Batista, who was to dominate politics for the most of the next 25 years, moving from a populist alliance which incorporated the Communists in the 1930s to a representative military dictatorship in the 1950s.

As a reaction to Batista's rule, in 1953 a group of young nationalists led by Fidel Castro made an unsuccessful attempt to overthrow the government. Despite its failure, Castro became the head of an anti-Batista movement which launched a guerrilla campaign in 1956 and overpowered the forces of the dictatorship by 1958. A process of socialist transformations and anti-American initiated by the new regime provoked a hostile reaction from the United States administration. Under attack from exiled groups supported by Washington, the Castro regime went into an increasingly closer alliance with the Soviet Union. This resulted in the missile crisis of October 1962, when the US government eventually prevailed, forcing the Soviets to withdraw their nuclear missiles from the island, but in exchange gave an undertaking that it would not invade Cuba. During the 1960s the revolutionary regime consolidated its political support, with nationalist groups, socialists and members of the pre-1959 Communist party becoming integrated into a new organization, which became the Communist party of Cuba. Throughout this period Cuba maintained an independent foreign policy, strongly supporting liberation movements in Latin America and Africa, and trying to remain neutral in the Sino-Soviet dispute. Policy failures at home and abroad meant that in the 1970s Cuba became once again more closely aligned with the Soviet Union, sending troops to Angola and Ethiopia, and becoming part of Comecon.

Economic and social trends

The economy has been traditionally dependent on sugar and foreign relations once revolved around access for Cuba's sugar to the North American markets. From the 1960s that dependency was replaced with one on the Communist bloc. Although the economy diversified after 1979, sugar remained central. Hopes of developing tourism as a source of hard currency earnings were blocked by shortages of essential inputs, as well as by lack of investment. The industrial sector developed fast after 1980, but in 1987 the lack of hard currency or credit to import spare parts led to a decline in industrial production.

The main achievements of the revolutionary experience have been in the areas of health, nutrition and education. In the early 1990s life expectancy, diet and schooling compared well with figures for the industrialized countries; the same applied to access to medical services. Dissent, however, was fueled in later years by the reappearance of shortages of essential consumer goods. Such shortages, had practically disappeared by the mid-1980s but were revived by major reductions in Soviet supplies and a drop in export earnings.

▲ **Fidel Castro in 1959.**

◀ **Santa Clara is liberated by Castro's forces, 1959.**

BARBADOS

BARBADOS

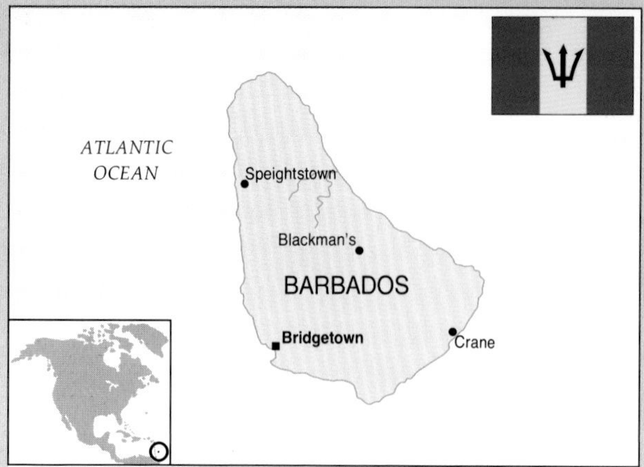

CHRONOLOGY

1924	Democratic League founded
1937	Social and industrial unrest
1944	Adams starts the Barbados Labor Party
1946	The Bushe Experiment
1951	Adult suffrage
1954	Ministerial government introduced
1955	Errol Barrow forms Democratic Labor Party
1958	Barbados joins West Indies Federation
1961	Full internal self-government
1966	Independence from Britain
1983	Barbados provides base for US military intervention in Grenada
1991	DLP returned to power under Erskine Sandiford

ESSENTIAL STATISTICS

Capital Bridgetown

Population (1989) 255,000

Area 430 sq km

Population per sq km (1989) 593

GNP per capita US$5,330

CONSTITUTIONAL DATA

Constitution Constitutional monarchy with two legislative houses (Senate; House of Assembly)

Date of independence 30 November 1966

Major international organizations UN; OAS; Commonwealth of Nations; ACP; I-ADB; CARICOM

Monetary unit 1 Barbados dollar (BDS$) = 100 cents

Official language English

Major religion Anglican

Heads of government since independence Errol Barrow (1966–76); Tom Adams (1976–85); Bernard St John (1985–86); Errol Barrow (1986–87); Erskine Sandiford (1987–)

Barbados, the most easterly of the Caribbean islands, has had the longest experience of self-government of any island in the Caribbean. The House of Assembly, composed of European settlers or local whites, was first established in 1639. The colonizers surrendered their patent to the British Crown in 1663 but the House of Assembly has endured. This long period of local control helps explain the sturdy independence of the Barbadian character. Never having been occupied by any other power than Britain, Barbadians have traditionally seen themselves as "little England" in the Caribbean.

For most of the first half of the 20th century, Barbados politics were almost exclusively controlled by the white planter class. However, the voice of the black middle and working classes demanded to be heard and was represented by such organizations as the Democratic League, founded by Dr Charles Duncan O'Neale in 1924. By the late 1930s, however, the League was defunct and mass party politics in Barbados did not really begin until 1944, when Grantley Adams added a political wing, the Barbados Labor party (BLP), to his pressure group, the Barbados Progressive League.

Despite the dominance of the sugar barons, the BLP won eight seats in the 1944 election, with the newly formed Congress party also picking up some seats. The BLP again won the majority of seats in the 1946 election. The governor, Sir Grattan Bushe, asked Adams to suggest names of elected members for appointment to the executive committee, which administered the island's affairs. This became known as the Bushe Experiment. By 1951 the BLP was well established as the main party in the House of Assembly. Formal ministerial government was introduced three years later, with Adams as the first premier. In 1955, Errol Barrow broke away from the BLP and founded the Democratic Labour party (DLP). The two parties have alternated in government since then. Both parties espoused a Liberal-democratic ideology, with the DLP to the "left".

Economic and social trends

Barbados entered the 1900s in a state of economic depression, the world price of sugar, its only cash crop, having fallen dramatically in the latter part of the 19th century.

The early years of the 20th century saw the first mass emigration of Barbadians to the United States and Panama, where they helped build the canal. Those who returned bought parts of the white-owned sugar estates and, by 1920, there were about 13,000 black peasant farmers. Sugar again went into decline in the 1920s and 1930s, the situation being made worse by the world depression which began in 1929. In 1937, there were riots in Barbados caused by the poor economic and social conditions of the working class.

Barbados prospered again under the impact of World War II, with sugar prices reviving and a US naval base being established on the island. After the war, population growth became a serious problem, and the island had one of the world's highest population densities. Many Barbadians emigrated to England in the 1940s and 1950s, until the British government imposed controls. In the early 1950s a new tourist industry was born. By 1990 tourism was the largest earner of foreign exchange for the island. Barbadians are generally regarded as the most conservative of the Caribbean peoples. The education system has retained a heavy emphasis on classical subjects and the arts.

JAMAICA

JAMAICA

CHRONOLOGY

1938	Social unrest. People's National party formed (PNP)
1943	Jamaica Labor party founded
1944	Universal adult suffrage granted
1953	Full ministerial system established
1955	PNP comes to power for first time
1958	Jamaica joins West Indies Federation
1959	Internal self-government introduced
1961	Jamaicans vote in referendum to secede from the Federation
1962	Jamaica becomes independent in August
1972	Michael Manley wins first election and becomes prime minister
1980	JLP returns to power under new leader, Edward Seaga
1989	PNP back in power under Michael Manley

ESSENTIAL STATISTICS

Capital Kingston

Population (1989) 2,376,000

Area 10,991 sq km

Population per sq km (1989) 216.2

GNP per capita (1987) US$960

CONSTITUTIONAL DATA

Constitution Constitutional monarchy with two legislative houses (Senate; House of Representatives)

Date of independence 6 August 1962

Major international organizations UN; Commonwealth of Nations; OAS; ACP; IADB; CARICOM

Monetary unit 1 Jamaican dollar (J$) = 100 cents

Official language English

Major religion Protestant

Heads of government since independence (Prime minister) Sir A. Bustamante (1962–67); H. Shearer (1967–72); M. Manley (1972–80); E. Seaga (1980–89); M. Manley (1989–)

Jamaica has had a form of representative government since the 1880s, when the British government agreed that nine members could be elected to the Legislative Council. Pressure from those elected members led to more attention being paid to education and infrastructure. In 1894, the railway from Kingston to Montego Bay was completed. By the turn of the century, the area of land under cultivation had greatly increased, thanks to the tens of thousands of peasant farmers, ex-slaves, who had become the backbone of the economy.

Despite its relatively small population, Jamaica sent 10,000 men to fight on the Allied side in World War I. A series of misfortunes in the 1920s and 1930s – the failure of the banana crop through disease, the collapse of sugar prices and several hurricanes which did extensive damage – were the main causes of the economic depression which gripped the island at that time. As elsewhere in the Caribbean, this sparked off social unrest, culminating in the riots of 1938.

Alexander Bustamante, one of the main leaders of the rebellion, was arrested by the colonial authorities. On his release, he formed the first trade union to be recognized in Jamaica – the Bustamante Industrial Trade Union (BITU). The same year, his cousin, Norman Manley, founded the first political party, the People's National party (PNP). The two worked closely together until Bustamante was interned in 1940. He was released in 1942 and, a year later, started his own party, the Jamaica Labor party (JLP). Manley countered with his own union, the Trade Union Congress (TUC), and the rivalry for political and union leadership intensified between the two. Under Manley, the PNP espoused a Jamaican form of the British socialism of the day and attracted the intellectual middle class, the civil servants and some sections of the working class.

The JLP, despite its name, was a more "pragmatic" party, taking its cue from Bustamante's lack of interest in any formal ideology. It was supported by middle-class businessmen and small traders, as well as many trade unionists.

In 1951, a new governor, Sir Hugh Foot, arrived in Jamaica. A progressive thinker, he wanted to see the island advance constitutionally at a faster pace. By 1953, Jamaica had been granted a further measure of control over its own affairs, which gave responsibility to elected members under a ministerial system. The PNP finally overturned the JLP's hold on power in 1955 and by 1959 there was full internal self-government.

A referendum took Jamaica out of the West Indies Federation and in 1962 it proceeded to independence on its own. Since independence, control of the government has alternated between the JLP and the PNP, at roughly ten-year intervals.

Economic and social trends

Jamaica was historically dependent on sugar, bananas and citrus before the post-1945 boom in bauxite and tourism. Jamaica has over 1,000 factories – but heavy unemployment, which by 1990 stood at about 20 percent of the workforce.

A "back to Africa" movement inspired by Marcus Garvey in the 1920s and 1930s became, after World War II, the focal point around which the deprived elements in the country could rally. The movement took the name "Rastafarian", after the Ethiopian Emperor Haile Selassie whose title, before he was crowned Emperor in 1930, was Ras Tafari. After independence the Rastafarian cult, with its emphasis on marijuana smoking and "dreadlocks", remained strong.

TRINIDAD & TOBAGO

REPUBLIC OF TRINIDAD AND TOBAGO

CHRONOLOGY

1903	Destruction of the Red House by angry mob
1929	Cipriani revives Trinidad Workingmen's Association
1921	Wood Commission recommends electoral change
1936	Butler party formed
1937	Riots in oil belt
1939	Moyne Commission recommends further constitutional change
1945	Adult suffrage. Albert Gomes first enters parliament
1956	Dr Williams and the PNM gain power
1962	Trinidad and Tobago becomes independent
1976	Queen replaced by nonexecutive president
1980	Tobago House of Assembly established with local autonomy
1981	Dr Williams dies, succeeded by George Chambers
1986	NAR comes to power, overturning 30 years of PNM rule

ESSENTIAL STATISTICS

Capital Port of Spain

Population (1989) 1,285,000

Area 6,128.4 sq km

Population per sq km (1989) 250.6

GNP per capita (1987) US$4,220

CONSTITUTIONAL DATA

Constitution Multiparty republic with two legislative houses (Senate; House of Representatives)

Date of independence 31 August 1962

Major international organizations UN; Commonwealth of Nations; OAS; ACP; I-ADB; CARICOM

Monetary unit 1 Trinidad and Tobago dollar = 100 cents

Official language English

Major religion Roman Catholic

Heads of government since independence (Prime minister) Dr Eric Williams (1956–81); George Chambers (1981–86); Raymond Robinson (1986–)

In 1903 the Red House (seat of government) in Port of Spain was destroyed by angry citizens incensed by a lack of water. Nevertheless, political activity in Trinidad and Tobago remained largely unfocused and undirected until Capt. Arthur Cipriani revived the old Trinidad Workingmen's Association (TWA) and called the first strike in the city's history, bringing out the workers on the Port of Spain waterfront in a demand for better pay and working conditions.

Cipriani then agitated for full adult franchise and Britain sent out a Royal Commission in 1921, under Major E.L.F. Wood. Wood recommended that an elective element be introduced into the Legislative Council. In the 1925 election, Cipriani handsomely won the Port of Spain seat. He set about pursuing social and humanitarian causes, including the legal recognition of trade unions, the abolition of child labor, the introduction of old age pensions and compulsory primary education.

Cipriani later transformed the TWA into the Trinidad Labor Party but was considered by his colleagues not militant enough for the times. One of these was Tubal Uriah (Buzz) Butler, a Grenadian national, who decided to form his own British Empire Workers' and Citizens' Home Rule Party in 1936. He organized workers in the petroleum industry.

In 1937, Butler called a strike in the oil belt, which led to riots and general social disturbances that continued into 1938. Sought by the authorities, Butler went into hiding but emerged in 1939 to give evidence to another Royal Commission appointed by the British Government, the Moyne Commission. The Commission confessed itself appalled by the social and economic conditions then existing.

The Moyne report led to the introduction of adult suffrage in 1945, and in the 1946 election, Butler's party captured the largest number of seats in the Legislative Council, but was not invited to have any say in government. Albert Gomes, a Trinidadian of Portuguese descent, was regarded as the *de facto* head of the elected element in the Council. Further constitutional reforms in 1950 saw Gomes returned to the Legislative Council, this time under the banner of a conservative, business-oriented group. By 1956 politics was taking on a whole new tone, with the arrival on the scene of the Oxford-educated Dr Eric Williams and his People's National Movement (PNM) – the first properly organized and funded party in the country's history. Williams's intellectual approach struck a responsive chord in the electorate. He remained in power for 25 years thereafter, until his death in 1981. Williams was succeeded by George Chambers in 1981 and the PNM was finally defeated in the 1986 election by a coalition of opposition forces called the National Alliance for Reconstruction (NAR). The coalition was showing signs of disintegration by the late 1980s.

Economic and social trends

Trinidad and Tobago's economy has been dominated by oil since the early 1900s. Older agricultural industries, such as sugar, cocoa, coffee and citrus, still survive, however. Large natural gas deposits have been used in huge petrochemical plants. Gas also provides cheap electricity for a new steel industry. A thriving light manufacturing sector makes a variety of goods for sale in the region and abroad.

Trinidad and Tobago's annual Carnival festival lasts for two days and features large costumed bands parading through the streets, depicting historical, biblical or contemporary themes.

GUYANA
CO-OPERATIVE REPUBLIC OF GUYANA

CHRONOLOGY

1917	Indentured labor ends
1947	Dr Jagan first elected to Legislative Council
1948	Rioting at Enmore estate
1953	People's Progressive party (PPP) secures firm mandate but constitution suspended
1955	Burnham forms People's National Congress (PNC)
1962	General unrest and social disorder
1964	Burnham comes to power on basis of proportional representation
1966	Independence; name changed to Guyana
1980	Guyana substitutes monarchy for executive presidency
1985	Burnham dies
1989	Privatization commences

ESSENTIAL STATISTICS

Capital Georgetown

Population (1989) 754,000

Area 215,083 sq km

Population per sq km (1989) 3.5

GNP per capita (1987) US$380

CONSTITUTIONAL DATA

Constitution Unitary multiparty republic with one legislative house (National Assembly)

Date of independence 26 May 1966

Major international organizations UN; Commonwealth of Nations; ACP; I-ADB; CARICOM

Monetary unit 1 Guyana dollar (G$) = 100 cents

Official language English

Major religion Christian

Heads of government since independence (Prime minister) Forbes Burnham (1966–80); (President) Forbes Burnham (1980–85); Desmond Hoyte (1985–)

By 1917 British Guiana's population of 300,000 was made up of Indians, numbering about 126,000 Africans, about the same, with Portuguese, Chinese, British and native Amerindians making up the rest. The franchise, however, was strictly limited to the white-planter class. A British Guiana Commission produced a critical report, recommending constitutional changes that would give the local middle class greater say in the colony's affairs by increasing the elected element in the new Legislative Council established in 1928. But the colonial Governor still held sway, through the nonelected members whose appointment he controlled. One forward step was the extension of the franchise to women, though restrictions with regard to who could exercise it remained in force for both men and women. It was not until 1943 that the elected members in the legislature gained numerical ascendancy. Full adult franchise finally arrived in 1953. World War II forced the postponement of the 1943 election. It was eventually held in 1947 and resulted in the election of Dr Cheddi Jagan, talented son of an East Indian plantation foreman, who ran as an independent.

Jagan later formed the People's Progressive Party (PPP) and brought in an articulate African lawyer, Forbes Burnham, to be chairman and to signify "Afro-Asian solidarity". The 1953 election was held under the new constitution, which provided for a House of Assembly of 24 elected members. The PPP secured a firm mandate, winning 18 seats. But the policies of the PPP ministers, which included secularization of education and more power to the unions, alarmed the British government, which promptly suspended the new constitution, dismissed the ministers and sent in troops.

By 1955, Burnham, a ruthless and ambitious man, had parted company with Jagan and formed the People's National Congress (PNC). He took the PPP's African supporters with him and British Guianan politics was henceforth more or less divided along racial lines. In 1957 the constitution was restored. The country quickly descended into social chaos following the 1961 PPP victory, with widespread arson and looting in Georgetown, the capital. More disorder and strikes followed in 1963, after an abortive attempt in London to negotiate full independence. The PPP won the largest number of seats in 1964, based on votes received, but Burnham and the small, right-wing United Force (UF) party teamed up to form the government. Independence was quickly granted in 1966. The PNC has, on its own, won every election held under proportional representation since then.

Economic and social trends

The PNC pursued a determined nationalization policy under Burnham's leadership in the 1970s and early 1980s. Public ownership, however, produced few benefits. It drained the treasury, frightened off new investors and led to decreases in production. President Desmond Hoyte realized in the late 1980s that Guyana's only hope lay in reversing the nationalization strategy and privatization of large chunks of government-owned assets.

Bauxite, sugar and rice remained the three pillars of the economy; but strikes in 1989 damaged production of the first two. By 1990 an increasing number of Indians supported the PNC. Large-scale emigration to North America and Trinidad and Tobago acted as a safety valve for those who saw little hope in the country.

VENEZUELA

REPUBLIC OF VENEZUELA

CHRONOLOGY

1899	Castro marches on Caracas
1938	General Conteras begins his three-year plan
1945	Revolution brings the Acción Democrática party to power
1948	Military coup headed by Delgado Chalbaud and Pérez Jimńez
1950	Delgado Chalbaud is assassinated
1958	Military junta seize power
1989	Venezuela's foreign debt of US$32 billion is suspended

ESSENTIAL STATISTICS

Capital Caracas

Population (1989) 19,246,000

Area 912,050 sq km

Population per sq km (1989) 21.1

GNP per capita (1987) US$3,230

CONSTITUTIONAL DATA

Constitution Federal multiparty republic with two legislative houses (Senate; Chamber of Deputies)

Major international organizations UN; I-ADB; OPEC

Monetary unit 1 bolivar (b) = 100 céntimos

Official language Spanish

Major religion Roman Catholic

Heads of state (President) Gen. C. Castro (1879–1908); Gen. J. Vicente Gómez (1908–35); Gen. E. López Conteras (1935–41); Gen. I. Medina Angarita (1941–45); R. Betancourt (1945–48); R. Gallegos Freire (Feb–Nov 1948); Lt.-Col C. Delgado Chalbaud (1948–50); G. Suarez Flamerich (1950–52); Col. M. Pérez Jiménez (1952–57); Rear-Adm. W. Larrazábal Ugueto (Jan–Nov 1958); E. Sanabria (1958–59); R. Betancourt (1959–63); R. Leoni (1964–69); R. Caldera Rodriguez (1969–74); C. Pérez Rodriguez (1974–78); L. Herrera Campins (1979–83); J. Lusinchi (1984–89); C. Pérez Rodriguez (1989–)

Until 1821, Venezuela was dominated by Spain; Creoles – native born whites – owned all the land, which comprised most of the colony's wealth. It was two creoles, Simón Bolívar and Francisco de Miranda, who led the South American independence movement in 1810-25. In 1821, Venezuela, with Colombia and Ecuador, was part of the republic of Gran Colombia, but it seceded in 1830.

The year 1899 marked the beginning of a new era in Venezuelan history. A *caudillo* from the Andean state of Tachira, General Cipriano Castro, marched on Caracas and took control of the presidency. For the next 59 years, with only a short interlude in 1945-48, Venezuela was controlled by five "Andinos" from Tachira. Castro's rule was characterized by repression, economic mismanagement, internal unrest and foreign intervention. The latter came about due to Castro's refusal to pay reparations for foreign property damaged during insurrections within the country; a blockade of the Venezuelan coast in 1902-03 by British, German and Italian warships was the response. Castro departed for Europe in 1908 to receive medical treatment, and his powers were taken over by General Juan Vicente Gomez. Gomez was dictator of Venezuela until his death in 1935: he abolished elections, freedom of the press and organized political activity, and effectively suppressed all opposition.

This "stability" encouraged foreign investors to enter Venezuela – the Royal Dutch-Shell combine just before World War I, and immediately after, the US company Standard Oil. By 1928 only the United States surpassed Venezuela in oil production, and the latter had become the world's leading exporter. High petroleum revenues during the 1930s preserved

the country from financial collapse during the Great Depression, and paid for agricultural subsidies, greater government revenues and increased trade. Also expanded was the country's infrastructure. The entire foreign debt was repaid and the budget deficit reduced. In spite of this wealth, Venezuela still had a society divided sharply between rich and poor.

Gomez's successor, Eleazor López Conteras, restored civil liberties and political activity, though an interlude of dictatorship in 1937 kept the opposition from becoming too threatening. He made some concessions to the poor, building public hospitals and schools as part of a three-year plan that began in 1938. His successor, Isaias Medina Angarita, continued this initiative when he took over in 1941. Medina increased the nation's share of the oil revenues as they were reduced by the effects of World War II during 1941 and 1942. Increased revenues during the last two years of the war boosted the economy, but in spite of this, the Medina administration was abruptly overthrown in a revolution that placed the first popularly-supported party, the Acción Democrática, in government, under the leadership of Rómulo Betancourt. This civilian-military junta adopted a liberal-left constitution in 1947, and in the following year novelist Romulo Gallegos was made president. The new government immediately began implementing a series of widespread and profound reforms: the oil revenues were to be shared out half to the nations, half to the oil companies; unions were encouraged; support was given to social services and industrial and agricultural development. This was too much for the conservative groups, who staged a coup in 1948. From 1952 until 1957 the country was ruled by Major Marcos Pérez Jiménez, one of the coup leaders, and he reversed the reforms, outlawing unions, opposition parties and press freedom. His selfishness in increasing his own and his associates' wealth led to a military coup, after which a civilian-military junta ruled Venezuela again until Rómulo Betancourt was elected president.

In his second stint as president Betancourt adopted a more moderate line. Modernization programs, in agriculture and industry, were launched, and there was a drive to increase literacy. The 1960s saw large revenues from the oil and iron-ore industries, and this helped the reform programs that had been launched at the beginning of the decade. Venezuela acquired majority ownership of foreign banks and the natural gas industry, and nationalized iron-ore and petroleum industries.

The periods in office of Carlos Andrés Pérez Rodriguez (elected in 1973) and Luis Herrera Campins (elected 1978) were widely criticized for their management of the oil revenues. In 1983, the newly elected president Jaime Lusinchi, an Acción Democrática candidate, announced austerity measures, as well as a campaign against corruption. Uncertainty over international petroleum prices made reforms difficult, though the rescheduling of Venezuela's foreign debt in 1986 provided a needed boost, as did the suspension in 1989 of the repayment of Venezuela's US$32 billion foreign debt.

Economic stagnation was at the root of another more insidious problem, drug-trafficking in neighboring Colombia. In 1988 Colombia and Venezuela agreed to increase military presence along their mutual border, and to cooperate in attempting to suppress drug-trafficking in this region.

▼ **Oil installation on and around Lake Maracaibo.**

BRAZIL

FEDERAL REPUBLIC OF BRAZIL

CHRONOLOGY

1930 Coup d'état led by Getúlio Vargas

1932 Defeat of São Paulo rebellion

1942 Declaration of war on the Axis Powers

1964 Military takeover

1967 Initiation of Brazilian "economic miracle"

1968 Widespread terrorism, repression and torture begins

1985 Transition to civilian rule begins

ESSENTIAL STATISTICS

Capital Brasília

Population (1989) 147,404,000

Area 8,511,965 sq km

Population per sq km (1989) 17.3

GNP per capita US$2,020 (1987)

CONSTITUTIONAL DATA

Constitution Multiparty federal republic with two legislative houses (Federal Senate; Chamber of Deputies)

Major international organizations UN; OAS; I-ADB

Monetary unit 1 new cruzado (NCz$) = 100 centavos

Official language Portuguese

Major religion Roman Catholic

Heads of government (President) R. Alves (1903–07); A. Pena (1907–09); N. Peçanha (1909–11); H. de Fonseca (1911–14); W. Pereira Gomes (1914–18); D. Ribeiro (acting) (1918–19); E. Pessoa (1919–22); A. Bernardes (1922–26); W. Pereira de Souza (1926–30); Military junta (Oct 1930); G. Vargas (1930–45); J. Linhares (1945); Gen. E. Dutra (1946–50); G. Vargas (1951–54); J. Café Filho (1954–55); C. Luz (acting) (Nov 1955); N. de Oliveira Ramos (acting) (1955–56); J. Kubitschek (1956–60); J. Quadros (Jan–Aug 1961); J. Goulart (1961–64); Marshal H. Castelo Branco (1964–67); Marshal A. da Costa e Silva (1967–69); Gen. E. Garrastazú Médici (1969–74); Gen. E. Geisel (1974–79); Gen. J. Figueiredo (1979–85); J. Sarney (1985–89); F. Collor de Mello (1989–)

Brazil won its independence in 1822 from Portugal. Emperors Dom Pedro I (r.1822–31) and Dom Pedro II (r.1840–89) gave Brazil "a political center of gravity" and fostered an enduring centralized parliamentary system through which wealthy plantation- and mine-owners exercised considerable control over taxation and legislation. The constitution of 1891 established a federation of 20 states, the United States of Brazil, with a strong directly elected executive. The president and the army together inherited most of the prerogatives previously vested in the emperor. The 1891 constitution defined the army as a permanent institution entrusted not only with national defense, but also with safeguarding law, order and the working of the constitutional organs of government. It was enjoined to obey civilian governments only in so far as they observed the rule of law, a requirement that proved crucial to later Brazilian politics.

From 1894 to 1930 the presidency had incestuously alternated between candidates backed by Brazil's two most powerful states; but the rapidly industrializing "coffee state" of São Paulo's President Washington Luís Pereira de Sousa (1926–30) broke the rules of the game by rigging the election of another "Paulista" to the presidency in 1930, instead of installing the candidate backed by the state of Minas Gerais, Getúlio Vargas. Undaunted, Vargas capitalized on the growing disenchantment with Brazil's corrupt and oligarchic "first republic".

Vargas was installed in the presidency by a military coup amid the economic crisis induced by the 1930s' world depression. He had a capacity to hold together his heterogeneous liberal alliance of the "national bourgeoisie" with interests somewhat divergent from those of the hitherto dominant plantation-owners and foreign capitalists. In 1930 Vargas established an interventionist Ministry of Labor, Industry and Trade, reflecting a paternalistic concern to conciliate labor.

After destroying both right- and leftwing opposition in 1935–38, Vargas retracted his earlier commitment to hold elections in 1938 and appointed himself to another term as president (1938–43, later extended to 1945). He aligned Brazil with the United States against the Axis Powers in World War II. Brazil not only declared war on Germany and Italy in August 1942 and supplied key materials such as natural rubber and quartz to the Allies; it also allowed the United States to use vital Brazilian air and naval bases and participated (with distinction) in the liberation of occupied Europe in 1944–45. In return, the United States provided military training, Lend-Lease aid and capital for the construction of Brazil's integrated steel mill at Volta Redonda – the largest in Latin America.

By 1945 Vargas could no longer head off pressure for free elections. In 1943–44 opponents of fascism and dictatorship formed the *União Democrática National* (UDN), which quickly became the main party of Brazil's powerful landowning, commercial and pro-American interests. At the same time, Vargas launched a Brazilian Workers' party (PTB) and a middle-class Social Democratic party (PSD). He hoped to head an electoral coalition of the workers he had patronized and the expanding "national bourgeoisie", to defend Brazilian industry and workers against fresh inroads and domination of American capitalists. But the army (with US and UDN encouragement) forced Vargas into (temporary) retirement in October 1945.

General Dutra, the defense minister who ousted Vargas, easily won the December 1945 elections. As president, Dutra

aligned Brazil behind the US in the Cold War, outlawed the Communist party, purged the trade unions, squandered the large foreign exchange reserves amassed by Brazil during the war years, opened the doors to foreign capital and white immigration, and won US approval. But Dutra's attempt in 1948 to open up oil exploration and production in Brazil to foreign oil companies unleashed a wave of nationalist indignation. This nationalist backlash helped Vargas to win almost half the vote in the 1950 presidential election.

The final Vargas presidency (1951–54) walked a tightrope between nationalist concern to regulate foreign capital and Brazil's dependence on foreign capital to finance costly industrial and infrastructural projects. Moreover, clashes between the finance ministry's policy of wage restraint and labor minister João Goulart's campaign to raise minimum wages drastically led to a damaging dismissal of this Vargas protégé. Finally, amid allegations of involvement in financial scandals and in the course of an attempted assassination of his arch-critic, journalist Carlos Lacerda, Vargas shot himself on 24 August 1954.

The 1955 elections were won by PSD-candidate Dr Juscelino Kubitschek, the dynamic big-spending state governor of Minas Gerais, with PTB-leader João Goulart as running mate. Army attempts to prevent their investiture were thwarted. Kubitschek's presidency was characterized by unbounded confidence in Brazil's "great future": he built up infrastructure and manufacturing industries; planned a futuristic new federal capital (Brasília) in the middle of nowhere; undertook overamibitious development; defiantly rejected IMF prescriptions; and left a legacy of rocketing inflation, deficits and debt.

Jânio Quadros, the independent reformist "new broom" who decisively won the 1960 presidential election, inherited an economy in crisis. President Quadros began by denouncing the profligacy and massive misappropriation of funds under Kubitschek, implicating vice-president Goulart in alleged large-scale corruption. He announced an orthodox stabilization and austerity program and more controversial agrarian, fiscal and currency reforms, while securing emergency loans from the new Kennedy administration in the United States and cultivating closer relations with Eastern Europe, China and Castro's Cuba. But his austerity measures antagonized key elements in the labor movement and the propertied classes. After eight months in office, Quadros suddenly announced his resignation. He apparently expected to be recalled and given emergency powers, but he miscalculated.

Vice-president Goulart became president from September 1961 to March 1964. The propertied classes and the senior military resisted Goulart's presidency, considering him dangerously radical; but he was widely supported, and his opponents eventually accepted him, though only as a figurehead; they established a premiership and ministerial responsibility to Congress, to strip the president of much of his

▲ **Carnival floats in Rio de Janeiro.**

power. Goulart and successive prime ministers were often at loggerheads. The deadlock was not broken until January 1963 when a plebiscite restored full executive power to the President. Bold stabilization plans were initiated and US and IMF backing was secured by finance minister Santiago Dantas and economist Celso Furtado, as annual inflation exceeded 60 percent and employment failed to keep up with population growth. But when the austerity and US and IMF backing of the Dantas-Furtado "three-year plan" brought Goulart into conflict with his chief supporters he ditched the Plan and moved leftward. There was an unprecedented radical upsurge in late 1963 and early 1964 when Goulart was deposed by an almost bloodless coup and until 1985 Brazil was ruled by military dictators in alliance with civilian technocrats. General Castello Branco (1964–67) presided over a tough deflationary austerity program. This paved the way for the so-called "economic miracle" under Generals Costa e Silva (1967–69) and Médici (1969–74), when the economy grew by 10 percent a year. However, the high social costs of the initial austerity program (and of the subsequent boom) engendered mass protests and strikes, eliciting strong repression followed by widespread urban terrorism, political kidnapping and guerrilla warfare in 1969–73. This precipitated massive and ferocious repression by the security forces and off-duty "death squads". Mounting international criticism of the widespread killing and torture of "suspects", potential informants and radical critics by the security forces, combined with defeat of the main guerrilla movements in 1973, reduced rather than ended human rights violations under General Geisel (1974–79). In 1974, however, he permitted relatively free congressional elections, in which the liberal-conservative opposition party (the PMDB) trounced the government party (the PSD). In 1974-81 Brazil's economic boom was slowed down by the steeply rising cost of imported energy, resurgent inflation and industrial unrest, large trade deficits and the growing burden of external debt. General Figueiredo (1979–85) presided over economic recession, 100 percent a year inflation, widespread strikes and unrest, the first of several partial defaults on payments to Brazil's creditors and a gradual return to elective government.

President José Sarney (1985–89) had neither the will nor the authority to carry through his land reform plans (violently resisted by the landlord class). His 1986 "Cruzado Plan" to freeze prices and establish a new stable currency (the Cruzado) was a costly failure (by 1989 inflation had reached 1,765 percent a year). And he cynically exploited Brazil's debt crises as a xenophobic device to divert public attention from the more fundamental *internal* causes of Brazilian poverty: excessive concentration of income and wealth in the hands of the few; massive misappropriation of public funds; and routine violation of even the most basic human rights of the poor.

The presidential election of December 1989 was won by Fernando Collor de Mello, governor of Algoas, where his wealthy rightwing family enjoyed baronial power. Collor won on a platform of drastic economic liberalization, privatization and deregulation; commitments to promote social justice and to halt reckless capitalist despoliation of Amazonia; and the promise to reduce inflation to under ten percent per month in one year, by slashing public expenditure, public employment and import tariffs and implementing a draconian liquidity squeeze.

Economic and social trends

Brazil has the world's fifth largest territory, the sixth largest population and the eleventh largest economy. It is a multiracial society and racial melting-pot. Nevertheless racial inequalities have long existed and they undoubtedly involve racial discrimination and racial stereotyping. In 1976 the average income of whites was twice that of nonwhites. Historically, the Portuguese married and had sexual relations with nonwhites to far greater degrees than did other European imperial nations and Brazil has never had legal segregation of races or laws against miscegenation.

In 1925, only 23 percent of Brazil's population lived in towns and cities; the proportion rose to 75 percent in 1988. The city of São Paulo grew from 31,000 people in 1872 to 579,000 in 1920, 2.2 million in 1950 and 14 million in 1988, becoming one of the world's largest cities.

Up to the 1930s, Brazil's economy was in large measure characterized by successive boom-and-bust cycles based upon Brazil wood, sugar cane, gold, diamonds, rubber and coffee. Each cycle was tied to a single region and staple export. Each involved speculative and exploitative pursuit of quick profits

on relatively modest initial outlays. Capital and labor shifted, when necessary, from one staple export and region to another.

Nevertheless, to picture Brazilian development in the primary export phase simply as a succession of boom-and-bust cycles is misleading. Booms subsided, but, as staple followed staple and as the main focus shifted from region to region, the older staples continued to be produced and there was cumulative economic diversification. Furthermore, beside its successive staple exports, Brazil also came to produce major quantities of cotton, coconuts, cacao, tobacco, sisal, agricultural products, meat, hides and semi-precious stones. Brazil remains a major producer of most of these products. Since the 1940s, Brazil has also developed major new primary exports, notably manganese, iron ore, soya and wood pulp.

The coffee and rubber booms attracted large inflows of British capital. Foreign capital outstanding increased from about US$190 million in 1880 to $2.6 billion in 1930. In 1914 Britain's holdings were roughly sixty percent of the total; the largely British-built railroad network expanded from 3,400 km in 1880 to 32,000 km in 1930.

Brazilian industrialization began during the primary export phase. The expansion of primary exports, above all coffee, expanded domestic earnings and purchasing power, created an internal market for manufactured goods and financed imports of machinery and equipment. From the 1850s steadily increasing quantities of Brazilian cotton, hides, tobacco and sugar cane were transformed into cloth, clothing, footwear, edible sugar, cigarettes and alcoholic beverages for domestic consumption. Construction of physical infrastructure and cities increased demand for bricks and cement. Railroads aided market

▼ **Destruction of the Amazonian forests in the 1980s.**

integration. And World War I, by temporarily reducing the availability of imported manufactures, induced a 150-percent increase in Brazil's industrial output in 1914–19. The 1907 industrial census recorded 3,258 industrial enterprises with 150,841 workers; the 1920 census recorded 13,336 and 275,512, respectively. By 1930 Brazil had over one million industrial workers. Up to 1929, in the main, industry grew in proportion to the extent that growth of primary exports expanded domestic purchasing power and Brazil's capacity to pay for the requisite imports of machinery, equipment and fuel.

The world depression more than halved the prices fetched by Brazil's staple exports (especially coffee) in 1929–33, correspondingly reducing Brazil's capacity to pay for imports and to service its foreign debts. The Vargas regime repudiated most of Brazil's foreign debt in 1930–31, but this meant that Brazil was unable to borrow any fresh capital from abroad until the 1940s, when Brazil's US$1,050 million wartime trade surpluses were used to reach mutually acceptable settlements with Brazil's (mainly British) creditors and to buy back the (mainly British-owned) railways and other public utilities. In the 1930s, protective import tariffs and quotas and exchange controls were adopted to reduce imports. These defensive emergency measures had the effect of strongly stimulating and protecting Brazilian industry, initiating the "import-substituting industrialization" which propelled Brazilian economic development from the 1930s to the 1960s. Heavily protected markets meant inflated and often monopolistic prices and profits for Brazilian manufacturers, helping them to finance industrial investment largely without recourse to foreign capital in the 1930s–40s, and subsequently attracting massive foreign and multinational investment in Brazilian production facilities in the 1950s–70s.

Industrial output rose steadily until it grew at nine percent in the 1950s. From the 1950s, moreover, Brazil increasingly produced capital goods and consumer durables. During the 1964–85 so-called "Brazilian economic miracle", Brazil also became a major exporter of manufactured goods and began to move into information technology and aircraft construction. Brazil's economy grew at seven percent a year on average from 1945 to 1981.

It is often asked how a country with so many very poor people could have sustained such massive industrial growth. One answer is that, although most Brazilians have remained very poor, by 1960 there were at least 10–15 million (and by 1980 at least 25–30 million) prosperous upper- and middle-class Brazilians, representing a considerable retail market. Their real purchasing power was enhanced by the comparatively low real cost of living arising from the extremely low wages of the rest of the population. Secondly, large-scale industrialization has nearly always proceeded partly on the basis of "productive consumption": growing inputs into expanding productive processes. Thus the creation of the world's eighth largest automotive industry in Brazil in the 1950s–60s induced subsidiary booms among suppliers of car-components, as well as stimulating fuel refining and highway construction.

Twentieth-century Brazil's best-known cultural exports have been musical: the samba, the Bossa Nova, the popular compositions of Antonio Carlos Jobim and the classical compositions of Hector Villa-Lobos (1887–1959). Brazilian national culture and identity has been actively promoted by the "Brazilianization" of education since 1935 and compulsory schooling since 1937 resulting in a rise of adult literacy from 35 percent in 1900 to 70 percent in 1970. In successive censuses, over ninety percent of Brazilians have declared themselves to be Roman Catholic and Brazil is the world's largest Catholic country, leading many to believe that there ought to be a Brazilian Pope.

The state of São Paulo

The state of São Paulo has become the industrial powerhouse of 20th-century Brazil and the major industrial region of Latin America. It accounts for over forty percent of Brazil's GDP and exports and over half its industrial output. Its 33 million inhabitants, known as Paulistas, occupy an area the size of Britain and constitute a "state within a state". São Paulo came close to seceding from Brazil in the early 1930s and again in the mid 1960s. Most Paulistas are descended from the 1.5 million northern Italians who migrated to Brazil between 1860 and 1896 in response to the labor requirements of the concurrent coffee boom. The need to make São Paulo more attractive to Italians was an important factor in the final abolition of slavery in Brazil in 1888. Consequently, São Paulo developed a much more fluid, dynamic and enterprising society than other parts of Brazil.

The coffee boom also stimulated and financed the development of the cities of São Paulo, Santos, Campinas and Jundiai. The city of São Paulo grew from 31,000 inhabitants in 1872, to 240,000 in 1900, 1.3 million in 1940, and over 14 million in 1990. Rich coffee barons invested in industry, transport, the port of Santos and urban real estate, establishing a stranglehold on mushrooming industrial and commercial wealth. São Paulo's wealth is very unequally distributed and millions of its inhabitants live in slums. Urban overcrowding, poverty and pollution, aggravated by rapid inflows of labor from impoverished rural areas in various parts of Brazil, kept down wage costs and maintained the international competitiveness of São Paulo's industries. They also fueled the growth of crime and voodoo.

▶ **São Paulo city – financial success, social disaster.**

PARAGUAY

REPUBLIC OF PARAGUAY

CHRONOLOGY

1932 War breaks out with Bolivia over Gran Chaco region

1938 International arbitration settles borders in Paraguay's favor

1954 General Alfredo Stroessner assumes Presidency

1989 General Andrés Rodríguez ousts Stroessner after 34-year dictatorship

ESSENTIAL STATISTICS

Capital Asunción

Population (1989) 4,157,000

Area 406,752 sq km

Population per sq km (1989) 10.2

GNP per capita (1987) US$1,000

CONSTITUTIONAL DATA

Constitution Republic with two legislative houses (Senate; Chamber of Deputies)

Major international organizations UN; OAS; I-ADB

Monetary unit 1 Paraguayan Guaraní (G) = 100 céntimos

Official language Spanish

Major religion Roman Catholic

Heads of government since 1900 (President) E. Aceval (1899–1902); J. A. Escurra (1902–04); J. Gaona (1904–05); C. Báez (1905–06); B. Ferréira (1906–08); E. González Naverro (1908–10); M. Gondra (1910–11); A. Jara (Jan–Jul 1911); Liberato M. Rojas (Jul–Dec 1911); E. Schaerer (1912–16); M. Franco (1916–19); J. P. Montero (1919–20); M. Gondra (1920–21); Eligio Ayala (1921–28); J. P. Guggiari (1928–31); E. González Naverro (1931–32); Eusebio Ayala (1932–36); R. Franco (1936–37); F. Paiva (1937–39); J. F. Estigarribia (1939–40); H. Moríñigo (1940–48); F. Chávez (1949–54); (Dictator) A. Stroessner (1954–89); A. Rodríguez (1989–)

Paraguay has had a bloody history since its independence in 1852. Before that time, though formally part of the Spanish empire, Paraguay was ruled from Buenos Aires and independence was won only with Brazilian support. In 1868 the country, led by president Francisco Solano Lopez, suffered traumatic defeat at the hands of Brazil at the battle of Lomas Valentinas, and as a result of this and ensuing campaigns the population declined by two-thirds, with most of the male population dead. A liberal constitution was adopted in 1870 and Paraguay's borders were confirmed by the United States in 1877. Ten years later the major political groupings of 20th-century Paraguayan politics emerged – the Partido Colorado, which assumed power until 1904, and the Liberal party.

The major theme of Paraguayan affairs from the 1870s was a longstanding dispute over the Gran Chaco region to the west of the country, a poorly drained alluvial plain divided between Paraguay, Argentina and Bolivia. The discovery of oil nearby in the 1920s fueled interest in this region, which had previously been given over mainly to cattle-ranching; Bolivia also saw the region as offering a route to the Atlantic. After several Bolivian military incursions, the Paraguayan president Eligio Ayala claimed the region, and war broke out in June 1932. The ensuing conflict did not end until 1935 when General José Felix Estigarribia succeeded in driving the better-equipped Bolivians from the region. The borders were settled, in Paraguay's favor, by international arbitration in 1938. The hoped-for oil finds in the region failed to materialize.

The liberal government under Ayala continued until its removal by a short-lived revolution in 1936; in 1939 Estigarribia was elected president, and in 1940 assumed dictatorial powers but died a few days later in an air crash. His successor, Higinio Moríñigo, used these powers to destroy the Liberals, but in 1948, following a year of violence throughout the country, was himself removed by the Colorados. After a period of instability, Alfredo Stroessner took power with support from the army and Colorados in 1954.

Stroessner, working in close collaboration with the United States, undertook important investments in communications and industry, and in 1967 he granted civil liberties and set up a two-chamber parliament (always controlled by the Colorados). Relations with the United States worsened in the 1970s, and Stroessner relied increasingly on support from Brazil. In 1989 he was removed in a coup by General Andrés Rodríguez, who held elections which resulted in overwhelming victory for Rodríguez and the Colorados. Rodríguez undertook a program of economic reform, freeing the exchange rate. Nevertheless the country remained one of the poorest in the region.

Economic and social trends

The people of Paraguay are mainly of *mestizo* (mixed Indian and Spanish) descent. There are small groups of immigrants from other western European countries and East Asia, and significant groups of Mennonites and Hutterites (extreme Protestant Christians, mostly immigrants from Germany). The economy is primarily agricultural, with significant export earnings from timber and cattle. Despite varied mineral deposits, mining and industry remain underdeveloped. The country has suffered from high rates of emigration since the 1940s. The Catholic Church remains a strong influence, and has acted as a focus for human rights and land reform movements.

URUGUAY

ORIENTAL REPUBLIC OF URUGUAY

CHRONOLOGY

1903	José Batlle y Ordóñez inaugurated as president
1904	Last civil uprising
1929	Death of Batlle y Ordóñez
1933	Coup d'état led by Gabriel Terra
1968	Tupamaro guerrilla campaign opens
1973	Military coup d'état
1980	Plebiscite rejects military's proposed constitution
1985	Democracy is restored
1989	Plebiscite accepts amnesty for military's human rights record

ESSENTIAL STATISTICS

Capital Montevideo

Population (1989) 3,017,000

Area 176,215 sq km

Population per sq km (1989) 17.2

GNP per capita (1987) US$2,180

CONSTITUTIONAL DATA

Constitution Republic with two legislative houses (Senate; Chamber of Representatives)

Major international organizations UN; OAS; I-ADB

Monetary unit 1 Uruguayan new peso (NUr$) = 100 centésímos

Official language Spanish

Major religion Christian

Heads of government since 1900 (President) Juan Lindolf Cuestas (1897–1903); José Batlle y Ordóñez (1903–07); Claudio Williman (1907–11); José Batlle y Ordóñez (1911–15); Feliciano Viera (1915–19); Baltasar Brum (1919–23); José Serrato (1923–27); Juan Campisteguy (1927–31); Gabriel Terra (1931–38); Alfredo Baldomir (1938–43); Juan José Amézaga (1943–47); Tomás Berreta (1947); Luis Batlle Berres (1947–51); Andrés Martínez Trueba (1951–52); National Council of Government (1952–67); (President) Oscar Gestido (1967); Jorge Pacheco Areco (1967–72); Juan María Bordaberry (1972–76); Aparicio Méndez (1976–81); General Gregorio Alvarez (1981–85); Julio María Sanguinetti (1985–89); Luis Alberto Lacalle (1989–)

The history of Uruguay in the 20th century contrasts sharply with its early years. Originally confirmed in existence in 1828 as a buffer state between Brazil and Argentina, the republic was for many years victim to a succession of civil wars as rival bands known as Blancos and Colorados fought for territorial supremacy. Political and economic organization was primitive. The structure of a modern state began to appear only in the final decades of the 19th century with the growth of wool exports to Europe and north America, investment of British capital in railways, and the arrival of immigrants from Mediterranean Europe.

The transformation which gave Uruguay its utopian reputation as one of Latin America's most stable, prosperous and democratic countries (it won the reputation of being "South America's first welfare state"), at least until 1970, took shape in the first two decades of the 20th century. The inauguration of José Batlle y Ordóñez (a Colorado) as president in 1903 was of the greatest significance for the subsequent history of the country. In 1904 he suppressed the last revolt of disaffected Blanco landowners, and thus allowed livestock producers to modernize their cattle herds and take advantage of the commercial possibilities offered by the new trade in frozen meat to Europe.

Batlle, however, was no friend of the landowning class, whose exploitation of unimproved pasture discouraged arable agriculture, stimulated the migration of population from the countryside to the capital city, Montevideo (or neighboring countries), and placed an eventual limit on rural output and exports. He therefore used the prosperity given by the high level of exports to develop the economy of Montevideo, and gave Uruguay the distinctively urban character which it has retained, in spite of its continuing reliance on rural sector exports. Blancos and Colorados ceased to fight for territorial control, and as political parties they now began their enduring shared monopoly of political power.

Another legacy left by Batlle, whose second term ended in 1915 but whose vision of a model society continued to haunt some sectors of the Colorado party, was a belief in constitutional reform, the aim being to replace the one-person presidency with a nine-person presidential council. This was achieved first in the constitution of 1919, in which the office of president was retained, though with reduced powers. The system was criticized as reducing the decision-taking capacity of the executive, and promoting "co-participation", by which supporters of both parties shared in the distribution of jobs in the bureaucracy. Gabriel Terra, president from 1931, used the inadequacies of this system as a pretext for his coup in March 1933, though personal ambition and the desire of landowners to influence government policy during the world depression were stronger reasons.

Terra's semidictatorial regime ousted the followers of Batlle (known as *batllistas*), but the growth of manufacturing industry from the late 1930s made their restoration inevitable. Terra's 1934 constitution was replaced in 1943, when all political groups became free to participate, and in the late 1940s Uruguay entered a new era of prosperity. Manufacturing grew rapidly, world prices for meat and wool were high, and in 1948 the British-owned public utilities were nationalized. In 1947 Luis Batlle Berres, nephew of Batlle y Ordóñez, became president until 1951. His populist style and interventionist economic policies were instrumental in retaining trade-union support for the traditional parties, and weakened the ideological left wing. Luis Batlle was the dominant personality of the 1950s, but his ambitions for a second term of office were frustrated by the decision of his party (supported by the Blancos) to promote the reintroduction of the collegiate presidential council, in 1952. The National Council of Government was notorious for its ineffectuality and the institutionalized corruption of coparticipation, on a grander scale than in the 1920s.

During the 1950s export prices fell, rural output stagnated, and by 1956 the era of industrial growth based on import substitution was at an end. Uruguay's crisis had begun. In elections in 1958 the Blancos won their first ever victory (there were repeat triumphs in 1962, and later in 1989), but there was no serious economic or political reform. The two parties continued to dominate, using the resources of the state to construct their electoral clienteles.

The long stalemate reached its unhappy conclusion during 1966–73. A plebiscite in 1966 ended the second collegiate experiment. Pacheco, president from 1967 to 1972, introduced uncompromising policies using "emergency" powers which antagonized students and trade unions and stimulated the emergence of the well-organized, mainly middle-class, Tupamaro urban guerrilla movement. Successful kidnappings and the threat to the established order resulted in the intervention of the armed forces, first to defeat the Tupamaros in 1972, and then to displace the entire political system in the military coup of 1973. Bordaberry, the president, offered only token resistance and was retained as a figurehead until his own ambitions ended his usefulness in 1976. His successor, Méndez, was even more a nominal figure.

For a decade from 1973 all political activity was proscribed by the dictatorship. The practice of torture was routine, and although the number of killings and "disappearances" was less than in Argentina or Chile, the military's control over the country was absolute. The military ruled as a junta, and no single figure was allowed to accumulate power. However, the intention of the military to institutionalize its rule was upset when its proposed constitution was rejected in a plebiscite in 1980. By then its economic program, based on a strong financial sector, reduced wage levels, and the processing of agricultural raw materials for export, was also in difficulties.

In 1985 the military withdrew, largely on its own terms (in 1989 a plebiscite confirmed an amnesty for crimes against human rights), and democracy returned. Prosperity, however, did not. Although export performance in the late 1980s was respectable, the burden of a large external debt and the reluctance of the private sector to invest restricted the rate of economic growth and any improvement in living standards. Inflationary pressures remained strong. The capacity of the political system to reform the welfare system and the huge bureaucracy was limited by the continuing dominance of the traditional and factionalized Colorado and Blanco parties. The national crisis which began in the mid-1950s thus continued though, by 1989, the transition from military to civilian rule was nominally complete.

Social and cultural trends

For a society of just three million inhabitants, somewhat provincial in character and isolated geographically, Uruguay's cultural history is remarkably rich. This is a consequence of the preeminence of Montevideo (with half the population), near-universal literacy, a large middle-class population, and fairly high average incomes, as well as the proximity of the much larger cultural center of Buenos Aires, capital of Argentina. The influence of Europe in Uruguay's social and cultural life is strong, reflecting the absence of an indigenous population. Football is preeminently the national sport, though international success in this has dwindled since the heady days of the 1930s and 1950s when the national team twice won the World Cup. Uruguayan literature and painting achieved international renown through such figures as the novelist Juan Carlos Onetti and the artist Joaquín Torres-García. In the universities, law, accountancy and medicine were popular courses, but in contrast the country had a very restricted technological capacity.

▼ **Traditional leatherworking in Montevideo.**

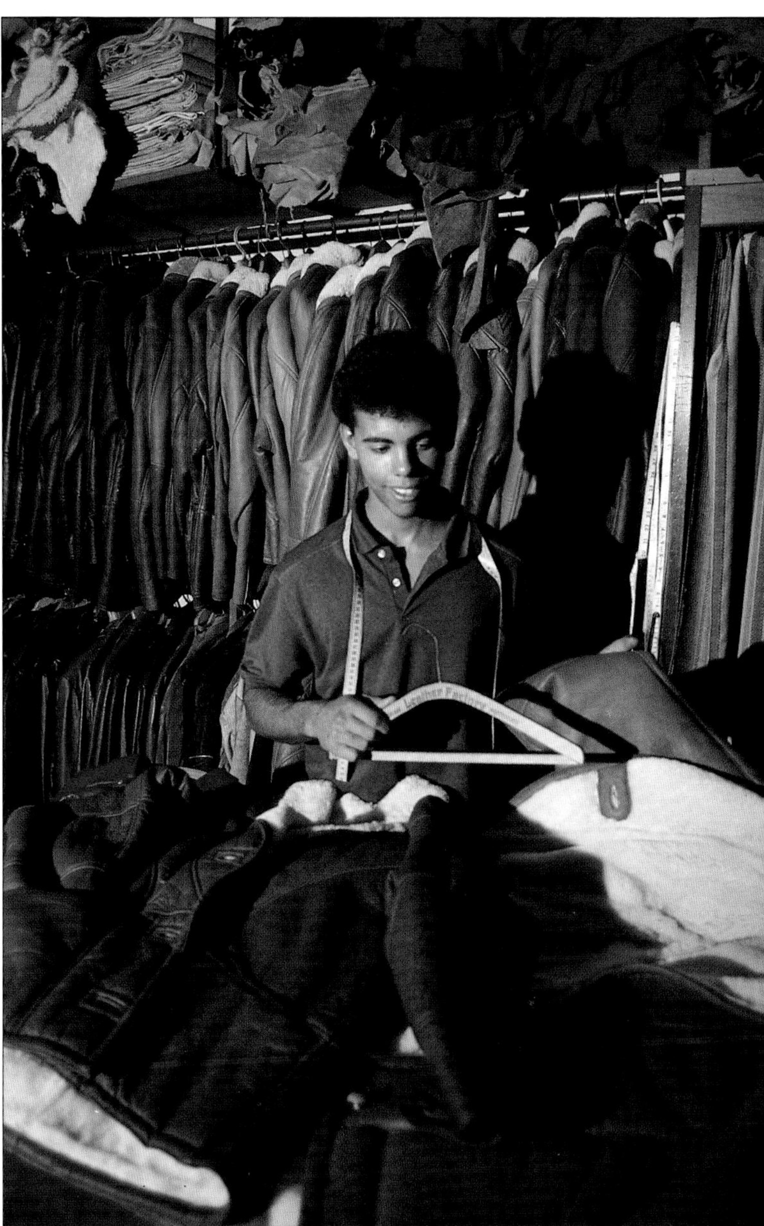

ARGENTINA

ARGENTINE REPUBLIC

CHRONOLOGY

1943	Moves towards fascism under the military after a coup
1955	Papal excommunication of Perón after mob violence
1976	"Dirty War" against the Left begins – over 9,000 "disappear"
1982	Falklands (Malvinas) War ends in defeat
1991	Controversial presidential pardons

ESSENTIAL STATISTICS

Capital Buenos Aires

Population (1989) 32,425,000

Area 2,780,092 sq km

Population per sq km (1989) 11.7

GNP per capita (1987) US$2,370

CONSTITUTIONAL DATA

Constitution Federal republic with two legislative houses (Senate; Chamber of Deputies)

Major international organizations UN; OAS; I-ADB

Monetary unit 1 austral (A) = 1,000 pesos

Official language Spanish

Major religion Roman Catholic

Heads of government since 1900 (President) J. Roca (1898–1904); M. Quintana (1904–06); F. Alcorta (1906–10); R.S. Peña (1910–14); V. de la Plaza (1914–16); H. Yrigoyen (1916–22); M.T. de Alvear (1922–28); H. Yrigoyen (1928–30); J.F. Uriburu (1930–31); A. Justo (1932–37); R.M. Ortiz (1937–42); R.S. Castillo (1942–43); P. Ramírez (1943–44); E. Farrell (1944–45); Colonel J. Perón (1946–55); E. Leonardi (1955); P.E. Aramburu (1955–58); A. Frondizi (1958–62); J.M. Guido (1962); A. Illia (1962–66); J.C. Onganía (1966–70); R.M. Levingston (1970–71); A. Lanusse (1971–73); Dr. H. Cámpora (1973); J. Perón (1973–74); M.E.M. de Perón (1974–76); J.R. Videla (1976–81); R. Viola (1981); L. Galtieri (1981–82); General R. Bignone (1982–83); Dr. R. Alfonsin (1983–89); Carlos Menem (1989–)

From 1862 to 1916 Argentina was corruptly controlled by a property-owning and mercantile "Oligarchy" presiding over economic booms which attracted wave upon wave of European investment and immigration. Half the six million immigrants settled in Argentina; almost a third of the population was foreign-born by 1914. World War I disrupted Argentina's trade-dependent economy, causing some loss of faith in the economic strategy of the increasingly disunited "Oligarchy". This helped Hipólito Yrigoyen's *Unión Cívica Radical* to win the 1916 presidential election. But the Radicals were almost as corrupt as their predecessors.

In September 1930 Yrigoyen was overthrown by the military. Until late 1931 Argentina was ruled by General Uriburu, allied to the nationalist right. But from 1932 to 1938 the presidency was held by General Agostín Justo, a moderate conservative. His main policies were to balance the budget, reduce imports and increase exports; by means of retrenchment, increased import-duties, exchange-control and devaluation. Profits from exchange-control funded farm support. The paramount concern was to maintain meat exports to Britain, in return for which British industrial exports and companies were granted preferential treatment in Argentina. This economic strategy helped to increase Argentina's GDP by 14 percent between 1929 and 1937, kept unemployment down, allowed manufacturing output to grow through the 1930s and led to massive trade surpluses in the 1940s.

Argentina remained neutral in World War II, but its nationalist-corporatist policies led to Axis sympathies and mounting friction with the United States and Britain. General Ramírez, president from 1943 to 1944, made attempts to draw closer to the United States. But US refusal to make "face-saving" concessions helped pro-Nazi officers, including war minister General Edelmiro Farrell and his aide Colonel Juan Perón, to take partial control of the military regime in October 1943. They finally deposed Ramírez in late January 1944.

The emerging Farrell–Perón regime had already begun to suppress political parties, gag the press, treble military expenditure and the size of the army, create a siege economy and impose Catholic and "patriotic" instruction in the schools. In March 1944 railroadmen staged the first of many workers' demonstrations in support of Perón. Encouraged, Perón began to decree compulsory conciliation and arbitration of industrial disputes, improving labor conditions for various groups of workers, to the alarm of the propertied classes and foreign companies.

In early 1945, Farrell belatedly decided to play along with US policy in Latin America and enter the war on the winning side. In October 1945, under mounting pressure, Farrell permitted the imprisonment of Perón. Immediately, however, Perón's charismatic 26-year-old wife-to-be Eva Duarte and their allies successfully called out massive workers' marches to set their tribune free and launch a new Labor party which won Perón 54 percent of the vote in the February 1946 presidential election.

Perón's presidency, lasting from June 1946 to September 1955, established the trappings of a modern welfare state. Perónist programs were funded by the foreign currency reserves accumulated by Argentina during World War II and by purchasing farm produce cheaply at home to be sold dearly abroad, until European agriculture fully recovered from the war. By 1949, however, the export bonanza was over, foreign currency reserves had largely been wasted on nationalizing run-down public utilities, and rocketing wages, public expenditure and money-supply were sucking in unsustainably high imports. Hardships caused by austerity measures threw into sharper relief mounting misappropriation and waste of public funds by Perónist agencies. In 1952 the 32-year-old "Evita" Perón died of cancer. Perónist vandalism and mob-rule culminated in 1954–55 in the burning of churches and papal excommunication of Perón's government, facilitating his overthrow by the army in September 1955.

After 1955 Argentina oscillated between military and civilian rule. The military were repeatedly greeted as "saviors" by the propertied classes and the foreign business community. Since the working class had been lastingly mobilized by Perónism and the trade unions, Argentine society was no longer controllable by sham constitutionalism and electoral fraud. But reliance on the military carried a high price: increasing military expenditure and especially ferocious suppression of Perónistas, trade unions and the left. General Galtieri tried to gain popularity by invading the Falkland Islands or Malvinas (islands off the Argentinean coast that had been under British rule since the early 19th century), in April 1982. National jubilation turned to humiliation as a British "Task Force" recaptured the islands in mid-1982.

Since 1955 Argentina has had three Radical civilian presidents. Their relaxed rule facilitated uncontrolled growth of public expenditure and wages, soaring budget and trade deficits, burgeoning internal and foreign debts and galloping inflation.

The Perónist movement was rehabilitated in 1972. Dr Hector Cámpora won the May 1973 presidential election. Perón himself, aged 78, won 62 percent of the vote in September 1973 and cracked down on terrorism and the left while the economy boomed, inflation abated and real wages rose. In 1974, however, the Perónist package was destroyed by a wages explosion, leftist terrorism, the death of Perón and the accession of his widow, former nightclub-dancer "Isabelita" Perón, but the military waited until June 1976 before deposing her.

The Perónists returned to power in July 1989 under Carlos Menem. President Menem inherited a US $63 billion foreign debt, 200 percent per month inflation and massive trade and budget deficits. However, contrary to expectations, Menem embarked on an orthodox austerity program and major programs of economic liberalization and privatization, jeopardizing many public sector jobs, and proposed that the Malvinas/Falkland issue be "put under an umbrella", to normalize relations with Britain. He also pardoned officers responsible for the "dirty war" of 1976–81.

▼ **Buenos Aires has a distinctly European air.**

CHILE
REPUBLIC OF CHILE

CHRONOLOGY

1927	Military dictatorship established under Colonel C. Ibáñez
1931	President Ibáñez overthrown
1938	Popular Front Government elected
1973	Government overthrown in violent military coup
1983	Protests break out against military government
1990	Return to democracy

ESSENTIAL STATISTICS

Capital Santiago

Population (1989) 12,961,000

Area 756,626 sq km

Population per sq km (1989) 17.1

GNP per capita (1987) US$1,310

CONSTITUTIONAL DATA

Constitution Multiparty republic with one legislative house containing two chambers (Chamber of Deputies; Chamber of Senate)

Major international organizations UN; OAS; I-ADB

Monetary unit 1 peso (Ch$) = 100 centavos

Official language Spanish

Major religion Roman Catholic

Heads of government (President) G. Riesco (1901–06); P. Montt (1906–10); R. Barros Luco (1910–15); J. L. Sanfuentes (1915–20); A. Alessandri Palma (1920–25); E. Figueros Larrain (1925–27); (Military dictator) C. Ibáñez (1927–31); (President) A. Alessandri Palma (1932–38); P. Aguirre Cerda (1938–41); J. Antonio Ríos (1942–46); G. González Videla (1946–52); C. Ibáñez (1952–58); J. Alessandri Rodríguez (1958–64); E. Frei Montalva (1964–70); S. Allende Gossens (1970–73); (Military dictator) A. Pinochet Ugarte (1973–90); P. Aylwin (1990–)

Chilean politics in the 20th century has revolved around the competition between class-based, political parties within a democratic tradition established in the mid 19th century. Attachment to the democratic system was usually strong enough to contain political competition within legal bounds. But in the 1920s economic recession and political stalemate led to military intervention in politics and in 1973 the constitutional system was unable to contain an acute social and economic crisis.

Political struggle in Chile has revolved around the distribution of the earnings of the export economy, based at first on nitrates and then copper. As export earnings fluctuated sharply, political life reflected their instability, and governments were rarely able to fulfill their electoral promises. The electorate did not often vote for the same government twice. Since 1900 Chile has been governed by Conservatives, Liberals, independents, military populists, Radicals, Socialists, Christian Democrats, and military dictators. All these governments faced one common problem: that of endemic inflation.

A cosy arrangement whereby a small political elite monopolized the profits from the export of nitrates ended in the 1920s when the economy was devastated by economic recession following World War I and the discovery of artificial substitutes for nitrates. A powerful labor movement, organized by the Chilean Communist party, threatened the social order. A reformist president, Arturo Alessandri, was blocked by a conservative Congress until the military intervened to push through reform legislation and a new constitution in 1925. Once in power the military were reluctant to leave, but the effects of the international depression of the interwar period swept them from power in a tide of social and political protest.

Chile has long looked to Europe for political models and ideologies. In 1938, imitating much of Europe, Chile elected a Popular Front government. This government enacted many overdue social reforms, but the real beneficiaries were the middle classes, whose political expression, the Radical party, controlled the presidency from 1938 to 1952. Chile remained formally neutral in World War II until declaring war on the Axis powers in February 1945, but cheap Chilean copper subsidized the Allied war effort and in practice there was close cooperation with the United States. The economy was again hit by the collapse of copper prices following the end of the Korean War, and the Radical governments had been unable to curb inflation. The election of former dictator Carlos Ibáñez showed popular discontent with the parties.

From 1958 to 1990 Chile was governed by four contrasting administrations. Jorge Alessandri, the son of the former president, was elected in 1958, and hoped to create a new capitalist class to undertake the modernization of the country. But the right wing in Chile was too accustomed to state protection to accept this role, and landowners refused to reform the archaic system of land tenure. In 1964 the Chilean electorate turned to the Christian Democratic party of Eduardo Frei.

The presidency of Frei saw profound reforms. The largely American-owned copper mines were taken into partial state ownership. Agrarian reform was implemented, and the peasantry was encouraged to form trade unions. There was a real income redistribution to the poor. But the employers opposed the reforms of the government, the unions felt that the reforms did not go far enough and were intended to undermine the hold of the Marxist parties, and acute internal

disputes weakened the Christian Democratic party.

In 1970, the electorate chose a Marxist, Salvador Allende, who offered a peaceful road to socialism. But from the start the government was weakened by the lack of internal unity, and by the indiscipline of Allende's own Socialist party. The United States encouraged the right wing to undermine the government. A worsening international economy added to the difficulties of the government, and mistaken economic policies led to spiraling inflation and declining production. Political violence grew in intensity. A military coup was not a great surprise, but what was unexpected was the brutality of the coup, the longevity of the ensuing dictatorial government of General Pinochet, and its adoption of free-market economic policies.

Pinochet ruled from 1973 to 1990. The basis of his power was in the loyal and disciplined armed forces, and the ruthless and brutal secret police. But he also enjoyed the support of the middle class which had suffered in the economic chaos of the Allende years. He came to enjoy more support as his economic policies, at first very harsh and recessive, after 1985 at last produced economic growth with low and controlled levels of inflation. But his attempt to secure election in 1988 as president for another eight years was defeated. In 1990 Chile returned to democracy with reforms of the 1980 constitution passed by referendum; and the electorate chose Patricio Aylwin, a Christian Democrat with socialist and social democratic backing promising sociopolitical change.

Economic and social trends

Chile's economy is based essentially upon the export of copper. Chilean industry is not competitive internationally, and it was only in the 1980s that the agricultural sector has became more efficient and an important exporter. The economy is very sensitive to fluctuations in international prices. Chile was devastated by the recession of the interwar years, and fluctuations in prices since then magnified the instability of the economy. A protected manufacturing sector was a further drain on the balance of payments, and only in the 1980s was Chilean agriculture able to contribute to exports.

The free market reforms of the Pinochet period sharply reduced the role of the state, cut back the inefficient manufacturing sector, and opened the country to international competition. The legacy of that period to the new democratic government in 1990 was a more efficient state, a healthy export economy and a vigorous private sector.

For a small country, Chile has a rich cultural life combining the European cultural traditions with a South American flavor. One of the greatest poets of this century, Pablo Neruda, a lifelong Communist, was a Nobel prize winner (1945), as was Chilean poet Gabriela Mistra (1971). One of the most famous novelists of the Latin American "boom" is Isabel Allende, a distant relative of president Salvador Allende.

▲ **Salvador Allende, democratically elected Marxist president.**

BOLIVIA

REPUBLIC OF BOLIVIA

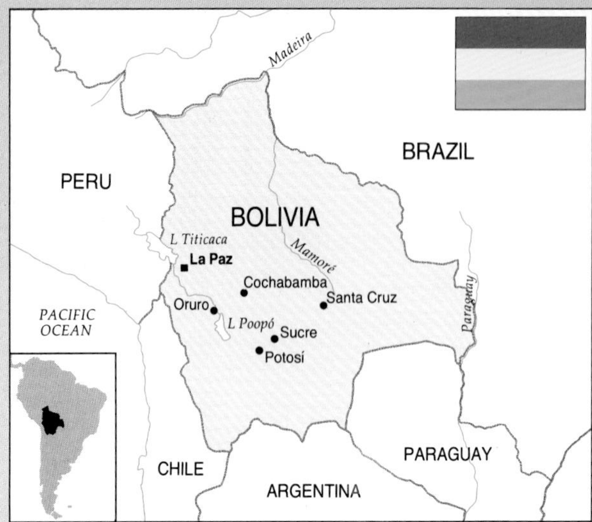

CHRONOLOGY

1932 Chaco War with Paraguay

1952 National revolution brings MNR to power under Víctor Paz Estenssoro

1980 Power is seized by a military junta linked to drug-traffickers

ESSENTIAL STATISTICS

Capital La Paz (Sucre: legal capital)

Population (1989) 7,193,000

Area 1,098,581 sq km

Population per sq km (1989) 6.5

GNP per capita (1987) US$570

CONSTITUTIONAL DATA

Constitution Unitary multiparty republic with two legislative houses (Chamber of Senators; Chamber of Deputies)

Major international organizations UN; OAS; I-ADB; Andean Pact

Monetary unit 1 boliviano (Bs) = 100 centavos

Official languages Spanish; Aymara; Quechua

Major religion Roman Catholic

Heads of government since 1900 (President) J. M. Pardo (1899–1904); I. Montes (1904–09); E. Villazón (1909–13); I. Montes (1913–17); M. Gutiérrez G. (1917–20); B. Saavedra (1920–25); H. Siles (1926–30); D. Salamanca (1931–34); J. L. Tejada (1934–36); D. Toro (1936–37); Lt.-Col. Busch (1937–39); Gen C. Quintanilla (1939–40); Gen. E. Penaranda (1940–43); D. G. Villaroel (1943–46); T. Manje G. (1946–47); E. Hertzog (1947–49); M. Urralangoita (1949–51); Military junta (1951–52); (President) V. Paz Estenssoro (1952–56); H. Siles (1956–60); V. Paz Estenssoro (1960–64); Gen. R. Barrientos (1964–69); L. A. Siles Salinas (Apr–Sep 1969); (Military dictator) A. Ovando (1969–70); J. J. Torres (1970–71); H. Banzer (1971–78); J. Pereda (Jul–Nov 1978); D. Padilla (1978–79); L. Gueiler (1979–80); Military juntas (1980–82); (President) H. Siles (1982–85); V. Paz Estenssoro (1985–89); J. Paz Zamora (1989–)

For the first half of the 20th century political power in Bolivia was exercised mainly by professional and landed groups in support of the economically dominant mineowning interests. The turn of the century saw the Liberals replacing the Conservatives in political control. In practice this made little difference, except in that it confirmed the primacy of La Paz over the colonial city of Sucre. The division of the Liberals and the growth of the Republican party which occurred in the 1920s led to the traditional two-party politics giving way to a less stable multiparty system.

Early in the century, as a result of clashes with its more powerful neighbors over ill-defined frontiers, Bolivia was reduced to its present territorial limits. After a sequence of border skirmishes, Bolivia went to war with Paraguay in 1932 over control of the Chaco, a wilderness region where discoveries of oil reserves were anticipated, and by 1935 had lost heavily. In the 1930s and 1940s, organized labor, based mainly on the mines, came to play a significant political role. Also, new, more popular-based political parties came into being, notably the National Revolutionary Movement (MNR), founded in 1941 with a reformist program.

The 1952 national revolution was the major turning-point of the 20th century for Bolivia, involving a popular upsurge, the collapse of the old oligarchic regime, and the replacement of the armed forces by popular militias. The revolution brought the MNR under Víctor Paz Estenssoro to power; mining and oil interests were nationalized; a sweeping land reform took place; and universal suffrage was established, giving the majority of the population a political voice for the first time. However, subsequently, under strong pressure from the United States, the militias were disbanded, the armed forces reconstituted, and the government moved to the right. Twelve years of government by the MNR leaders, Paz Estenssoro and Hernan Siles Zuazo, gave way in 1964 to 18 years of mainly rightwing military government with only short interludes of unstable civilian rule.

After 1982 Bolivia enjoyed unusual political stability under civilian, constitutional government, even though the military remained a strong force behind the scenes. This was partly because the right wing in Bolivian politics managed to retain access to power. A right wing coalition under Paz Estenssoro of the MNR and Banzer's supporters took over from a discredited leftwing government in 1985. In implementing a stabilization strategy to deal with hyperinflation, the Paz Estenssoro government broke the power of the leftist labor confederation, the COB, through a program of mass redundancy in the mining and other state-run industries.

Economic and social trends

Economic development in the 20th century was shaped by the rise and fall of a single commodity – tin. Exploitation was made possible by the construction of railways between 1890 and 1910 which linked the mines with coastal ports in Chile and Peru. Output increased rapidly during World War I from an annual average of 15 million tonnes (1900–09) to 33 million tonnes (1920–29). But, a classic example of an open, commodity-dependent economy, Bolivia felt the full force of the 1929 stock market crash. Between 1929 and 1932 prices fell to only 40 percent of their previous round value. With government income based on export taxes, the fiscal shock was immediate. In 1931

Bolivia defaulted on its foreign debt. Despite foreign competition in the early years of the century, Bolivian mining is exceptional in that control stayed in the hands of Bolivian companies. However, compared to Malaya and Indonesia, Bolivia was a high-cost producer. Inadequate investment from the 1930s onward and falling ore grades further undermined international competitivity, though periods of high world prices helped to mask the fact.

Beyond stabilizing an economy threatened by hyperinflation (the highest in the world in the early 1980s) the economic priorities of post-revolutionary governments were, first, for Bolivia to get more for its exports by building up a domestic industry to process ores; and second to diversify its export base (and to develop the eastern lowlands) by encouraging cash-crop agriculture. But investment in metallurgy and agriculture was at the expense of mining itself, and by the 1970s it achieved the growth of an unmanageable foreign debt. The fall in tin prices in the 1980s again revealed the vulnerability of the economy. Balance-of-payments difficulties led to default once again, and underscored the failure of successive stabilization plans to bring inflation under control. The radical package of 1985, helped by the resumption of foreign aid and cocaine earnings did succeed in stopping inflation. But it did little to stimulate growth in what was still Latin America's poorest economy.

The people of Bolivia are mostly of Indian or Spanish descent, or *mestizo* (of mixed descent). Literary and artistic output in Bolivia has had to contend with the lack of an educated public, lack of wealthy patrons, or both.

▼ **Indians outside a church in La Paz.**

PERU

REPUBLIC OF PERU

CHRONOLOGY

1924	American Popular Revolutionary Alliance (APRA) is formed under Victor Haya de la Torre
1930	General Sánchez Cerro seizes power in rebellion
1948	Military coup led by General Odria
1968	Military coup ushers in period of radical social reform
1985	Peru defies foreign banks by limiting debt payments

ESSENTIAL STATISTICS

Capital Lima

Population (1989) 21,792,000

Area 1,285,216 sq km

Population per sq km (1989) 17.0

GNP per capita (1987) US$1,430

CONSTITUTIONAL DATA

Constitution Unitary multiparty republic with two legislative houses (Senate; Chamber of Deputies)

Major international organizations UN; OAS; I-ADB; Andean Group

Monetary unit 1 Inti (I/.) = 1,000 soles

Official languages Spanish; Quechua

Major religion Roman Catholic

Heads of government since 1900 (President) E. de Romaña (1899–1903); M. Candamo (1903–04); J. Pardo (1904–08); A. B. Leguía (1908–12); G. Billinghurst (1912–14); O. R. Benavides (1914–15); J. Pardo (1915–19); A. B. Leguía (1919–30); Col. Sánchez Cerro (1930–31); D. Samánez Ocampo (Mar–Dec 1931); Col. L. Sánchez Cerro (1931–33); Gen. O. Benavides (1933–39); M. Prado y Ugarteche (1939–45); J. L. Bustamante (1945–48); Gen. M. A. Odría (1948–56); M. Prado y Ugarteche (1956–62); Military junta (1962–63); (President) F. Belaúnde Terry (1963–68); Military junta (under J. Velasco) (1968–75); (President) Gen. F. Morales Bermúdez (1975–80); F. Belaúnde Terry (1980–85); A. García Pérez (1985–90); A. K. Fujimori (1990–)

During the first half of the 20th century, Peruvian political institutions failed to match social change by incorporating mass-based political movements into the formal political life of the country. There were long periods of dictatorial rule in defense of the status quo, with only brief periods of civilian government elected on a narrow franchise. Up until the 1960s there was no attempt to introduce much-needed social reform. The Leguía dictatorship (1919–30) led to a degree of modernization, but fell victim to a coup, just at the moment that Peru felt the full impact of the world slump. Social change, however, prompted the establishment of new political forces. One was APRA, established by Víctor Raúl Haya de la Torre in exile in 1924. Its ideology was anti-oligarchic, nationalist and reformist, influenced by the Mexican revolution. The other was the Socialist party, founded in 1928 by José Carlos Mariátegui, which was later renamed the Communist party. Haya's populist skills, combined with Mariátegui's early death, led to APRA's rapid expansion and near-victory in the 1931 elections. The "Aprista" challenge, involving a number of abortive attempts at insurrection, led to the party's being driven underground during much of the following 30 years. As a consequence of the need to remain clandestine and Haya's distrust of Communism, the party shifted to the right.

In the 1950s and 1960s a new party emerged with a reformist agenda: Fernando Belaúnde's Acción Popular (AP). Briefly it seemed that a two-party system was taking root in Peru, but APRA's systematic blocking of AP's mild reform program led to a leftist military takeover in 1968. Fearing that Peru was heading for a Cuba-style revolution, the Velasco government took up a number of Belaúnde's reforms and pushed them much harder. It nationalized key United States-owned mining and oil companies, undertook a radical land reform and brought in a degree of worker participation in industry. Velasco himself was ousted in 1975 by army officers. A phased transition back to constitutional rule began in 1976.

Despite economic difficulties and the emergence of an extreme and uncompromising guerrilla organization, Sendero Luminoso, the 1980s brought an unprecedented period of unbroken constitutional government. Erosion of living standards, however, made for electoral volatility. Both the center-right Belaúnde government (1980–85) and the center-left García government (1985–90) began on a wave of popularity only to see their support rapidly ebb as a result of unpopular economic measures. Disenchanted by the existing options, Peruvians backed a complete outsider, Alberto Fujimori, in the 1990 elections. Neither Belaúnde nor García were able to quell Sendero.

Economic and social trends

Peru had suffered disastrously in the War of the Pacific with Chile (1879–83), but the economy gradually recovered on the basis of the expansion of the mining industry, the development of coastal cash-crop agriculture for export (primarily sugar and cotton), and the discovery of oil on the coast close to the Ecuadorean frontier. Actively encouraged by Leguía in the 1920s, the most dynamic sectors of the economy came to be controlled by American companies. Export-led growth was seriously interrupted by the impact of the 1929 crash and Peru was forced to default on its foreign debt. Helped by its agricultural exports, Peru recovered more rapidly than its neighbors from the crash, and the economy prospered during World War

II and the Korean War in response to the strong demand for strategic minerals. By the 1950s, however, it was no longer so easy to respond to periodic increases in world demand by producing more. Peru's capacity to increase production of raw materials hinged on substantial investments being made in new mines, the development of the newly discovered oil reserves in the jungle, and irrigation to extend the frontiers of coastal agriculture.

Increased state intervention in economic development in the early 1970s coincided with a period of cheap foreign credit, with the result that Peru's foreign debt multiplied eightfold over the course of the decade. In 1976 Peru was the first Latin American country to experience a debt crisis. Under pressure from creditors and the International Monetary Fund (IMF), it was forced to abandon Velasco's ambitious model of state capitalist growth in favor of economic stabilization and liberalization. Successive stabilization packages, however, were conspicuously unsuccessful in bringing inflation under control while they helped reduce the pace of growth. Peru experienced a prime example of "stagflation" in the 1980s, except for a brief period from 1985 to 1987. This was when the newly elected García government decided to jettison orthodox economic management in favor of price controls and the nonpayment of debt, producing a short-lived growth spurt with relatively low inflation. García's hostile attitude towards the foreign banking community prompted an effective international boycott on lending to Peru. This made it almost impossible to get loans when they were desperately needed in 1988 and 1989, which, in turn, made it even more difficult to get resurgent inflation back under control. The García government ended in 1990 with hyperinflation imminent, and with GDP per capita at its lowest level since the 1950s.

Cultural trends

In Peru, more than in most other countries of Latin America, culture is hard to divorce from politics. The intellectual tradition was profoundly influenced by the writings of both Haya de la Torre and Mariátegui. Both gave great weight in their political writing to the question of national integration, and to the role of a basically Indian peasantry. Both also stressed the importance of understanding Peru's Indian past and the present in elaborating any strategy of this political change. One of the greatest Andean poets this century, César Vallejo, worked away from the modernist tradition and from avant-garde experimentalism towards an overtly political stance. Probably the most successful novelist in the indigenist tradition, which took up themes and styles peculiar to Peru, up until the 1960s, Ciro Alegría, was also a follower of Haya de la Torre. José María Arguedas, more radical in his views, developed the indigenist genre, blending it with notions of contemporary class struggle. He scorned the attempts by white Peruvians to ape the alien culture of Europe and the United States, seeing true cultural values in indigenous culture itself. Politics also loomed large in the novels of Peru's most commercially successful and internationally renowned author, Mario Vargas Llosa, though in a less ideologically committed way.

▲ The Inca site of Machu Picchu, rediscovered in 1911.

ECUADOR

REPUBLIC OF ECUADOR

CHRONOLOGY

1912 Ex-president Alfaro murdered

1922 Riots break out in Guayaquil

1925 Military coup

1949 Earthquake kills 7,000

1981 President Roldós killed in aircraft accident

ESSENTIAL STATISTICS

Capital Quito

Population (1989) 10,490,000

Area 269,178 sq km

Population per sq km (1989) 39

GNP per capita (1987) US$1,040

CONSTITUTIONAL DATA

Constitution Unitary multiparty republic with one legislative house (National Congress)

Major international organizations UN; OAS; I-ADB; OPEC

Monetary unit 1 ucre (sl.) = 100 centavos

Official language Spanish

Major religion Roman Catholic

Heads of government since 1900 (President) E. Alfaro (1897–1901); L. Plaza (1901–05); L. García (1905–06); E. Alfaro (1906–11); L. Plaza (1912–16); A. Baquerizo (1916–20); J. Luis Tamayo (1920–24); G. Córdoba (1924–25); Military junta (1925–26); I. Ayora (1926–31); A. Baquerizo (1931–32); J. D. Martínez (1932–33); A. Montalvo (1933–34); J. M. Velasco (1934–35); F. Páez (1935–37); C. A. Enriquez (1937–38); A. Mosquero N. ((1938–39); C. Arroyo del Rio (1940–44); J. M. Velasco (1944–47); C. J. Arosemena Tola (1947–48); G. Plaza Lasso (1948–52); J. M. Velasco (1952–56); C. Ponce Enriquez (1956–60); J. M. Velasco (1960–61); C. J. Arosemena Mouroy (1961–63); Military junta (1963–66); (President) C. Yerovi Indaburú (1966); O. Arosemena Gómez (1966–68); J. M. Velasco (1968–72); G. Rodríguez Lara (1972–76); Military junta (1976–79); (President) J. Roldós (1979–81); O. Hurtado (1981–84); L. Febres (1984–88); R. Borga (1989–)

Ecuador entered the 20th century strongly marked by the autocratic rule of Gabriel García Moreno (1860-75). He had realized that the only unifying factor in the then young nation was the Roman Catholic Church, and that strong leadership would be needed to create the nationalism needed to hold Ecuador together. He based his leadership on these two ideas: education, welfare and many other areas of government were turned over to the Catholic Church, and opposition was firmly suppressed. His assassination in 1875 brought a gradual inrease in Liberal influence in Ecuador, spreading into the more conservative sierra from the coast. Liberalization was consolidated under the leadership of General Eloy Alfaro, who served two periods as president (1897-1901 and 1906-11). He dismantled the administrative structure that had been created under Moreno, and gradually disengaged the Church from the State, instituting freedom of religion, civil marriage and burial and divorce. In spite of these reforms, the new Liberal regime proved similar in character to the previous one, and had little impact on the lives of the poor indians and peasants who worked in the great estates.

Alfaro was overthrown in August 1911 by a coalition of conservatives and dissident Liberals, but attempted to return a few months later, after the new president had died in office. He was rejected by the Liberals and arrested, to be murdered in January 1912 by a lynch mob that had broken into the prison in which he was being held.

Henceforth a Liberal clique dominated Ecuador, its various factions constantly engaged in infighting within the government, though real power lay in the hands of the rich bankers and merchants of Guayaquil. In the economic boom that followed World War I, the latter endeavored to take complete control of the agriculture of the coastal plains. But disaster struck in the early 1920s, in the form of a currency devaluation and a cacao crop blighted by fungal disease, bringing a depression and unrest in urban areas. Riots in Guayaquil ended in the deaths of hundreds at the hands of the army in November 1922. As a result of the unsettled situation, the army staged a coup on in 1925, putting a stop to the activities of the wealthy clique in Guayaquil, on whom blame had been laid for the turbulence. The following 23 years were to prove just as unsettled, however, with the internal strife in Ecuador exacerbating its difficulties in coming to terms with new economic realities. The high prices fetched for raw materials during World War II aided economic recovery from the lingering effects of the Great Depression of the 1930s.

Ecuador had not succeeded in settling its large Amazonian territory, and in July 1941 the army of neighboring Peru invaded the area after a series of border clashes with Ecuador. The inefficacy of the Ecuadorian defense and the humiliation of further territorial loss (Ecuador had, at different times, lost territory to each of its neighbors, Colombia and Peru) caused the downfall of president Carlos Arroyo del Río. Ecuador was forced to relinquish the Amazonian region at a peace conference in Rio de Janeiro in 1942, as the major allies were too preoccupied with the war effort to allow the local dispute to disrupt the supply of raw materials. Subsequent Ecuadorian governments contested the "Protocol of Rio", though despite their having a good legal and moral argument only harm came of the protest which was often used by nationalists to distract attention from internal problems.

The postwar period until 1972 was dominated by the popular figure of José María Velasco Ibarra, who was president of Ecuador five times. His terms of office tended to be turbulent periods, with abrupt reversals of policy, equivocal economic programs and suspensions of civil liberties.

Velasco's last period of rule ended in 1972, when a military junta took over for seven years until transferring power to a constitutionally elected civilian president, Jaime Roldós Aguilera. He pledged to redistribute more fairly the oil wealth that had come with the boom of the 1970s after the 1973 OPEC price hike, which had helped to improve the infrastructure of Ecuador, although causing severe inflation. He was killed in an aircraft crash in 1981, and was replaced by Osvaldo Hurtando Larrea of the Christian Democrat party.

In this year hostilities broke out with Peru over a border dispute in the Cordillera del Condor area. Eventually the area was awarded to Peru under the 1942 Rio Protocol, though Ecuador did not recognize this protocol. There were further clashes during 1982–83, as well as skirmishes between Ecuadoran and Colombian troops in the border zone. Hurtado's attempts to resolve the dispute with Peru amicably ended with the dismissal in 1982 of the defense minister and the resignation of the leader of the armed forces.

Austerity measures introduced in 1983 encountered strong resistance from trade unions, bringing the resignation of three ministers and their later replacement in the elections of 1984, when León Febres Cordero was elected president. He had campaigned on a platform of "breads, roofs and jobs", and his plans for economic recovery by way of austerity measures and the expansion of the private sector met resistance from the leftwing-dominated Congress.

Congress and the Government quarreled again later in 1984, over the appointments procedure for the Supreme Court, but fears of a coup brought a solution by the end of the year, when Congress allowed the government to appoint the new Supreme Court justices.

A more serious incident occurred in 1986 when the Chief of Staff of the Armed Forces and Commander of the Air Force Lt-Gen. Frank Vargas Pazzos was dismissed by President Febres Cordero. The former refused to accept his dismissal, and barricaded himself in at the Mantas military base with several hundred supporters. The siege ended peacefully after a minister and the army commander whom Vargas had accused of embezzlement resigned. While in detention at another base, Vargas made a further move, provoking troops loyal to the president to attack, capturing him and many of his supporters. In January 1987 supporters of Vargas, in retaliation, abducted president Febres Cordero, and held him for eleven hours before exchanging him for Vargas, who was then granted amnesty. Though this amnesty had been approved by Congress, the president refused to implement it, and Vargas had to go into hiding until a ransom had been raised.

▼ **Ecuadoran peasants at a political gathering.**

COLOMBIA

REPUBLIC OF COLOMBIA

CHRONOLOGY

1902 War of the Thousand Days ends

1934 Leticia returned to Colombia following conflict with Peru

1948 *La violencia* devastates the capital

1984 Assassination of Minister of Justice leads to clampdown on drugs mafia

1989 Drugs mafia murder leading Presidental candidate

ESSENTIAL STATISTICS

Capital Bogotá

Population (1989) 32,317,000

Area 1,141,748 sq km

Population per sq km (1989) 28.3

GNP per capita (1989) US$1,280

CONSTITUTIONAL DATA

Constitution Unitary, multiparty republic with two legislative houses (Senate; House of Representatives)

Major international organizations UN; OAS; I-ADB; Andean Group

Monetary unit 1 peso (Col$) = 100 centavos

Official language Spanish

Major religion Roman catholic

Heads of government since 1900 (President) J. Marroquín (1900–04); R. Reyes (1904–09); R. Gonzalez Valencia (1909–10); C. E. Restrepo (1910–14); J. V. Concha (1914–18); M. F. Súarez (1918–21); H. Holguín (1921–22); P. Nel Ospina (1922–26); M. A. Méndez (1926–30); E. O. Herrero (1930–34); A. López (1934–38); E. Santos (1938–42); A. López (1942–45); A. Lleras (1945–46); M. Ospina (1946–50); L. Gomez (1950–53); (Military dictator) G. Rojas (1953–57); Military junta (1957–58); (President) A. Lleras (1958–62); G. Leon (1962–66); C. Lleras (1966–70); M. Pastrana (1970–74); A. López (1974–78); J. C. Turbay (1978–82); B. Betancur (1982–86); V. Barco (1986–90); C. Gaviria (1990–)

Colombia is one of the few Latin American countries where the political parties of the 19th century, Liberals and Conservatives, managed to survive and adapt to the changed political and social conditions of the 20th century. Even so, Colombian politics has frequently been violent. From the 1880s through to 1930, government remained in the hands of the Conservatives, despite an unsuccessful Liberal revolt of 1899–1902, the "war of the thousand days" which prompted Panama to secede in 1904 and emphasized the Liberals' military weakness. From 1910 to 1930 Colombia enjoyed an unprecedented degree of political stability, based on a series of power-sharing arrangements which gave the Liberals access to political power. Conservative dominance was ensured by a restricted suffrage and the frequent recourse to fraud in elections. The Conservatives were well entrenched in a country which was still overwhelmingly rural, and in which they tended to exercise control over the land and administration at the local level, as well as receiving decided support from the Church.

The Liberal victory of 1930 came about because the Conservatives were divided, but the Liberals had also begun by this time to develop their own popular base among an emerging middle and working class. In the 1930s this was reflected by the enactment of modern social legislation. Division among the Liberals brought the Conservatives back to power in 1946. The assassination of the Liberal populist leader, Jorge Eliecer Gaitán, prompted the return to bloody political strife between the parties, known simply as "*la violencia*". This ended only in 1957 with the creation of the National Front, another power-sharing agreement under which Conservatives and Liberals were to alternate in office for four consecutive presidential terms (16 years). Since 1974, when there was a return to controlled competitive politics, the Liberals have tended to dominate the political scene, with the Conservatives only winning (as they did in 1982 with Belisario Betancur) when the Liberals have been divided.

The persistence of the two-party system and its flexibility, allowing collaboration and the coopting of opposition from outside, helped generate a sense of political exclusivity. This was one of the main reasons for the persistence of leftwing guerrilla organizations in Colombia since the 1960s. By far the most important of these was the Communist party *Fuerzas Armadas Revolucionarias de Colombia* (FARC) with some thirty fronts fighting over many different parts of the country, helped by a terrain ideal for fighting guerrilla warfare. Others include the M-19, the *Ejército de Liberación Nacional* (ELN) and the *Ejército Popular de Liberación* (EPL). A major achievement of recent governments was the negotiation of peace settlements with the FARC and the M-19, under which guerrillas were encouraged to take part in parliamentary and municipal politics. However, dissident factions of the FARC remained under arms, as did the ELN and EPL. The Barco government (1986–90) also took active steps to suppress the cocaine mafias, prompted by the assassination of Liberal presidential hopeful, Luis Carlos Galán, in August 1989. As well as unleashing a military clampdown (in which key drug baron Gonzalo Rodríguez Gacha lost his life), Barco authorized the extradition of Colombian drugs suspects wanted in the United States. His successor, César Gaviria, also a Liberal, seemed set to continue the offensive. Meanwhile, Colombia continued to be one of Latin America's most violent countries.

Economic and social trends

Largely an economic backwater in the 19th century, the Colombian economy was transformed in the 20th century by the development of the market for coffee. By the 1930s Colombia was exporting three million bags, having become the world's second largest exporter after Brazil. Unusually for Latin America the industry began and remained in national hands, though foreign companies were active in the railroad construction which the coffee boom helped to stimulate. At the same time, by the 1920s Colombia had become an important exporter of bananas and crude oil.

The improved balance-of-payments situation in the 1920s facilitated foreign borrowing, and this in turn led to greater spending on public works and the modernization of the economy. By the 1930s industrialization was just beginning, with the growth of textiles, food processing and other industries geared toward domestic consumption. Although the coffee industry remained dominant until the 1960s, World War II saw a significant increase in industrial development, which, with protection and government assistance, continued to diversify in the 1950s and 1960s. By the 1960s Colombia had developed both chemical and metallurgical industries. Industrialization benefited from the ready accessibility of cheap hydroelectric power.

In the 1970s, Colombia avoided the extreme levels of indebtedness incurred by most other Latin American countries. In addition, successive governments maintained a cooperative disposition toward the private sector and toward foreign investment. However, Colombia suffered (as elsewhere) from the restrictions in new commercial lending in the wake of the Mexican default in 1982. During the 1980s Colombia reduced its export dependence on coffee by developing new export lines, notably coal, crude oil and nontraditional products like cut flowers. Total exports rose from US$3 billion in 1983 to well over US$6 billion in 1989. There was a remarkable degree of continuity and stability in policy-making, aided by relatively low rates of inflation. The Colombian economy also benefited from sales of cocaine, having become the main intermediary between coca-producing countries like Bolivia and Peru, and the main market in the United States. Only a relatively small amount of the profit returns to Colombia.

In Colombia, as elsewhere in Latin America, the modernist literary genre, predominant in the early part of the century, was disdainful toward local influences and sought inspiration primarily in Europe. European influences were also dominant in the plastic arts. Writers, however, starting with José Eustasio Rivera, came to "discover" social realities beyond the sophistication of the major cities. Rural isolation and violence came to be central and critical preoccupations in the Colombian novel from the 1950s on, notably in the work of Gabriel García Márquez. García Márquez's *Hundred Years of Solitude* and his *Autumn of the Patriarch* are landmarks in the history of the Latin American novel, helping to earn their author a Nobel prize in 1982.

▲ Processing the coca plant in making cocaine, a source of Colombian wealth.

GAZETTEER

ANTIGUA AND BARBUDA

(Antigua and Barbuda) **Capital** St John's **Population** (1989) 78,400 **Area** 442 sq km **GNP per capita**(1988) US$ 2800 **International organizations** UN, Commonwealth, OAS, ACP, CARICOM.

Antigua and Barbuda are two islands that formed part of the Leeward Islands Federation under the British Crown from 1871 until 1956 when Antigua became a separate Crown Colony. After a spell as part of the West Indies Federation (1958–62) it became an Associated State of the UK in 1967. During the 1970s the islands debated the advantages and disadvantages of full independence; Barbuda, a long-standing dependency of Antigua, sought to secede for a time. The islands finally won independence in November 1981 under an ALP (Labor party) prime minister, Vere Bird. The head of state is the British monarch, represented by a governor-general. There is a legislature comprising a Senate and a House of Representatives. The islands are more prosperous than many in the eastern Caribbean, and the economy relies heavily on tourism. Earlier in the century cotton and sugar were important crops, but the sugar industry closed down in the early 1970s. In 1985 the NDP (Democratic party) replaced the crumbling original opposition party; in the 1989 elections the NDP won 31 percent of the votes, but only one seat. Amid opposition accusations of government corruption after an arms smuggling scandal in 1990, as well as criticism about the number of Bird family members in government, P.M. Bird took more executive power on himself.

BAHAMAS

(The Commonwealth of the Bahamas) **Capital** Nassau **Population** (1989) 249,000 **Area** 13,939 sq km **GNP per capita** (1988) US$10,570 **International organizations** UN, Commonwealth, OAS, ACP, I-ADB, CARICOM.

The Bahamas archipelago were in British hands, with short interruptions, from the 17th century. Their position off the coast of the United States meant that they prospered during periods of disruption there, notably during the Civil War of the 1860s, and in the 1920s when the Prohibition laws in the United States brought a powerful bootlegging operation to the islands. The islands did not, however, begin to prosper permanently until after World War II, when tourism began to develop. In 1967 the Progressive Liberal party (PLP) led by Lynden Pindling undertook to end racial segregation and worked toward independence, which was achieved in 1973. During the 1980s the islands faced problems of drug trafficking; relations with the USA were damaged, but restored by 1989. In 1990 economic decline forced austerity measures. Attempts at economic diversification did not flourish; but in 1990 a Venezuelan company brought a major Bahamian oil refinery, and hopes grew for economic expansion.

BELIZE

(Belize) **Capital** Belmopan **Population** (1989) 185,000 **Area** 22,965 sq km **GNP per capita** (1988) US$1460 **International organizations** UN, Commonwealth, ACP, OAS, CARICOM.

Belize, an ancient center of Mayan civilization, was settled by the British in the 18th century and in 1862 became a British Crown Colony, as British Honduras. Guatemala asserted a territorial claim over the colony, and continued to do so into modern times. The colony was taken toward independence in 1964 by George Price, but Guatemalan hostility caused the final implementation to be delayed until 1981. A British force remained after independence to protect the new state, which was not admitted to the Organization of American States (OAS). The economy remained primarily agricultural, with an emphasis on sugar and citrus fruits. It was in the early 1980s a major marijuana producer, and on account of this the government came under severe pressure from the USA. In 1984 UDP (United Democratic Party) leader M. Esquival became prime minister; but he was defeated in 1989, and George Price regained the office. Economic diversification was hindered by the effects on the tourist trade of drug-trafficking and the dispute with Guatemala.

COSTA RICA

(Republic of Costa Rica) **Capital** San José **Population** (1989) 2,941,000 **Area** 51,100 sq km **GNP per capita** (1988) US$ 1760 **International organizations** UN, OAS, I-ADB, CACM.

In 1838 Costa Rica became independent of the Central American federation. In 1890 a free election brought José Joaquin Rodriguez to the presidency, and initiated a tradition of thorough democracy in the country, despite attempted revolutions in 1917 and 1948. A longstanding border dispute with Panama was resolved in 1941. The same year, Costa Rica followed the United States in declaring war on Japan, Germany and Italy. A new constitution was promulgated by José Figueres in 1949, whereby the army was abolished. In the 1970s and 1980s, Costa Rica received refugees from Nicaragua, and the country was used as a base for guerrilla activities. Coffee has long been an important crop, and fruit-growing developed into an important element in the economy from the 1940s. The exploitation of bauxite deposits encouraged the establishment of an aluminum smelter in the 1980s. In 1983 Costa Rica declared itself neutral, and maintained this, refusing to be used as a base for the attack of other nations.In 1987 President Arias drew up a peace plan for the region, signed by the five central American countries; he was awarded the 1987 Nobel Peace Prize; his work on a Nicaraguan cease-fire helped to pave the way for free elections in Nicaragua. At home, economic stringency provoked unrest in the late 1980s, but inflation and unemployment were reduced; and rescheduling, and other strategies to deal with the national debt, were very successful. However, a drug scandal implicating government members led to the electoral defeat in 1990 of the PLN (National Liberation party) by the PUSC (Christian party for Social Unity) under Rafael Calderon.

DOMINICA

(Commonwealth of Dominica) **Capital** Roseau **Population** (1989) 82,800 **Area** 750 sq km **GNP per capita** (1988) US$1,650 **International organizations** UN, Commonwealth, OAS, ACP, CARICOM.

Until 1940 Dominica was joined administratively to the Leeward Islands, then transferred to the Windwards as a separate colony. In 1958 it joined the West Indies Federation. After this was dissolved in 1962, there were discussions for alternative forms of federation. The West Indies Act of 1967 gave Dominica the status of association with Britain and under the 1967 constitution it became fully self-governing, gaining full independence in 1978. Cabinet crises followed in 1979, caused by the involvement of the government in an alleged invasion of Barbados. In the 1980 elections, Eugenia Charles became the Caribbean's first female prime minister. Under her conservative administration Dominica made marked advances toward economic recovery, with considerable decreases in unemployment and inflation. She was reelected in 1985. Several coup attempts in 1981 were overshadowed by successive natural disasters: in 1979 Hurricane David severely damaged the island, virtually wiping out the nation's agriculture by carrying away most of the topsoil. The following year the island was battered by Hurricane Allen, and in 1989 by Hurricane Hugo.

DOMINICAN REPUBLIC

(Dominican Republic) **Capital** Santo Domingo **Population** (1989) 7,012,000 **Area** 48,443 sq km **GNP per capita** (1988) US$680 **International organizations** UN, OAS, I-ADB.

In 1916 the United States took complete control of the politically and economically weak Dominican Republic. This period of US occupation (1916–24) saw the improvement of infrastructure and services. Horacio Vásquez became president in 1924, but the incompetence and corruptness of his rule led to a revolution, aided by Rafael Trujillo, who controlled the US-trained and organized army. Trujillo instituted an extremely harsh dictatorial regime. However, his rule brought prosperity, resulting in the repayment of the national debt. He was assassinated in 1961, and successors took a more democratic course. The hegemony of the right wing was maintained, however, and this caused an uprising in 1965, which was repressed by the USA, who then occupied the country (1965–66). The nation then fell into the grip of rightwing business interests, and the economy gradually declined, until in 1986, with the reelection of Joaquin Balaguer (president 1966–78), sugar prices (upon which the country depended) improved, the trade deficit was improved and international loans were restructured. In 1988, however, the economy declined sharply, with general strike, riots, rapid inflation answered by the aging Balaguer with austerity measures and a continuing overspend on public works. In 1991 he resigned as party leader.

EL SALVADOR

(Republic of El Salvador) **Capital** San Salvador **Population** (1989) 5,138,000 **Area** 21,041 sq km **GNP per capita** (1988) US$950 **International organizations** UN, Commonwealth, OAS, I-ADB, CACM.

In El Salvador the vital coffee crop, and politics, were controlled by a rich elite; each president chose his successor. When Pío Romero Bosque (president 1927–31) decided to leave this choice to the legislature, a military revolt ensued and Gen. Maximiliano Hernández Martínez became president, imposing a harsh regime until 1944, when a revolt provoked his resignation. During the 1950s a military junta ruled the country, and cotton became another important crop. Lieut. Col. José María Lemus, the last president of the junta, was deposed in 1960, and within months a second junta took control, under which many economic reforms were enacted, and opposition parties allowed. Emotions stimulated by a soccer game with Honduras in 1969 provoked the "Soccer War". El Salvador finally agreed to withdraw unconditionally, although peace was not restored until 1980. The harsh regime of the Partido de Conciliación Nacional (PCN) during the 1970s saw economic conditions also grow harsh. By 1980, the country had plunged into a civil war of guerrilla violence. A junta then attempted to reform the system and commit itself to human rights (the ousted president Romero's government had been accused of torture and murder of political prisoners), but violence continued to spread, and pressure grew in the USA to withdraw its support. José Napoleón Duarte, the only civilian member of the junta, was appointed president; Álvaro Magaña Borjo became president in 1982 after Duarte was forced to resign. Fresh elections in 1984 brought back Duarte, who began a dialog with the guerrilla leaders, and tried to reduce human rights abuses, the latter being a condition to which US aid was attached. Still the civil war did not abate. In 1989 Alfredo Cristiani of ARENA (Nationalist Republican party) was elected president in an atmosphere of high tension, with postelection murders of two leading politicians. Cristiani introduced austerity measures as the bloodshed increased; his term of office was marked by massive violations of human rights. 1990–91 saw the launch of an attempt to negotiate peace under UN auspices.

GRENADA

(Grenada) **Capital** Saint George's **Population** (1989) 96,600 **Area** 345 sq km **GNP per capita** (1988) US$1265 **International organizations** UN, Commonwealth, OAS, I-ADB, CARICOM.

In 1967 Grenada, previously British, became a self-governing state, becoming fully independent in 1974 as a constitutional monarchy under the British monarch. In 1967 the Grenada United Labour Party took office under Eric M. Gairy. In the 1976 election its majority was reduced by an opposition coalition (NJM), which in 1979 staged a bloodless coup, and proclaimed a People's Revolutionary Government (PRV) with their leader Maurice Bishop as prime minister. In a

▲ **Soldiers in El Salvador in the late 1980s.**

▲ **A group of women in Haiti.**

military coup in 1983 Bishop was killed and a US-led invasion followed, which overthrew the coup leaders and returned power to the governor-general. When constitutional government was restored in 1984, a general election brought in the New National Party (NNP) under H.A. Blaize, who tried to revive tourism, which, with agriculture, was a major source of income. An acute budgetary crisis caused internal divisions in the NNP in 1987, and growing opposition. This continued until Blaize's death in 1989. His successor, Ben Jones, gave way in 1990 to Nicholas Braithwaite, of the NDC (National Democratic Congress); two defections to the NDC gave Braithwaite an absolute majority. Grenada is governed as a constitutional monarchy with the British monarch represented by a governor-general as the nominal head of state; but the elected government wields executive power.

GUATEMALA

(Republic of Guatemala) **Capital** Guatemala **Population** (1989) 8,935,000 **Area** 108,889 sq km **GNP per capita** (1988) US$880 **International organizations** UN, OAS, I-ADB, CACM.

In Guatemala after independence (1838), a conservative rule continued until 1871, followed by a Liberal regime that lasted until 1944, when a leftwing regime came to power. Liberal leader Barrios instigated sweeping reforms, reducing the power of the upper classes, developing the economy and stimulating the coffee industry. These progressive moves continued under Barrios' politically repressive successors. In 1931 Jorge Ubico ushered in a period of one-sided economic development, with continued military repression. Discontent grew, and in 1944 a general strike forced Ubico to resign, and a leftist revolutionary junta took over, bringing better conditions for workers and a welfare system. Communist influence in the country grew, in 1954 provoking an invasion by Col. Carlos Castillo Armas, aided by the CIA and foreign recruits. He became president, attempting to mitigate the Communists' social reforms, but he was assassinated in 1957. His successor, Ydígoras Fuentes, was removed in a coup in 1963, and it was not until 1966 that a free election was held, this time won by Julio César Méndez Montenegro, but his need to pander to the military undermined his authority. Further regimes continued this repressive policy, and popular reformers were stymied by electoral manipulation. The discovery of oil brought further unrest and a dispute with Britain over the independence of Belize, under whose territory the oilfield extended. A coup in 1982 installed a junta headed by Gen. Ríos Montt, who pledged to rid the country of corruption, end the guerrilla war and disband the death squads, but he failed, and was overthrown a year later by Gen. O. Mejía Victores, who promised to restore democracy. In 1984 elections were held, in which parties of the center took a third of the vote. The new civilian government under President V. Cerezo pledged to improve human rights and the economic situation and keep down the counterinsurgency of the right wing, which was the guarantee of the continuation of vital US aid. Strikes and terrorism continued rampant, and relations with the USA deteriorated. In 1989–90, Jorge Serrano Elias of the MAS (Solidarity Action movement) was elected president and promised a cabinet of national unity.

HAITI

(Republic of Haiti) **Capital** Port-au-Prince **Population** (1989) 5,520,000 **Area** 27,400 sq km **GNP per capita** (1988) US$360 **International organizations** UN, OAS, I-ADB.

In the early 20th century the USA had economic predominance over Haiti, and US Marines occupied the country between 1915 and 1934. Hostilities between Haiti and the Dominican Republic marked the postoccupation period, as well as further political instability. In 1957 François "Papa Doc" Duvalier was elected president, and soon after had himself elected president for life, organizing a private militia, the Tontons Macoutes, to maintain his regime. He presided over a period of economic decline, instability and poor relations with the Dominican Republic, but remained dictator of Haiti until his death in 1971, when he was replaced by his son Jean-Claude "Baby Doc" Duvalier. The period of the latter's rule was equally one of repression and economic decline, and popular demonstrations during 1986 caused him to go into exile, to be replaced by a five-man military-civilian council, which elected Leslie Manigat as president in January 1988. In June of that year, a coup by Gen. Henri Namphy reinstated a military regime, but his lack of support among the military forced him to make concessions to Duvalier supporters. He was deposed in September by Sgt. Joseph Hébreux who installed Brig. Gen. Prosper Avril as president. He too lacked sufficient authority to control the country, and economic decline continued. In 1990 Avril resigned and went into exile; the next president, appointed by Gen. H. Abraham, was a judge, Ertha Pascal-Trouillot; her regime was marked by sporadic violence and assassination attempts and at general elections in 1991 leftwing priest and theologian Jean-Bertrand Aristide was chosen president. He promised to eradicate corruption, and renounced his presidential salary. He declared himself nonparty and appointed a baker, agronomist and friend, René Preval, as prime minister, and a whole cabinet of similar individuals.

HONDURAS

(Republic of Honduras) **Capital** Tegucigalpa **Population** (1989) 4,530,000 **Area** 112,088 sq km **GNP per capita** (1988) US$850 **International organizations** UN, OAS, I-ADB, CACM.

During the first decade of the 20th century, Miguel Davila was president of Honduras, but his unpopularity caused enough unrest for US Marines to be sent in to protect US banana investments. Political unrest continued after World War I. In 1932 Gen. Tiburcio Carías Andino was elected president (until 1949), and he was a virtual dictator. Honduras backed the Allies in World War II as it had in World War I. Wartime economic problems led to unrest and the government barely survived, though Carías was forced to resign the presidency. The next three presidents (1949–63) made some headway in modernizing the infrastructure and administration of the country. The last of them, Ramón Villeda Morales, was deposed in 1963 by Col. Osvaldo López Arellano. In 1969 the Soccer War with El Salvador, triggered by events during a soccer game, caused further economic degradation. Until 1971, with the election of Ramón Ernesto Cruz, Honduras was ruled by a military government. Cruz's rule was short as he was removed by López Arellano in 1972, but López Arellano himself resigned in 1975 after an international bribery scandal, and was replaced by Col. Juan Alberto Melgar Castro, under whose rule (until 1978) Honduras prospered. A bloodless coup in 1978 brought in Gen. Policarpo Paz García, who continued Melgar's policies. Despite internal calm and the election in 1981 of the first civilian government for 17 years, Honduras was disturbed by the bloody revolutions in Nicaragua (1979) and El Salvador (1980). US bases in Honduras were used for antigovernment Contras in Nicaragua, and the Honduran government banned this in 1984. Honduras walked a tightrope in its relations with the USA and Nicaragua until 1988 when a centrist party more acceptable to the USA took power in Nicaragua.

SAINT KITTS-NEVIS

(Federation of Saint Kitts [or, Saint Christopher] and Nevis) **Capital** Basseterre **Population** (1989) 44,100 **Area** 267 sq km **GNP per capita** (1988) US$2,770 **International organizations** UN, Commonwealth, OAS, ACP, CARICOM, OECS.

Saint Kitts-Nevis consists of two islands in the Lesser Antilles. Since independence in 1983, the federation has been a member of the Commonwealth of Nations, with the British monarch as head of state. The islands of Saint Kitts, Nevis and Anguilla were united by a federal act in 1882 and became an independent state in association with Britain in 1967. The islands were granted full self-government, with Britain retaining responsibility for defense and foreign affairs. In 1967 Anguilla complained of domination by the Saint Kitts administration and proclaimed independence. Negotiations between the Anguillan leaders, Britain and Caribbean Commonwealth states were unsuccessful and in 1969, Britain sent police and troops to the island, and a temporary British commissioner was installed. The troops were withdrawn later the same year, but a team of military engineers remained. In 1971 Anguilla was placed directly under British control, and in 1976 was granted a constitution, though it remained technically a part of Saint Kitts-Nevis Anguilla. It was formally severed from the federation in 1980. Sugar is the main export and Britain and the USA are the main trading partners of the islands. After a traumatic drop in sugar prices in the mid 1980s, the government promised its intention to diversify the economy.

SAINT LUCIA

(Saint Lucia) **Capital** Castries **Population** (1989) 150,000 **Area** 617 sq km **GNP per capita** (1988) US$1,540 **International organizations** UN, Commonwealth, OAS, ACP, CARICOM.

Saint Lucia is an island in the Lesser Antilles, the second largest of the Windward Islands, a parliamentary democracy within the Commonwealth of Nations. After French and British territorial rivalry it was ceded to Britain in 1814, becoming a crown colony. The French influence on the island continued however, for instance in the preponderance of the Roman Catholic Church. Representative government was obtained by the constitution of 1924. In 1958 Saint Lucia joined the West Indies Federation, and in 1960 it became autonomous within the Federation, also achieving a greater degree of internal self-government. After the demise of the Federation in 1962, and after discussions for alternative forms of federation, in 1967 Saint Lucia assumed the status of association with Britain. Independence was achieved in 1979, and in the first elections the leftwing Labour Party became the government, and set about forming close political ties with the socialist regimes of the Caribbean, establishing relations with Cuba and joining the nonaligned movement. Attempts at a mixed economy proved unsatisfactory in dealing with the problems of the new country, especially after Hurricane Allen wiped out the banana crop in 1980. Factions within the Labour Party caused political instability, and the tourist trade slumped to half its preindependence level. In 1982 the more conservative United Workers' Party was voted in on a platform of inviting foreign investment and decentralizing government administration. The agricultural sector was gradually rebuilt after the hurricane and tourism increased, aided by the end of a recession in the USA. Sugarcane was the chief crop until 1964 when most of the cane fields were converted to banana cultivation, and bananas are now the chief export of Saint Lucia. Most exports go to Britain, USA and other Caribbean islands.

SAINT VINCENT AND THE GRENADINES

(Saint Vincent and the Grenadines) **Capital** Kingstown **Population** (1989) 114,000 **Area** 389 sq km **GNP per capita** (1988) US$1,100 **International organizations** UN, Commonwealth, OAS, ACP, CARICOM.

Saint Vincent entered the 20th century as a British colony. In 1958 it joined the Federation of the West Indies, and in 1960 was granted a new constitution, becoming in 1969 a state in association with Britain. In 1979 Saint Vincent achieved independence, and in the first elections held that year the Labour Party under M. Cato became the government. Cato was critical of the revolution in Grenada, and supported the US invasion in 1983. He developed close political ties with the centrist governments of Trinidad and Tobago and Barbados. In this year also the Soufrière volcano erupted, damaging agriculture and the tourist trade, and in the following year Hurricane Allen virtually wiped out the banana crop. Recession in the USA and the falling value of the pound badly affected both tourism and the export of bananas. A general election in 1984 brought the New Democratic Party (NDP) to power under J. Mitchell, who began a program of reorganizing agriculture and lowering unemployment, which was more than 30 percent, by encouraging the construction industry. Economic improvements abroad improved the tourist and export trades. Mitchell supported the 1987 movement for East Caribbean unity; and he moved toward a closer union with Dominica, Grenada and Saint Lucia. He was reelected in 1989.

SURINAM

(Republic of Surinam) **Capital** Paramaribo **Population** (1989) 405,000 **Area** 163,820 sq km **GNP per capita** (1988) US$2,450 **International organizations** UN, OAS, ACP, I-ADB.

Surinam gets its name probably from the Surinen tribe, who were expelled by the Caribs before the 15th century, when other indigenous tribes were the Arawak and the Warrow. It was first visited by Europeans in the early 16th century, and the coastal part appropriated but not settled by the Spanish in 1593. A settlement was founded there in 1651 by an Englishman, and developed economically mainly by Dutch Jews. In 1667 Britain ceded Surinam to the Dutch in exchange for what is now New York. It reverted to British possession twice (1799–1802 and 1804–16) between then and 1816, when the Treaty of Paris (Nov 1815) finally confirmed it as a Dutch colony, Dutch Guiana. After the abolition of slavery in 1863 workers from China, Java and India arrived in the colony to support the agriculturally based economy. After achieving partial autonomy in 1950, it became self-governing in 1954. Independence was gained in 1975 under President Ferrier, previously the governor. After a military coup in February 1980 President Ferrier appointed a civilian administration under Dr. A. Chin A Sen; a National Military Council (NMC) was established; a further coup in August of the same year brought about Ferrier's replacement by Chin A Sen, and the replacement of several NMC members. An attempt to contain the power of the army was thwarted by the formation of the RPF (Revolutionary People's Front), led by the army. In 1982 Lt. Col. D. Bouterse took over from Chin A Sen with another coup, imposing direct military rule. But Bouterse consistently broke his promises of economic and social reform, the people demanded a return to the constitution, and the army tightened its grip. A period of civil disturbance followed, with strikes and constant demands for free elections; in 1984 Bouterse dissolved the cabinet and created an interim government, declaring his intention to restore civilian rule. The Dutch government were critical of Bouterse's regime, and stopped aid in 1985; in this year a new National Assembly and cabinet were installed. In 1986 antigovernment guerrilla fighting started, on the part of the SLA (Surinam Liberation Army). In 1987 a new constitution was approved by the National Assembly and by general referendum. Guerrilla activity continued, with sabotage causing the closure of a bauxite processing plant; the Dutch ambassador was accused of supporting the guerrillas and recalled. despite this, a general election was held that year; it was won by the new opposition alliance, the FDO; and in 1988 the National Assembly elected Ramsewak Shankar president. Full relations were restored with the Netherlands. In 1989 the Kouru Accord (a truce) between government and SLA was opposed by the army, and Bouterse resigned as Commander-in-Chief. More guerrilla activity sprang up, this time from an anti-SLA Amerindian group. In 1990 President Shankar was deposed in a bloodless military coup, and J. Kraag installed as provisional president; but Bouterse held power; and in 1991 he achieved a peace accord with the SLA. A general election was held in 1991, and won by a Labor/ethnic coalition, the NDF.

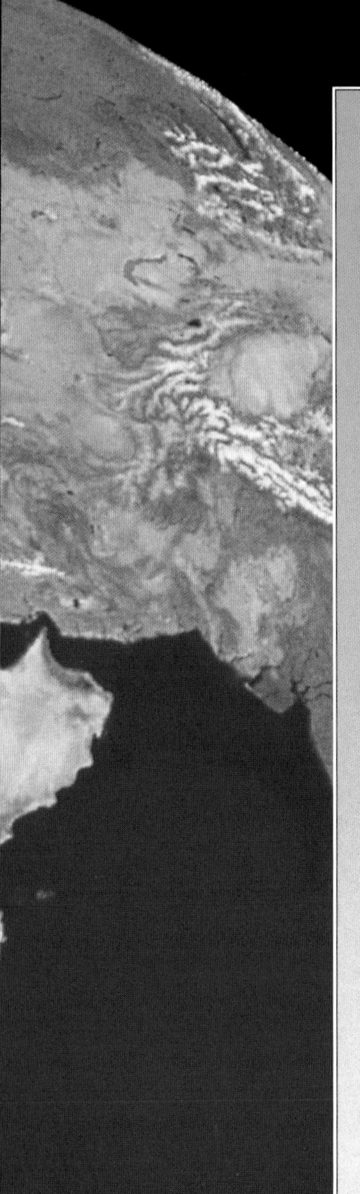

AFRICA

INDEX OF COUNTRIES

ALGERIA

DEMOCRATIC AND POPULAR REPUBLIC OF ALGERIA

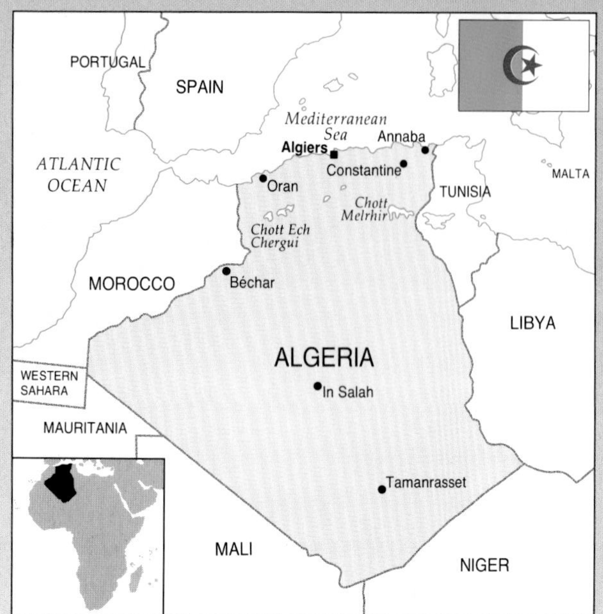

CHRONOLOGY

1900	Fiscal statute for Algeria
1944	Association of the Friends of the Manifesto and Liberty (Nationalist movement)
1945	Rebellion at Sétif and Guelma
1954	Front de Libération Nationale (FLN) formed; revolt begins
1955	Massacre at Skikda
1961	Referendum on self-determination
1962	Ceasefire and negotiated independence
1965	Military coup removes Ben Bella
1987	Trial of Islamic fundamentalists at Medea
1989	National referendum on single-party rule

ESSENTIAL STATISTICS

Capital Algiers

Population (1990) 24,700,000

Area 2,381,741 sq km

Population per sq km (1989) 10.3

GNP per capita (1987) US$2,760

CONSTITUTIONAL DATA

Constitution Multiparty republic with one legislative house (The National People's Assembly)

Date of independence 5 July 1962

Major international organizations UN; LAS; OAU; IDB; OPEC

Monetary unit 1 Algerian dinar (DA) = 100 centimes

Official language Arabic

Major religion Sunnite Muslim

Heads of government since independence (President) Ahmed Ben Bella (1962–65); Houari Boumédienne (1965–78); Chadli Bendjedid (1979–)

France began to conquer Algeria in 1830 and hardened its grip in the years leading to World War I. French government defined and created Algeria as the old indigenous ruling families died out. Algeria became a part of France with French laws, administration and control and was never a colony. A governor exercised overall authority, but an electorate made up of the white population and the few Muslims who had abjured their religious and cultural roots to become a political elite voted for deputies in the National Assembly in Paris. The Algerian budget, usually voted *en bloc* in Paris, was allocated by the army of civil servants familiar throughout the French state. When war came in 1914 thousands of Algerians joined the military and fought at Verdun and the battles of northeast France. By 1918 more than a third of the male population worked and served in continental France. Yet political rewards did not follow, and anger and resentment grew among the indigenous peoples while the white settlers, or *pieds noirs* ("black feet"), seemed largely oblivious to the changes taking place.

Attempts to extend full rights of citizenship to more Muslims failed against the opposition of the representatives of the settlers, and the first stirrings of Algerian nationalism were felt among the thousands of workers who lived and trained or studied in Paris and other large cities of metropolitan France. These were drawn to Messali Hadj and his newspaper the *Etoile Nord-Africaine* (established in 1926) and to the Algerian People's Party (PPA) in 1937. Others, like Ferhat Abbas, argued for some federal relationship for Algeria within the French state. These beginnings were interrupted by World War II. Algeria was part of Vichy France between 1940 and 1942, but the Anglo-American landings in November 1942 brought a significant development for the Muslims. Contact with the Americans and President Roosevelt's advisor, Robert Murphy, in particular, encouraged them to expect self-determination after the war. The white settlers were also initially hostile toward the Free French leader General Charles de Gaulle. Disappointed expectations led to a murderous rebellion on 8 May 1945, following the arrest of the nationalist leader Hadj. White settlers were massacred in Sétif and Guelma, and their subsequent avenging retribution left untold numbers of Muslims dead and injured.

Algeria then subsided into the inertia of prewar times. Local democracy of a kind came with the creation of a Muslim-elected assembly, for about nine million people, alongside the scarcely one million white settlers who elected their representatives. National representation in Paris was on the old basis of restricted citizenship. The simmering demands for reform were frustrated until 1954 when, encouraged by the French capitulation in Indo-China, the revolutionary *Front de Libération Nationale* (FLN) was formed. The revolt broke on 1 November 1954.

The War of Algerian Independence lasted for eight years, involved terrorist attacks, indiscriminate assassination and torture, and absorbed 500,000 French troops, many of these Muslims, the Hakis. The war raged through the cities, the French military gradually breaking the nationalists' organization by martial law in the 1957 Battle of Algiers, by frontier defenses (the Morice Line along the Tunisian border) and by rapid helicopter warfare, use of paratroops and counter-insurgency operations. However, the nationalists won the political battles. Governments in Paris found no way of ending the war; indeed in May 1958 street demonstrations by whites brought down the government in Paris and called for the return of Charles de Gaulle who was expected to save France from ruin. But he proved no friend of *Algérie-française*, and in September 1959 suggested the possibility of self-determination and full political participation by *all* Algerians. His perseverance brought success in talks at Evian, and in 1962 Algeria became independent.

The FLN remained the ruling party and opposition was not permitted, original leaders were ousted and many assassinated. During the 1980s Islamic fundamentalism gained much popular support. A trial of Islamic activists in 1987 brought riots and such forceful demands that single-party rule was ended by referendum in 1989. By 1990 the Islamic Salvation Front (FIS) had won control of most municipal councils and the regime had been obliged to modify its former revolutionary Marxism. Despite this, Algeria has acquired great standing as an honorable and responsible neutral state.

Economic and social trends

Algeria has faced an exceptionally high birth rate through the 20th century and its population has been increasingly concentrated in the cities. Industrialization has partly answered this serious social problem. There are several Algerian peoples, Arabs, Berbers and others. The resources have never been available to secure employment and wealth for this population.

Few primary materials provide stable exports beyond some oil and substantial natural gas supplies, which are sold to West European states, mainly France. Poor infrastructure, only 4,000 kilometers (2,500 mi) of railroads, and industry mean that Algeria must import modern equipment, much of which comes from Germany. Control over the economy was sustained by programs of nationalization of the oil and gas resources in 1971 and a National Charter in 1976. The lack of food for the rapidly expanding population was to be remedied by agricultural reforms in the 1970s and again in the National Plan of 1985–89 which placed emphasis on irrigation and water supplies.

The diverse origins of the peoples and the lack of historical unity meant that Algerian culture is both recent and political. Nationalism produced a strong and lively literature and poetry, much of this written in French which remains the basic language of all Algerian peoples. The writings echoed the bitter experience of the War of Independence. After 1962 new themes dominated, the role of women (important during the war as carriers and agents) and the effect of urbanization and exile in France. Muslim artists and writers took up more austere themes and in the 1970s and 1980s these were developed in a flowering of Algerian film. The Algerian National Institute of Cinema funded and inspired this achievement, represented in many Mediterranean festivals.

▼ **Islamic religious and political leaders in Algeria in 1990.**

TUNISIA

REPUBLIC OF TUNISIA

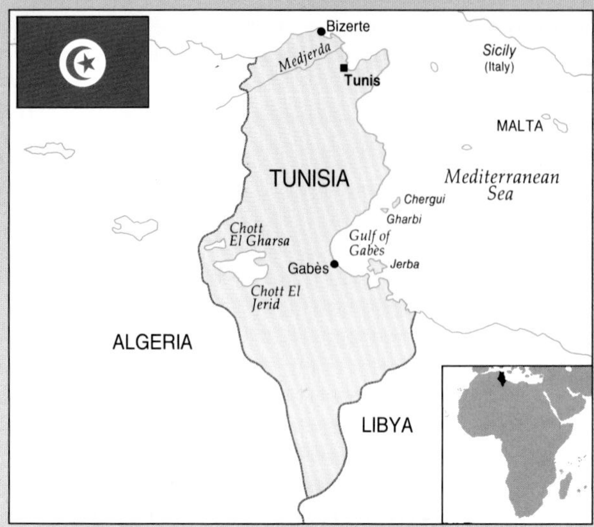

CHRONOLOGY

1883 Tunisia declared as French protectorate

1934 Habib Bourguiba forms Neo-Destour party

1956 Tunisia becomes independent

1957 Bey dynasty deposed – Bourguiba becomes President

1966 Oil is produced for the first time

1969 Tunisia is hit by disastrous floods

1987 Peaceful coup overthrows Bourguiba – led by Zine al-Abidine Ben Ali

1989 Tunisia is founder member of the Union of Arab Maghreb states (UMA)

ESSENTIAL STATISTICS

Capital Tunis

Population (1989) 7,973,000

Area 154,530 sq km

Population per sq km (1989) 51.6

GNP per capita (1987) US$1,210

CONSTITUTIONAL DATA

Constitution Multiparty republic with one legislative house (Chamber of Deputies)

Date of independence 20 March 1956

Major international organizations UN; LAS; OAU; IDB

Monetary unit 1 dinar (D) = 1,000 millimes

Official language Arabic

Major religion Sunnite Muslim

Heads of government since independence (President) Habib Bourguiba (1956–87); Zine al-Abidine Ben Ali (1987–)

Under the guise of a protectorate, France governed Tunisia until 1956 but much of Tunisian society and land ownership remained intact. Nominally the traditional ruler, the Bey, continued to rule. Within 20 years, French government generated a nationalist movement of Young Tunisians; World War I gave a vigorous impulse to nationalism and was followed by the founding of the Destour (Constitution) party. In 1934 Habib Bourguiba led a breakaway movement, forming the Neo-Destour. It dominated the nationalist movement and, as the Socialist Destourian Party, was the ruling party in 1990.

Controlling the Neo-Destour party and supported by a vigorous trade union movement, Bourguiba kept up moderate but unrelenting pressure on the French government. During World War II he resisted Mussolini's blandishments, supporting Gaullist France and returning to the struggle for independence after the war. The government of French premier Mendès-France gave Tunisia internal autonomy in 1954; the country became independent in March 1956.

A year later the Bey was deposed and Bourguiba became president. During the struggle for independence he had adopted Islamic positions as an adjunct to Tunisian nationalism. But once in power he introduced an exceptionally progressive personal code, establishing rights for women. For a time a well-organized birth control program was put in place, which Bourguiba justified by an interpretation of Islam; he also discouraged fasting in Ramadan because of the struggle for Tunisian development. A major effort was made in the field of education, which largely followed a French pattern, was secular and adopted French as the language of instruction after the first two years.

The economy was less successful. From 1961 to 1969 it was directed by Ahmed ben Salah, who established a system of state control and cooperative farming which strangled enterprise and nearly destroyed agriculture. It was brought to an end after disastrous floods in 1969 and private ownership of land was restored. Tourism was successfully developed. Oil was first produced in 1966 and has kept up with domestic consumption.

The regime was consistently moderate and pro-Western. Once independence was established Bourguiba readily expressed his admiration for French liberal values. As early as 1965 he spoke in favour of a dialog in the Arab-Israeli dispute.

But as Bourguiba aged the system he had created decayed. The party and the machinery of government became increasingly bureaucratic, privileged and corrupt. Opposition was expressed, in Tunisia as elsewhere, in the growth of Islamic activism which Bourguiba saw as a threat to his secular republic.

In October 1987 he appointed Zine al-Abidine Ben Ali as prime minister. Alarmed by the crescendo of Islamic violence and Bourguiba's intransigence, Ben Ali carried out a peaceful coup, with the backing of the army and much of the political elite, establishing himself as president in November 1987. He quickly set about reforming the economy in accordance with an IMF and World Bank program. He organized elections in 1989; but Islamic fundamentalists could stand only as independents and the elections were managed to ensure the dominance of the party. Looking across to the growing integration of the European Community, the Tunisian government joined with Algeria, Morocco, Mauritania and Libya to establish the Union of Maghreb states in February 1989.

MOROCCO

KINGDOM OF MOROCCO

CHRONOLOGY

1912	Morocco split in three – French protectorate; Spanish Zone; International zone of Tangier
1921	Abdel Krim contests European rule until 1926
1953	Sultan Mohammed is deposed
1956	Morocco gains independence led by Sultan Mohammed – International Zone of Tangier abolished
1975	Moroccan administration of Spanish Sahara agreed – resisted by Saharan guerillas
1988	Moroccan government agrees to settlement by referendum

ESSENTIAL STATISTICS

Capital Rabat

Population (1989) 24,530,000

Area 458,730 sq km

Population per sq km (1989) 53.5

GNP per capita (1987) US$620

CONSTITUTIONAL DATA

Constitution Constitutional monarchy with one legislative house (house of Representatives)

Date of independence 2 March 1956

Major international organizations UN; LAS; IDB

Monetary unit 1 Moroccan dirham (DH) = 100 Moroccan francs

Official language Arabic

Major religion Sunnite Muslim

Heads of government since independence (King) Mohammed V (1927–1961); Hassan II (1961–)

In 1900 Morocco was an independent kingdom of the 17th-century Sharifian dynasty, which still held power unchanged in 1990. Spain had established outposts in Tangier, Ceuta and Melilla without trying to push inland. Morocco then became the object of rivalry between France, established in Algeria, and Germany, at a time when the indigenous dynastic succession passed through a period of weakness. Two Moroccan diplomatic crises, in 1905 and 1911, were followed by the French occupation of Morocco, except for a Spanish zone established in the north. The French occupation was extended across the country by Marshall Lyautey, who believed that the French "Protectorate" had more than legal meaning. He respected and did his best to preserve Moroccan Islamic society.

In the interwar years European rule was contested first by an outstanding tribal leader, Abdel Krim, who established a republic in the Rif mountains and fought the Spanish until he was defeated, with French help, in 1926. In French Morocco the nationalist movement of the 1930s was an elite movement in which the Sultan, Mohammed ben Youssef, participated.

Morocco was drawn into World War II when Allied troops landed there in January 1943. As elsewhere the impact of the war was to increase nationalist feeling. The war also brought a significant meeting between President Roosevelt and Sultan Mohammed during the Casablanca conference of 1943. After the war nationalist feeling and organization increased. It was resisted by the French but helped by a more lax administration in Spanish Morocco. The French administration tried to subvert Sultan Mohammed by building up one of the southern tribal leaders, al-Glaoui and, in 1953, deposing Sultan Mohammed. The attempt was a bad miscalculation. It increased Mohammed's prestige as a nationalist leader. In 1956 both France and Spain conceded; Morocco regained its independence under Mohammed V until February 1961.

His son, King Hassan, at first lacked his prestige but inherited his political shrewdness. He continued the distinctive political system of constitutional monarchy. Morocco has maintained parliamentary government since independence, although parliament was often suspended, elections postponed and, when they occurred, always managed. Political parties were harassed and contained, but never suppressed and the King retained support across the political spectrum.

Morocco adopted a Western stance in foreign policy even though King Mohammed took a radical line in African politics in the 1950s. King Hassan was, above all, able to tap Moroccan nationalism and so acquire freedom of maneuver in foreign and domestic politics. Moroccan troops were sent to Syria in 1973 and to the October war against Israel.

But the most important nationalist cause has been the fight for former Spanish Sahara which won support from left to right. King Hassan in 1975 secured the agreement of the Spanish government to a Moroccan administration of the Sahara, with Mauritania as a minor partner. But the Moroccan occupation was resisted by Saharan guerrillas (Polisario), supported by Algeria and Libya. Moroccan troops fought a successful war, constructing a secure sand wall which was easy to defend. The diplomatic battle was more demanding but began to resolve itself when Algeria withdrew its support for the Polisario. In 1988 the Moroccan government agreed to a settlement by referendum; a timetable for this was approved by the UN in April 1991.

LIBYA

SOCIALIST PEOPLE'S LIBYAN ARAB JAMAHIRIYA

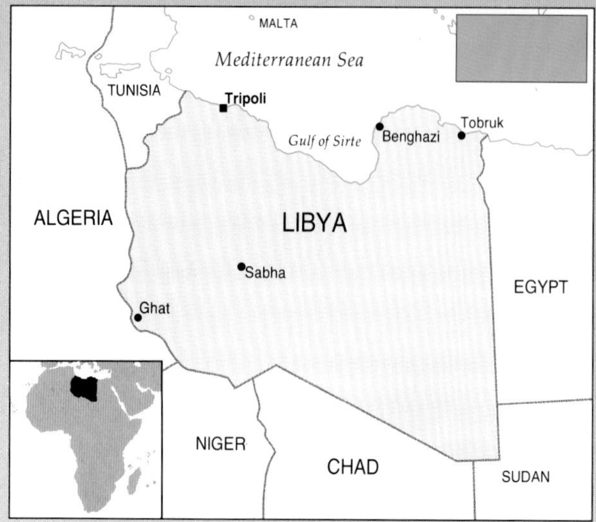

CHRONOLOGY

1911	Italian troops invade Tripoli
1951	Libya achieves independence
1955	First oil concessions granted
1969	Successful coup led by Captain Muammar Qadhafi
1971	Tehran–Tripoli agreement
1977	Libya is at war with Egypt
1986	US planes attack Tripoli
1989	Libya joins the Union of Arab Maghreb states (UMA)

ESSENTIAL STATISTICS

Capital Tripoli

Population (1987) 4,080,000

Area 1,757,000 sq km

Population per sq km (1989) 2.3

GNP per capita (1987) US$5,500

CONSTITUTIONAL DATA

Constitution Socialist state with one policy-making body (General People's Congress)

Date of independence 24 December 1951

Major international organizations UN; LAS; OAU; IDB; OPEC

Monetary unit 1 Libyan dinar (LD) = 1,000 dirhams

Official language Arabic

Major religion Sunnite Muslim

Heads of government since independence (Prime minister) Mahmud Muntasser (1951–54); Mohammed Saqizly (Feb–Apr 1954); Mustafa Halim (1954–57); Abdul Majid Coobar (1957–60); Mohammed bin-Othman al Said (1960–63); Mohieddine Fekini (1963–64); Mahmud Muntasser (1964–65); Husain Maziq (1965–67); Abdel Kader al Badri (Jul–Oct 1967); Abdel Hamid Bakkouche (1967–68); Wanis al Geddafi (1968–69); (President) Col. Muammar Muhammad Qadhafi (1969–)

In 1900 Libya was under nominal Ottoman rule, since Tripoli had been seized for Suleiman the Magnificent in 1551. The decay of the Ottoman Empire provoked, in Libya, a vigorous Islamic movement, fundamentalist and reformist, known (from its founder) as the Sanusiya movement. In its first phase, at the end of the 19th century, it was religious and evangelical, appealing to all Muslims and to pagans in Sudan and central Africa. The weakness of the Ottomans also provided an opportunity for Italy. In 1911 Italian troops invaded Tripoli and began a long period of conquest. By that time the Sanusiya movement had become a political and nationalist movement based on a network of *zawiya* (lodges), strongest in Cyrenaica but spread across the country. It led the resistance to the Italians until independence was achieved in 1951.

With difficulty Italy held on to Libya through World War I and then used military force to consolidate its rule. The country remained unattractive to Italians and large-scale colonization, settling some tens of thousands of Italians, was given massive government support. When Italy went to war in June 1940 Libya became a battleground where British and Commonwealth forces fought first the Italians and then the German army. Their victory, when Tripoli was captured in January 1943, was one of the turning points of the war.

World War II created a limited alliance between the British and the Sanusiya movement, now led by Emir Idris. He pressed his country's claims for independence. But amongst the Great Powers old imperial instincts merged with Cold War rivalry. The country was administered by the British in Cyrenaica and Tripolitania and the French in the southern province of Fezzan, but the Soviet Union sought a trusteeship over the country and the United States was intent on retaining the Wheelus air base. As diplomacy faltered Italy regained respectability as a western power in the Cold War, and a plan for an Anglo-Italian trusteeship was presented to the United Nations, where it was defeated. Under the aegis of the United Nations, Libya then became independent in December 1951.

The United Kingdom of Libya was a poor country with few known natural resources. The coastal strip between the platform of the Sahara and the Mediterranean is only one hundred kilometers (60 miles) wide at its maximum; along 650 kilometers (400 miles) of the Gulf of Sirte it meets the sea. In 1969 population was estimated at one person per square kilometer or 70 persons per square kilometer of arable land; in 1973 population was 2.26 million.

The discovery of oil transformed the economy and the politics of the country, and Libya then played a key role in changing the balance between oil-producing countries and international companies. The first oil concessions were granted in 1955 and Esso struck oil at Bir Zelten in 1959. Oil brought wealth, which proved subversive. Minor corruption became gross and the government of King Idris lacked the political energy and the administrative skill to absorb oil wealth into economic development. The overthrow of the Idris regime was widely expected but few if any identified the young officers who would carry out the coup. They were led by Captain Muammar Qadhafi and their coup succeeded against minimal opposition on 1 September 1969.

In 1970 a combination of events enabled Libya to obtain a higher price for oil. The vigor of a new regime was only one factor. There was a short-term tanker shortage which gave a

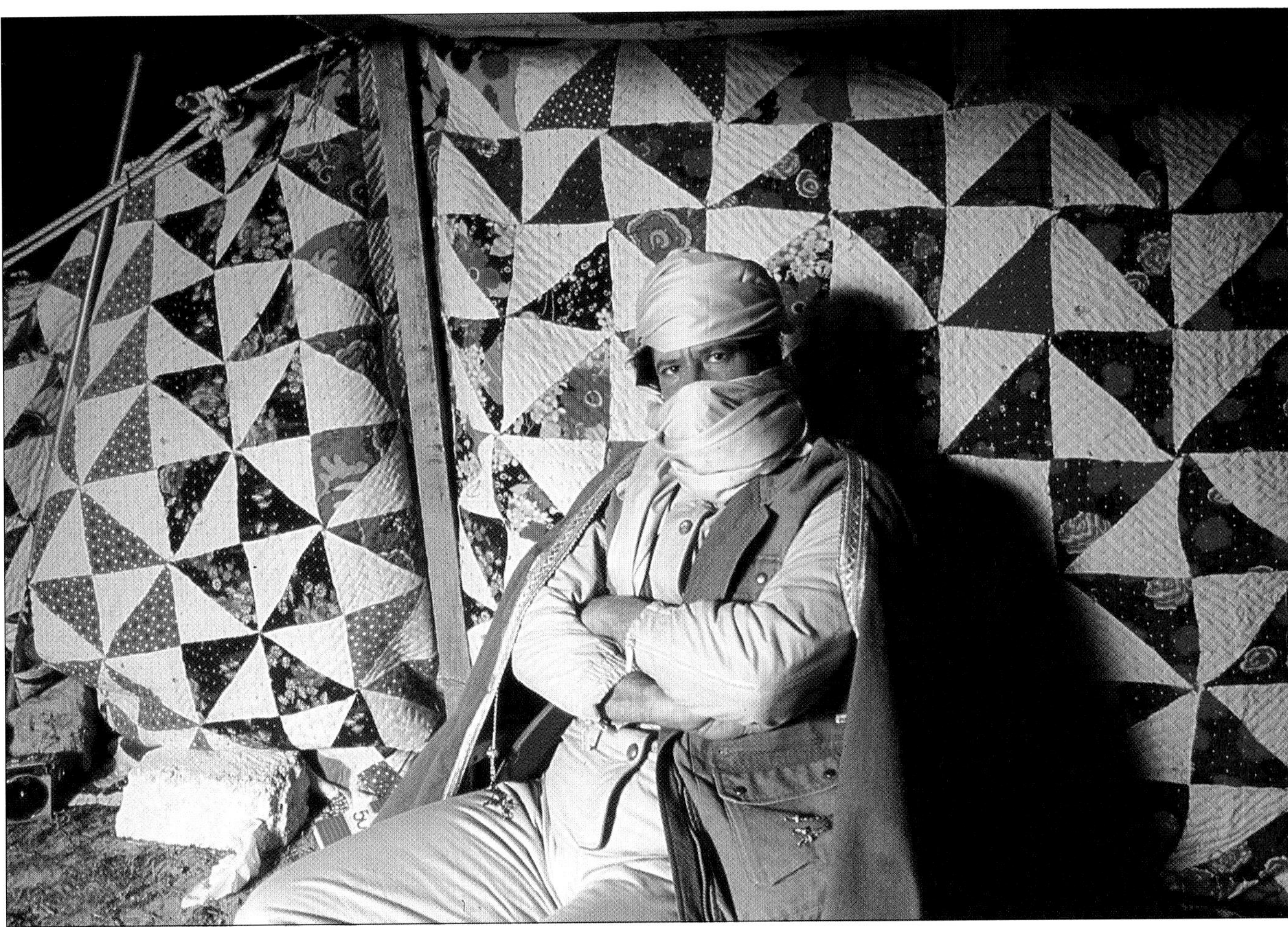

premium to Mediterranean oil. The concessionaires in Libya included small companies, more vulnerable than the majors. Occidental, the most vulnerable, yielded to Libyan pressure for higher prices and was followed by other companies. In this way OPEC scored its first major success, leading to the Tehran-Tripoli agreements of 1971 and heralding the spectacular price rise of 1973–74.

Qadhafi was inspired in his youth by Nasser's Arab socialism and also anticipated in his politics later movements of political Islam. He developed a political theory, eventually embodied in a "Green Book" which sought a middle road between capitalism and socialism. He professed ideas which may have been derived from Rousseau. The system of government which he installed included a series of popular committees, leading to a National Assembly, intended to embody a popular will without recourse to representation. The state was seen as an artificial creation impeding Arab unity.

Ideals and rhetoric failed to correspond to reality. Libya, although given the title Socialist People's Libyan Arab Jamahiriya, was in fact governed by a tight military security system, with East German help. Remaining poor except in the single resource of oil, Libya failed to develop economically. Oil revenues were spent on construction and on "turnkey" projects which failed to grow into a significant industrial sector. A major water project designed to tap an underground aquifer at Kufra and channel the water through a "man-made river" to coastal areas absorbed oil revenues but is unlikely, by itself, to develop a prosperous agriculture.

Qadhafi's genuine ambitions for Arab unity were dealt a fatal blow when a plan for union with Egypt (and Sudan) was overthrown by President Sadat shortly before the 1973 war. Like the early Sanusiyas Qadhafi wanted to extend Libyan influence southward into Africa. He led his army into inglorious support for Idi Amin in Uganda and an unsuccessful attempt to dominate Chad. Seeking revolutionary credentials which the 1969 coup failed to provide, he supported terrorism, not only by Palestinian groups but also by Filipino Muslims and the Irish Republican Army. They achieved little but alarmed the West and were given such importance by the Reagan administration in the United States that American planes attacked Tripoli in April 1986, although Qadhafi survived unhurt.

Thereafter Libya and Qadhafi resumed a low profile. The drop in the price of oil in the late 1980s and OPEC's adoption of quotas for its members had a severe impact on oil revenues. Lacking the massive reserves and flexible production capacity of the Gulf producers, Libya carried little weight in OPEC. The volatility of its foreign policy and disregard for the state interests of other Arab countries made permanent alliances impossible. In 1989 Libya joined the Arab Maghreb Union, but did so more from weakness than from strength.

▲ **President Qadhafi in a Bedouin tent.**

EGYPT

ARAB REPUBLIC OF EGYPT

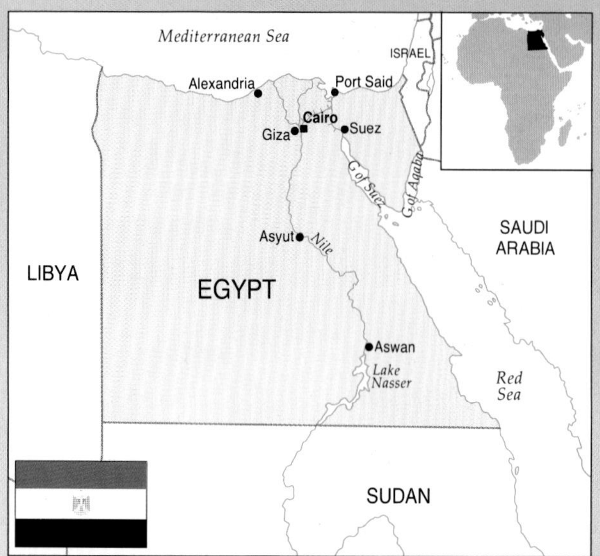

CHRONOLOGY

1914	British protectorate established
1922	Proclamation of independence
1937	Egypt joins League of Nations
1952	Riots break out in Cairo – military coup d'état
1954	British withdrawal from Egypt
1956	US cancels financial support of Aswan Dam and precipitate the Suez crisis
1962	Egypt is drawn into Yemen War
1967	Egypt demands withdrawal of United Nations forces; leads to Six-Day War
1981	Assassination of President Sadat
1990	Egypt hosts Emergency Arab League Conference

ESSENTIAL STATISTICS

Capital Cairo

Population (1989) 51,748,000

Area 997,739 sq km

Population per sq km (1989) 51.9

GNP per capita (1987) US$710

CONSTITUTIONAL DATA

Constitution Republic with one legislative house (People's Assembly)

Date of independence 28 February 1922

Major international organizations UN; LAS; OAU; IDB

Monetary unit 1 Egyptian pound (LE) = 100 piastres

Official language Arabic

Major religion Sunnite Muslim

Heads of government since independence (King) Fuad I (1922–36); Farouk (1936–52); Regency for Ahmad Fuad II (1952–53); (President) Gen. M. Naguib (1953–54); Col. G. Nasser (1954–70); A. Sadat (1970–81); S. Talib (acting) (Oct 1981); Lt.-Gen. H. Mubarak (1981–)

At the beginning of the 20th century Egypt was nominally a province of the Ottoman Empire, though in practice occupied by the British since 1882. On Turkey's entry into World War I it was made a British protectorate. By that time Egyptian nationalism was a well established political movement. It was followed by a broader movement which both enlisted popular support and was inspired by nationalist writing, secular and Muslim. World War I increased both Egyptian expectations for national self-determination, based on United States president Wilson's Fourteen Points, and British concern for the security of the empire. Popular demonstrations in 1919 led to a draft agreement but conservative and military opposition in Britain prevented its implementation. Instead the British government made a unilateral declaration of Egyptian independence, reserving to itself military security and the question of Sudan.

During World War II Egypt was controlled completely by Britain, which dictated who should form the government as the country became a major British base. Egypt itself declared war on Germany and Japan in February 1945, and also took the lead in establishing the Arab League, with its headquarters in Cairo. In 1946 Britain negotiated an agreement for the withdrawal of its forces but disagreement about the future of Sudan meant this was never ratified. The onset of the Cold War and the independence of India in 1947 then gave Egypt a renewed strategic importance for Britain. The ferment of nationalism increased. It was spread through Egyptian society by the impact of the war, which had created an artificial economic activity with consequent inflation. Nationalism was expressed in a vigorous literature and articulate daily press. It had a strong Muslim dimension; Sheikh Hasan al-Banna, who had founded the Muslim Brotherhood in 1928, was shot, presumably by the security forces.

British troops were attacked, often with the complicity or connivance of Egyptian police. Soon an explosion occurred, sparked by British forces occupying the police barracks at Ismailia. On this "black Saturday" in January 1952 serious riots broke out in Cairo; buildings were burned and foreign residents murdered. Meanwhile a group of army officers were plotting, in the Officers Club, a coup which they carried out on the night of 22–23 July 1952. The new government was headed by General Mohammed Naguib; its strongest member was Gamal Abdel Nasser who replaced Naguib in 1954.

Native-born Egyptians now ruled Egypt for the first time in 2,000 years. Consolidating their power they suppressed political parties, including the Communists and the Muslim Brotherhood, but failed repeatedly to establish an effective party of their own. Their ideology was diffuse but embraced two clear objectives: to drive the British out and to break the landowning class. Land reform was carried out immediately. Then, to negotiate with the British, the new government abandoned any Egyptian claim to Sudan. At the same time the British downgraded the base at Suez. The way was open for the agreement, reached in 1954, for British withdrawal.

The United States, as well as Britain, had cautiously welcomed the new regime and were ready to support, directly and indirectly through the World Bank, the construction of a high dam at Aswan. The dam was designed to give storage over a period of drought years, to irrigate and to generate electricity. It was a prestige project for the new government. But increasingly Egypt's international posture appeared dangerous to the

United States and Britain and even more so to France. Attempts to mediate negotiations with Israel were abortive; instead the Israelis launched a raid on the Gaza Strip and, in response, Egypt bought arms from the Soviet Union. President Nasser actively opposed British plans to construct an anti-Soviet Baghdad Pact, which was to include Jordan, and Cairo radio kept up its anti-British propaganda. Egypt recognized the Communist government of China. The British and the Americans had second thoughts about financing the Aswan Dam. The American decision to cancel was conveyed abruptly to the Egyptian ambassador in July 1956. In response President Nasser nationalized the Suez Canal and opened a new period in Egypt's history.

After the initial shock US president Eisenhower saw the nationalization as a commercial action. The British saw their security threatened; the French regarded Nasser's support as a decisive factor in the Algerian revolution. They brought in the Israelis and organized an expedition against Egypt, with the pretence that Britain and France were engaging in a police action to separate Egypt and Israel. The expedition was militarily successful but American pressure forced a withdrawal and a United Nations force (UNEF) was sent to Egypt.

The Suez affair gave Nasser great prestige as leader of progressive forces in the Arab world. Syria joined Egypt to form a United Arab Republic in 1958, although this union fell apart in 1961. Egyptian and foreign interests in banking, commerce and industry were nationalized and for a number of years the economy grew. But in 1962 Nasser, by supporting a republican coup in Yemen, was drawn into a debilitating Yemeni war. The economy declined, repression increased and Egypt's prestige in the Arab world was eroded. To regain the initiative Nasser opened a provocative crisis against Israel, in May 1967, by demanding the withdrawal of UNEF and closing the straits of Aqaba. Israel launched a preemptive attack defeating Egypt and occupying Sinai. Nasser died in 1970.

His successor Anwar Sadat gave Egypt a new direction. In a series of moves which were partly premeditated, partly improvised, he made first peace and then war with Israel and, in doing so, moved Egypt's client relationship from the Soviet Union to the United States. He gave primacy to Egyptian rather than Arab interests, and tried to liberalize the country economically and politically.

▼ **The Naguib–Nasser government of 1954.**

In contrast to the 1967 war, that of 1973 was carefully planned in alliance with Syria and with the support of the oil-producing states of the Gulf. Sadat kept Soviet support while the United States supplied Israel, provoking a shortlived oil embargo by the Arab oil producers. The crisis provoked, as Sadat hoped, intensive American diplomacy which resulted in disengagement between Egyptian and Israeli troops, sealed at the Geneva conference in 1973 and followed by further disengagement in 1975. Sadat then initiated his own peace process by securing an invitation to Jerusalem and recognizing Israel. The US president Carter mediated the peace process at Camp David. Sadat secured his major objectives: the support of the United States and a peace treaty with Israel. The cost was to be shunned by most of the Arab world, the Arab League itself moving to Tunis.

Islamic activism became an increasingly important political force. The mosque and university of Al Azhar remained the most important center of Islamic law and theology. It never spoke with a single voice but was a center of theological debate. Sadat encouraged hardline clergy (*ulema*) as a means of defense against the left and Nasserists. For the same reason the Muslim Brothers were allowed greater freedom of activity. But in the end this policy proved fatal for Sadat and dangerous for Egypt. Muslim calls for Islamic law (*shari'a*) to be made the law of the land alarmed the Christian Coptic church and put the Coptic community – 10 percent of the population – on the defensive. The Muslim Brotherhood was no longer a single coordinated organization: splinter groups, notably the *Takfir wa Hijra* and the *Jihad* group, broke away, making the movement more difficult for the security forces to control. In October 1981 Sadat was assassinated as he took the salute at an army parade. The aftermath showed the strength of Islamic opposition as groups of activists were rounded up, many of them well-armed.

Sadat's successor, Hosni Mubarak, appeared a colorless figure; his reputation increased steadily throughout the decade. The economy survived rather than succeeded. By slow stages the weight of the bureaucracy was reduced and subsidies held down. The population grew rapidly and Egypt remained a net food importer. There was little indigenous capital ready to invest in manufacturing. In terms of wages and productivity its labor force was at a disadvantage compared to the Far East, except in other Arab countries. Workers' remittances became an important source of foreign exchange, together with income from the Suez Canal and oil exports – all of which were dependent on the prosperity of the oil industry.

Politically the country remained relatively free. The press had long experience of self-censorship and expression was freer than in most other Arab countries. It was characteristic of Egyptian intellectual life that the novelist Naguib Mahfouz was awarded the Nobel Prize for Literature in 1988, but not all his books circulated freely in Egypt. Political parties were allowed within strict limits. Elections to the National Assembly were managed to ensure the dominance of the ruling National Democratic party but also to allow opposition freedom to express itself. Islamic opposition was the most vocal and there remained also a risk of sporadic communal violence between Muslims and Copts.

Egypt repaired its relationship with the Soviet Union but remained a client of the United States and, after Israel, was the most important recipient of American aid. By supporting Iraq in the war with Iran it won its way back into the Arab world, being readmitted to the Arab League in May 1989. Its new role was to be a mediator in the Arab world and a go-between in the American-sponsored peace process with Israel. In 1990, as none could have foreseen in 1980, Cairo hosted an emergency Arab League Conference to organize the Arab containment of an expansionist Iraq.

Suez Canal

The Suez Canal, linking the Mediterranean with the Red Sea, became practical when steamships made fast navigation in the Red Sea possible. It was built under an 1856 concession to Ferdinand de Lesseps' Suez Canal Company, with a predominantly French shareholding, and was opened in 1869. In 1882 Britain occupied Egypt and thereby secured physical control of the Canal, and became also the most important user. In 1888 the Constantinople Convention proclaimed the right of free transit to every vessel of commerce or war, in wartime as well as peace. By 1945 the Canal carried some seven percent of world trade; with the increase in oil transport the proportion increased to 15 percent.

The government of President Nasser, brought to power by the revolution of 1952, negotiated the withdrawal of British forces in 1954. Egypt then controlled the Canal and naturally considered the possibility of its nationalization. The Egyptian view was that the Canal was part of Egyptian sovereign territory, that it had been built with the sweat and blood of Egyptians and that the shareholders had been well rewarded. The trigger for nationalization, in July 1956, was the American and British decision not to finance the Aswan Dam, a decision taken without thought that nationalization might follow, possibly because the concession was due to expire in 1968.

Nationalization provoked an Anglo-French-Israeli invasion of Egypt, blocking the Canal, which had remained open. The expedition failed and withdrew. The Canal reopened in April 1957 and the amount of Egyptian compensation to be paid to the Suez Canal Company was agreed in May 1958. In 1957 Egypt's receipts from the Canal were LE24.5 million compared to LE2.3 million in 1955.

The Canal was again closed in the 1967 Arab–Israeli war and its ownership was the subject of conflict in the 1973 war. Between these interruptions it was widened and deepened. Although the transport of oil was partly shifted to supertankers, the growth in both oil and other trade ensured the continued prosperity of the canal. By the 1980s it was, with oil export and foreign remittances, Egypt's main source of foreign exchange with revenues of about US$1 billion annually. At the outbreak of the Gulf crisis in August 1990 it provided the transit route for warships and military transport to the Gulf.

▲ **The Suez Canal is one of the world's busiest waterways.**

SUDAN

REPUBLIC OF THE SUDAN

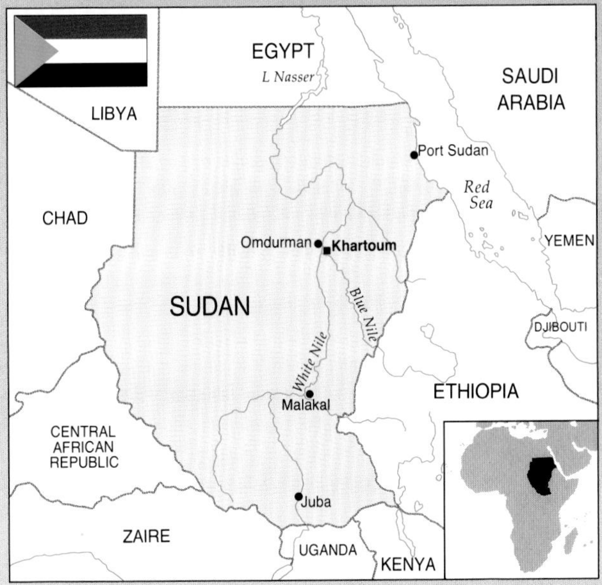

CHRONOLOGY

1930	Introduction of "Southern policy" to isolate African and Christian south from Arab and Muslim north
1956	Independence achieved
1958	Military coup overthrows liberal democratic government
1962	Sustained civil war breaks out in the south
1964	The October Revolution overthrows the military regime
1969	Second military coup installs radical military regime
1977	Military regime reconciled with old political parties
1983	Civil war breaks out again in the south
1985	Popular revolt overthrows the military regime
1989	Third military coup establishes regime strongly influenced by Islamic fundamentalists

ESSENTIAL STATISTICS

Capital Khartoum

Population (1989) 27,268,000

Area 2,503,890 sq km

Population per sq km (1989) 10.9

GNP per capita (1987) US$330

CONSTITUTIONAL DATA

Constitution Military regime

Date of independence 1 January 1956

Major international organizations UN; LAS; OAU; ACP; IDB

Monetary unit 1 Sudanese pound (LSd) = 100 piastres

Official language Arabic

Major religion Sunnite Muslim

Heads of government since independence Five-Man Council of State (1956–58); (President) Gen. I. Abboud (1958–64); I. al-Azhari (1965–69); Col. G. al-Nimeiri (1969–85); S. al-Dhahab (1985–86); Five-Man Supreme Council (1986–89); (President) Brig. Gen. O. al-Beshir (1989–)

Sudan, the largest country in Africa, has been noted for instability since independence in 1956, though many of the seeds were laid earlier. In 1898 Anglo-Egyptian forces reconquered Sudan, which from then until 1956 was effectively controlled by Britain. The rise of Egyptian nationalism after World War I encouraged a similar but smaller movement in Sudan in 1924. This was suppressed by Britain, which also then took the opportunity to reduce further the role of Egypt in the country.

Britain encouraged the isolation of the African and Christian southern third of the country from the Arab and Muslim north. But Sudanese nationalism was not so easily dismissed and by the late 1930s emerged again, establishing a moderate voice, the Graduates' Congress. Even before the end of World War II Britain was forced to embark on constitutional concessions; these led the country to independence in 1956, in spite of Egypt's continual hopes that it would join in a new union of the Nile valley. After independence the main features of the political history were the unstable alternatives of liberal-democratic governments and military regimes, and a long period of civil war in the south.

Liberal democracy meant in practice the emergence of two major political parties, centered around the two largest Islamic sects in northern Sudan. The Umma party was based on the Mahdist sect; while the Unionists were centered around the Khatmiyya. At successive elections it proved impossible for a clear majority to emerge, and the elected governments always took the form of unstable coalitions. In consequence military intervention was seen by many Sudanese as preferable, at least in the short term.

However, military regimes lacked widespread authority or coercive capacity. President Nimeiri collaborated briefly with the Communists and later with Islamic fundamentalists; while President Beshir cooperated with the same Islamic fundamentalists once more.

Meanwhile the southern Sudan was generally peripheral to political developments, which were dominated by the capital Khartoum, and increasingly it became alienated. Resentment in this region after independence led to prolonged civil war in 1962. For a decade the south was at peace and granted regional autonomy. However by the early 1980s Nimeiri's economic and political problems led to a decline in north-south relations and the resumption of civil war in 1983. This second war was more widespread and violent than the earlier one, and repeated attempts at negotiation ended in failure.

During this century Sudan's population rose from approximately two million to over twenty million, of whom two-thirds are in the north and one-third southerners. The country remained predominantly rural, though there was increasing urbanization, especially after independence. In the countryside there is still widespread pastoralism away from the Nile, while along the river settled agriculture has always predominated.

The main effort at economic development under British rule centered on the giant cotton-growing Gezira scheme. There was little attempt to alter this after independence until the 1970s, when President Nimeiri sought to make Sudan the breadbasket of the Arab world; and later oil was found. But hopes collapsed in the 1980s. By 1990 Sudan had a debt-ridden and shattered economy – problems exacerbated by frequent poor rains and warfare that contributed to repeated famine from 1984.

ETHIOPIA

PEOPLE'S DEMOCRATIC REPUBLIC OF ETHIOPIA

CHRONOLOGY

1913	Death of Menilik: Lij Iyasu becomes emperor
1916	Coup d'état: Zawditu empress; Ras Tafari (later Haile Selassie I) regent and heir
1930	Haile Selassie becomes emperor
1935	Italian invasion
1941	Liberation from Italian rule
1952	Federation of Eritrea with Ethiopia
1960	Attempted coup d'état
1962	Incorporation of Eritrea into Ethiopia
1974	Outbreak of revolution; Haile Selassie is overthrown
1977	Somali invasion of Ethiopia
1984	Famine kills almost half a million people
1991	Mengistu government overthrown

ESSENTIAL STATISTICS

Capital Addis Ababa

Population (1989) 48,898,000

Area 1,223,500 sq km

Population per sq km (1989) 40.0

GNP per capita (1987) US$120

CONSTITUTIONAL DATA

Constitution None (provisional government)

Major international organizations UN; OAU; ACP

Monetary unit 1 Ethiopian Birr (Br) = 100 cents

Official language Amharic

Major religion Ethiopian Orthodox

Heads of government since 1900 (Emperor) Menilik II (1889–1913); Lij Iyasu (1913–16); (Empress) Zawditu (1916–30); (Emperor) Haile Selassie (1930–74); (Head of Military Council) Aman Andom (Sep–Nov 1974); Tafari Banti (1974–77); Mengistu Haile Mariam (1977–87); (President) Mengistu Haile Mariam (1987–91); Meles Zenawi (interim 1991–)

After the Ethiopian emperor Menilik defeated the invading Italians at the battle of Adwa in 1896, Ethiopia became the sole indigenous African state to retain its independence into the 20th century. During the next few years, agreements between Ethiopia and surrounding colonial powers established the country's boundaries in broadly the form in which they remained in 1990. The sole but significant exception was the northern coastal region of Eritrea, much of which had long formed part of Ethiopia, but which now was an Italian colony. Menilik sought to unite his culturally diverse empire and introduce the trappings of the modern state, but succumbed from 1908 to a premature senility which gravely sapped the strength of central government. He was eventually succeeded by his grandson, Lij Iyasu, who was himself overthrown by a coup in 1916. Iyasu is sometimes credited with trying to reconcile Ethiopia's Muslims with the Christian government; but he offended powerful interests and was replaced by Menilik's daughter Zawditu as empress, with her cousin Ras Tafari (later to assume the name of Haile Selassie) as regent and heir. Zawditu is still the only woman to have been head of state in Africa in the 20th century. Shortly after his accession as emperor in 1930, Haile Selassie was confronted by the determination of Fascist Italy to wipe out the shame of Adwa and incorporate Ethiopia into the Italian empire. In a bloody eight-month war in 1935–36, Mussolini succeeded with the aid of overwhelming air power and poison gas in defeating the Ethiopian armies. Haile Selassie was forced into exile, but Ethiopian resistance fighters maintained their opposition to Italian rule until Italy's entry into World War II led to an invasion of the country, by British-led forces. By 1941, the Italians were defeated and Haile Selassie was restored.

Haile Selassie ruled Ethiopia for the next 33 years, acquiring in the process a worldwide reputation. Abroad he maintained an alliance with the United States and he emerged after 1960 as the doyen of independent African leaders. At home he pursued a policy of concentrating power in his own hands, at the expense of a regional aristocracy whose autonomy had already been undermined by Italian rule. The expansion of modern education, and the creation of a large central army and civil bureaucracy, all served this goal but at the same time led to the emergence of an intelligentsia to whom the aging emperor seemed increasingly anachronistic. In 1952, the federation of the former Italian colony of Eritrea with Ethiopia by decision of the UN gave previously landlocked Ethiopia access to the Red Sea. However, Haile Selassie's counterproductive determination to subordinate Eritrea to central control, culminating in its full incorporation into the empire in 1962, led to the formation of a guerrilla opposition seeking Eritrean independence.

In the final years of the empire, the task of controlling the growing forces of regional, intellectual and urban dissent became too much even for a politician with Haile Selassie's remarkable manipulative skills. Early in 1974, the combination of rising oil prices and revelations of a disastrous famine in Wollo region helped prompt a series of riots and demonstrations which soon became out of hand. Progressive concessions by the regime were shadowed by the formation of an organization within the armed forces, the PMAC or Derg, which by 12 September 1974 felt strong enough to overthrow the emperor. Internally divided, the Derg was itself uncertain as to what to do. Its first chairman, general Aman Andom, was killed in

November 1974, and power passed increasingly to one of the new vice-chairmen, Major Mengistu Haile-Mariam. A series of revolutionary measures included the nationalization of foreign companies and of urban and rural land. Resistance to the Derg, both in the countryside and among disaffected intellectuals in the towns, was met by brutal reprisals. In February 1977, Mengistu killed the Derg chairman, Tafari Banti, and took the position himself. Ethiopia's longstanding alliance with the United States was abandoned, and was replaced by links with the Soviet Union. The attempt from 1978 onward to build a new regime on the Soviet model was nonetheless a failure. The launching of a Marxist-Leninist ruling party in September 1984 coincided with a catastrophic famine in which about half a million people died. The regime's rigid centralism promoted regional opposition movements. In May 1991 the Mengistu regime collapsed and was replaced by an interim government led by a coalition of opposition forces.

Economic and social trends

Agricultural highland Ethiopia, home of one of black Africa's most ancient civilizations, has embraced Orthodox Christianity for some 1500 years, whereas most of the surrounding lowlands are inhabited by Muslim pastoralists. This distinction breeds basic conflicts which are partly overlaid by processes of national integration, and factional and ethnic disputes within each group. The principal language, Amharic, has become the language of government, and most leading officials are of Christian origin. These include members of Ethiopia's largest ethnic group, the Oromo, who are divided between the two main religions. Christian Tigreans, Muslim Afars and Somalis, and many more, make up a mosaic of peoples.

Economically, Ethiopia is among the poorest countries in the world. Rugged mountainous terrain hampers communications, and many Ethiopians, especially in the northern highlands, are at the mercy of uncertain rainfall and threats of famine. Periodic droughts have occurred throughout Ethiopian history. Catastrophic famines occurred in 1973–74 and 1984–85 and by the late 1980s the country depended even in good years on external food aid. Government attempts to resettle the "surplus" highland population in lowland areas have largely been unsuccessful. The major export crop, coffee, is grown mostly in the south and west of the country.

▼ Ethiopian families in a refugee camp, 1984.

KENYA

REPUBLIC OF KENYA

CHRONOLOGY

1895	British Protectorate declared over Kenya
1902	Kenya–Uganda railway opened
1952	State of emergency declared to cope with "Mau Mau"
1961	First national elections, leading to formation of minority Kenya African Democratic Union (KADU) government
1963	Independence is achieved
1964	Republic established, with Jomo Kenyatta as executive President
1966	Formation of Kenya People's Union (KPU)
1969	KPU banned
1978	Death of Jomo Kenyatta, Daniel Arap Moi elected
1982	One-party state inaugurated; attempted coup
1989	Internal and external pressure to maintain human rights and more liberal political regime

ESSENTIAL STATISTICS

Capital Nairobi

Population (1989) 23,883,000

Area 582,646 sq km

Population per sq km (1989) 41.8

GNP per capita US$340

CONSTITUTIONAL DATA

Constitution Unitary single-party republic with one legislative house (National Assembly)

Date of independence 12 December 1963

Major international organizations UN; Commonwealth of Nations; OAU; ACP

Monetary unit 1 Kenya shilling (K sh) = 100 cents

Official languages Swahili; English

Major religion Christian

Heads of government since independence (President) Jomo Kenyatta (1964–78); Daniel Arap Moi (1978–)

The British government ruled Kenya from 1895, when it took over from the Imperial British East Africa Company. The settler community, though small, was well organized and had interests which clashed with those of the African population; this clash diverted attention from the ultimately more important relationships between the indigenous people. The Asians constituted a second immigrant group. Asians came to occupy the lower middle ranks of the civil service and business, thereby blocking African advancement. Africans failed to improve their lot by constitutional means and after World War II a frustrated nationalism took a militant and, in central Kenya, a violent form in the emergence of the Kikuyu protest movement known as "Mau Mau". The British declared a state of emergency in October 1952, which lasted for seven years.

During the emergency period, the colonial regime restricted African political organization to the local (in practice communal or "tribal") level. However, African nationalism eventually triumphed and the London Constitutional Conference of January 1960 provided for African majority rule. The lifting of the ban on the formation of Kenya-wide African political parties enabled two main parties to emerge. These were the Kenya African National Union (KANU), which was dominated by the Kikuyu and the Luo "tribes" and the Kenya African Democratic Union (KADU). A minority KADU Government was formed following the first national election in February 1961. This gave way to a KADU/KANU coalition government in April 1962 and a KANU government following further elections in May 1963, led by Jomo Kenyatta.

Communal fears and suspicions over land and power were allayed in 1963 when a quasi-federal constitution was adopted. This gave Kenya internal self-government and provided for a central government and seven regional governments enjoying a high degree of autonomy. The independence constitution of December 1963, which was modeled on Westminster lines, was still strongly weighted toward regionalism. In 1964–65 a process of constitutional amendment established a republic with an executive president in December 1964 and led to the severe curtailment of the powers of the regions.

For a brief period Kenya became effectively a one-party state, but this ended when Oginga Odinga, the national vice-president since December 1964, left KANU (of which he was vice-president) and formed the Kenya People's Union (KPU) in April 1966. He, and other MPs who defected from KANU, sought to exploit popular grievances over land and unemployment. The government reacted by passing a fifth constitutional amendment requiring the resignation of MPs who left their parliamentary party. In the "little general election" held in June 1966 to fill the vacant seats, the KPU received a majority of votes in a low poll, but only 9 of the 29 seats contested (KANU won the other 20). The party's extra-parliamentary activities were severely restricted prior to its banning in late 1969.

The KANU government worked through civil service provincial and district commissioners who were given wide powers both to maintain law and order and to act as "agents for development". In 1969 the county councils lost to the civil service their major responsibilities for primary education, rural health and minor roads.

The checks on the arbitrary use of executive power were inadequate: the Senate, supposedly the champion of minority and district interests, was abolished; the government was

authorized to detain a person without trial, and, under a provision in the Penal Code, to ban publications. In 1990–91 President Moi resisted intense domestic pressure to end KANU's legal monopoly of power arguing that multiparty politics would be ethnically divisive. A constitutional one-party state was inaugurated in June 1982.

Economic and social trends

Most Kenyans are farmers and live in the fertile, southwestern part of the country, where pressure on available land is severe; much of the rest of the country is underpopulated, being arid or semi-arid and suitable for livestock production rather than the growth of crops. The Swynnerton Plan (1954) introduced individual land tenure for African farmers and allowed them to cultivate profitable export crops – a right hitherto reserved for white settlers. The 1961 Million Acre Settlement Scheme settled 34,000 families over the next ten years. Following independence, the government continued to favor the large farming sector, which was now increasingly in Kenyan hands. Commericalized peasant agriculture was not neglected and smallholders growing tea, coffee, sugar-cane and other crops received nearly the full value of the net export price.

The independent government pursued a liberal foreign investment policy, though in June 1986 it said that Kenyans should hold a controlling interest in joint ventures with foreign businesses. The sixth national development plan (1989–93) renewed the government's commitment to the equitable distribution of growth, which was to be achieved through agriculture, a revitalized industrialized sector, and small-scale enterprises. The 1990–91 budget emphasized the need for tight fiscal discipline and export-led growth in order to meet the nation's large budgetary and external deficits.

▲ **Mau Mau guerillas await questioning in the 1950s.**

UGANDA

REPUBLIC OF UGANDA

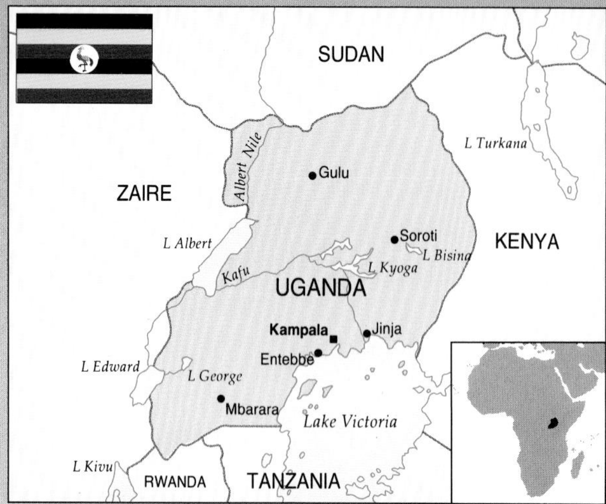

CHRONOLOGY

1894	Britain declare a Protectorate over Buganda; protection subsequently extended to the whole of Uganda
1962	Uganda becomes independent; Buganda achieves federal status
1964	The Uganda People's Congress (UPC) abrogates its alliance with Kabaka Yekka (KY) and forms a government on its own
1971	Military coup: Obote removed by General Idi Amin
1972	Amin expels the Asian community and seizes its economic assets
1980	"Rigged" general elections restore Obote to power; outbreak of civil war
1986	The National Resistance Army (NRA) led by Yoweri Museveni seize power
1990	Return to civilian rule is deferred for 5 years

ESSENTIAL STATISTICS

Capital Kampala

Population (1989) 16,452,000

Area 241,040 sq km

Population per sq km (1989) 83.5

GNP per capita (1987) US$260

CONSTITUTIONAL DATA

Constitution Quasi military regime with one interim legislative body (National Resistance Council)

Date of independence 9 October 1962

Major international organizations UN; Commonwealth of Nations; OAU; ACP; IDB

Monetary unit 1 Uganda shilling (Ush) = 100 cents

Official languages English; Swahili

Major religion Christian

Heads of government since independence (Prime minster) A.M. Obote (1962–66); (President) A.M. Obote (1966–71); General I. Amin (1971–79); Y. Lule (Apr–Jun 1979); G. Binaisa (1979–80); A.M. Obote (1980–85); General T. Okello (1985–86); Y. Museveni (1986–)

In 1894 Britain declared a protectorate over the long-established Kingdom of Buganda and subsequently extended its protection to cover the whole of Uganda. As independence approached after World War II, Buganda, which had wealth, size, unity and educated manpower on its side, sought to preserve the special position conferred by its 1900 Agreement with Britain. Its king (the Kabaka) was deported to Britain in 1954, but was allowed to return as a constitutional monarch the next year. Buganda achieved federal status under the independence (Westminster-type) constitution of 1962 and the Kabaka became president of Uganda.

In the run-up to independence, two main political parties emerged: the Democratic party (DP), backed by the powerful Roman Catholic missions, and the Uganda People's Congress (UPC), which drew its main support from the western kingdoms and the north. Political power tended to be diffused among the kingdoms and districts. The formation of a DP government in 1961 prompted the Baganda people to form Kabaka-Yekka (KY) – "the Kabaka only" – a movement designed to maintain Buganda's identity. In February 1962 KY became the junior partner in a UPC-KY coalition government. However, defections from both KY and DP enabled the UPC to dispense with this alliance in August 1964 and to form a government on its own, under Milton Obote.

The UPC was still subject to internal division. In February 1966 Obote became executive president, deprived Buganda of its federal status, and ordered government troops to sack the royal palace; the Kabaka went into exile. In September 1967 Uganda became a republic under a unitary constitution which converted the kingdoms into districts (four in Buganda).

In January 1971 Obote was removed by General Idi Amin, who instituted a personal, brutal and idiosyncratic dictatorship. Amin showed considerable political shrewdness in his use of patron-client networks to win local support and in popular moves such as baiting the British and expelling the Asian community in 1972. Asian economic assets were transferred to military personnel and other Ugandan Africans, often with minimal business experience or acumen.

The overthrow of Amin in April 1979 by the Tanzanian army and a small Ugandan exile force paved the way for the return to power of ex-president Obote and a revived UPC. However, rigged general elections in December 1980 led to the outbreak of civil war. Obote's cabinet was inexperienced and inept and the administration was understaffed, unable to control the armed forces or the UPC officials.

In 1985 Obote was removed by a rebel army, but the successor military regime was itself toppled in January of the next year by a large, well-organized and disciplined guerrilla force led by Yoweri Museveni, a former defense minister from Ankole in southwestern Uganda. Museveni, in a bid to secure national reconciliation, ruled with a broad-based cabinet reflecting the regional and "tribal" divisions of the country. Kampala, the capital, returned to something approaching normality and most of the sporadic rebel resistance in the north and east ended by mid-1991. Museveni was pragmatic in his choice of policy and pledged to restore democracy. Early in 1990 the National Resistance Council (NRC), the country's reconstituted legislature, extended the life of his government for five years beyond 1990, the year scheduled for the return to democratic civilian rule.

TANZANIA

UNITED REPUBLIC OF TANZANIA

CHRONOLOGY

1905	Maji-Maji rebellion against German rule
1961	Tanganyika achieves independence
1962	Tanganyika becomes a republic, with Julius Nyerere as executive President
1963	Zanzibar achieves independence
1964	United Republic of Tanganyika and Zanzibar formed
1965	One-party state legalized in mainland Tanzania
1967	Arusha Declaration is issued
1979	Tanzanian army invades Uganda
1983	Attempt to overthrow Nyerere's government
1985	Resignation of Nyerere, succeeded by Ali Hassan Mwinyi
1990	Nyerere hands over to Mwinyi as chairman of CCM

ESSENTIAL STATISTICS

Capital Dodoma

Population (1989) 23, 729,000

Area 945,037 sq km

Population per sq km (1989) 26.8

GNP per capita (1987) US$220

CONSTITUTIONAL DATA

Constitution Unitary single-party republic with one legislative house (National Assembly)

Date of independence 9 December 1961

Major international organizations UN; Commonwealth of Nations; OAU; ACP; SADCC

Monetary unit 1 Tanzanian shilling (T sh)= 100 cents

Official languages Swahili; English

Major religions Christian; Muslim

Heads of government since independence (Executive President) Julius Kambarage Nyerere (1962–85); Ali Hassan Mwinyi (1985–)

The modern country of Tanzania combines the historically separate regions of Tanganyika and Zanzibar. In 1885, Tanganyika was absorbed into Germany's East African Protectorate. During World War I the British took over the region and became the administering authority under, first, the mandate of the League of Nations and then, from December 1946, the Trusteeship Agreement of the United Nations. Nationalism increased after World War II and centered on the Tanganyika African National Union (TANU), a political party founded in July 1954 and led by Julius Nyerere, a Roman Catholic schoolteacher. Internal self-government was achieved in May 1961 and independence, as a monarchy within the Commonwealth, on 9 December 1961. Tanganyika became a republic a year later.

Nyerere, the executive president, survived a serious army mutiny in January 1964. The mutiny underlined the need to bring the trade unions under effective government control; this was achieved by legislation in 1964. Security considerations also prompted the union of Tanganyika and Zanzibar in April 1964, with Zanzibar retaining substantial autonomy within the union; the name Tanzania was adopted in 1965. Zanzibar, linked historically and culturally for hundreds of years with the Middle East and north Africa, had come under British protection in 1890, achieved independence in December 1963 and experienced a revolution ending Arab hegemony in January 1964. In February 1977 TANU and the Afro-Shirazi party of Zanzibar merged to form a new party, *Chama cha Mapinduzi* (CCM), the Revolutionary party. Demands for greater autonomy and even for secession surfaced in Zanzibar during the 1980s, coinciding with the downturn in its economy.

The Arusha declaration of January 1967 launched Tanzania on a socialist path; emphasis was placed on egalitarianism, public ownership and control, and self-reliance. Decentralization measures introduced in 1972 benefited regional and district directors of development and other bureaucrats rather than elected representatives. Elected authorities and marketing cooperatives were revived in the 1980s.

Poor economic performance was one reason for the attempt, in January 1983, to overthrow the government. In 1985 Nyerere resigned as state president and was succeeded by Ali Hassan Mwinyi, president of Zanzibar and national vice-president of Tanzania. Until August 1990 Nyerere remained chairman of CCM, a fact which enabled him to have some say in government policy. President Mwinyi replaced him and in October 1990 was reelected president for a second five-year term.

Tanzania is a predominantly rural society: coffee and cotton are the main exports of mainland Tanzania, while Zanzibar is almost entirely dependent on clove and coconut products.

The break-up in 1977 of the East African Community worsened Tanzania's relations with Kenya and Uganda, the other member states. War with Uganda in 1979 had a crippling effect on an economy already hard hit by oil price increases, the incidence of drought and flooding, the cost of servicing a massive foreign debt, the dismal performance of the state enterprise sector, and the failure of communal agricultural production. For years, the government refused to accept the formula of the International Monetary Fund (IMF) for economic recovery, but agreement was finally reached in August 1986. Under Mwinyi the country moved steadily towards the adoption of a market economy and more liberal investment laws.

MALAWI

REPUBLIC OF MALAWI

CHRONOLOGY

1891	Britain declares a protectorate known as Nyasaland
1915	Anticolonial uprising put down with force
1944	Nyasaland African Congress (NAC) formed
1959	NAC banned, Banda detained, state of emergency declared and Malawi Congress Party (MCP) formed
1964	Nyasaland becomes the independent state of Malawi
1966	Malawi becomes a Republic and one-party state
1971	Banda voted life-president
1973	Controls on foreign journalists imposed
1981	Suppression of dissidents and opposition parties begins
1983	Leader of Malawi Freedom Movement convicted of treason; Leader of Socialist League of Malawi murdered in Zimbabwe
1990	Student demonstrations against Banda's consort Mama Tamanda Kadzamira

ESSENTIAL STATISTICS

Capital Lilongwe

Population (1989) 8,515,000

Area 118,484 sq km

Population per sq km (1989) 90.3

GNP per capita (1987) US$160

CONSTITUTIONAL DATA

Constitution Single-party Republic with one legislative house

Date of independence 6 July 1964

Major international organizations UN; OAU; ACP; Commonwealth of Nations; SADCC

Monetary unit 1 Malaŵi kwacha (MK) = 100 tambala

Official language English

Major religion Christian

Head of government since independence Hastings K. Banda (Prime minister 1964–66; President 1966–)

Malawi is a slender, landlocked country lying in the southern reaches of the Great Rift Valley. As Nyasaland, it was declared a British protectorate in 1891 and remained under colonial rule until independence in 1964. Fertile and with considerable irrigation potential from the lakes and rivers that make up a fifth of its total area, Nyasaland saw the creation of an agrarian settler economy which took over most of the land from the native population. Despite widespread grievances and an abortive uprising in 1915, the response among Africans was not wholeheartedly militant. The political organization which gave expression to reform was the Nyasaland African Congress (NAC), formed in 1944.

In 1953 the Central African Federation was formed, uniting settler interests in Nyasaland, Northern and Southern Rhodesia. Opposition to the Federation gave impetus and focus to the activities of the NAC which was also served by growing discontent in the countryside. Not all leaders associated themselves with rural revolt but it galvanized and radicalized nationalist politics nevertheless. In the mid-1950s a younger and more combative leadership emerged, demanding independence and the vote. In 1958 the physician Dr Hastings Kamuzu Banda, who had spent almost 40 years abroad, returned to unite and lead the NAC. A struggle was waged in which the colonial state was only able to retain control through force and at great expense. The British were reluctant to pursue this option and, in the light of growing support for the newly-formed Malawi Congress party, they facilitated a transition to self-government in 1963 and full independence in 1964.

From these militant roots emerged one of Africa's most conservative political regimes. This can partly be explained by Malawi's economic dependence on South Africa to which it has remained supportive and loyal, and partly by Banda's own development objectives. He soon secured for himself a seemingly unassailable position, rapidly extinguishing any opposition. This was achieved through the one-party state, his life-presidentship and an appointed National Assembly and cabinet which he constantly reshuffled.

By 1990 Banda was over 90 years of age, yet refused to allow anyone to emerge as a potential successor. Although he was physically healthy, his frailty left as the most powerful people on the political stage, Malawi's "official hostess", Mama Cecilia Tamanda Kadzamira and her uncle, John Tembo. Despite their unpopularity, particularly with the army, they controlled access to the president and his patronage.

Economic and social trends

Undoubtedly a poor country, Malawi has had some development success since 1964. Although industrialization was given priority at independence, the manufacturing base remained limited, mineral resources underexploited and by 1990 the agrarian economy still accounted for nearly forty percent of GDP. Agriculture employed three-quarters of the working population, the majority as food crop producers in the ailing subsistence sector. Maize was the staple, other main food crops being cassava, millet, sorghum, groundnuts and rice. Exports included groundnuts, coffee, cotton, sugar, tea and tobacco. The last two were the most important in economic growth.

By 1990 Malawi had close economic ties with South Africa. It also sent migrants to the South African mines, though their numbers declined later in the 1980s.

MOZAMBIQUE

PEOPLE'S REPUBLIC OF MOZAMBIQUE

CHRONOLOGY

1891	Portugal confirmed as colonial power
1962	Frente de Libertaçào de Moçambique (FRELIMO) formed under Eduardo Mondlane
1975	Independence as The People's Republic of Mozambique
1976	Formation of the Resistência Nacional Moçambicana (RENAMO)
1981	South African destabilization begins
1982	War between FRELIMO and RENAMO begins
1986	President Samora Machel killed in air crash – succeeded by Joaquim Alberto Chissano
1987	Emergency appeals for relief of victims of drought, famine and war
1989	Negotiations between FRELIMO and RENAMO
1990	Chissano presents a liberal draft constitution and RENAMO put under international pressure to accept

ESSENTIAL STATISTICS

Capital Maputo

Population (1989) 15,293,000

Area 812,379 sq km

Population per sq km (1989) 19.1

GNP per capita (1987) US$150

CONSTITUTIONAL DATA

Constitution Single-party republic with single legislative house (People's Assembly)

Date of independence 25 June 1975

Major international organizations UN; OAU; ACP; SADCC

Monetary unit 1 metical (mt) = 100 centavos

Official language Portuguese

Major religion Traditional beliefs

Heads of government since independence (President) S. Machel (1975–86); Joaquim Alberto Chissano (1986–)

Mozambique was settled by Portuguese who earned a grisly reputation as colonists from the late 19th century. Resentment festered and eventually broke into open revolt in 1960 with the Mueda massacre, in which 600 Africans were killed by colonial forces. In 1974 a coup in Lisbon heralded changes at home and in Africa. It was engineered by the armed forces, battle-fatigued and demoralized by successive defeats by nationalists in Africa.

When Mozambique achieved independence in 1975, the majority of settlers fled, sabotaging what they left behind. The Portuguese abandoned a country into which little had been ploughed back, where illiteracy stood at 95 percent and there were only 80 doctors for a malnourished population which had never known a vaccination program. Differences within the nationalist movement were largely resolved before independence when the Frente de Libertaçào de Moçambique (FRELIMO) took over under Samora Machel.

A socialist party from 1977, FRELIMO committed itself to reconstruction through centralized planning and popular democracy within a one-party state. However, energy and enthusiasm were not matched by resources and personnel. Policy blunders together with drought led to famine and disillusionment. Mozambique's socialism threatened its white neighbors who engaged in destabilization tactics. Rhodesia formed the Resistência Nacional Moçambicana (RENAMO) from dissident exiles, and South Africa took over support for it after Zimbabwe won independence in 1980. After 1983 RENAMO grew, fighting a debilitating war and picking up people disaffected from FRELIMO in the process.

Destabilization continued although Machel signed the Nkomati accord with South Africa in 1984. In the later 1980s Western pressure and the diplomatic skills of Chissano, who took over after Machel's death in 1986, meant that relations between Maputo, the Mozambican capital, and Pretoria improved. Furthermore, negotiations toward an internal settlement with RENAMO and a multiparty system began.

Economic and social trends

Under colonial rule Portuguese trade with Mozambique was strictly bilateral until the 1960s. Thereafter Mozambique was opened up to foreign capital. This enabled South Africa to extend its economic power within the region, particularly through its investment in the Cahora Bassa hydroelectric project. It already used Mozambican migrants in the gold mines, the Portuguese exporting much of the Mozambican workforce annually for labor outside the country. After independence trade shifted away from Portugal and South Africa.

At home overambitious investments were made in state farms and industry to the detriment of subsistence and small-scale agriculture. At the Fourth Party Congress in 1983, FRELIMO belatedly acknowledged it had neglected the peasantry but policy reversals were rendered ineffective by war, drought, floods, a shortage of foreign exchange and skills. Famine and economic collapse reached crisis proportions, and refugees streamed across Mozambique's borders.

A turn to the West for assistance required a shift away from socialist planning. In 1984 Mozambique joined the International Monetary Fund and World Bank. Agricultural and commercial production grew but living standards of the urban poor plunged: riots broke out in Maputo in 1990.

ZAMBIA

REPUBLIC OF ZAMBIA

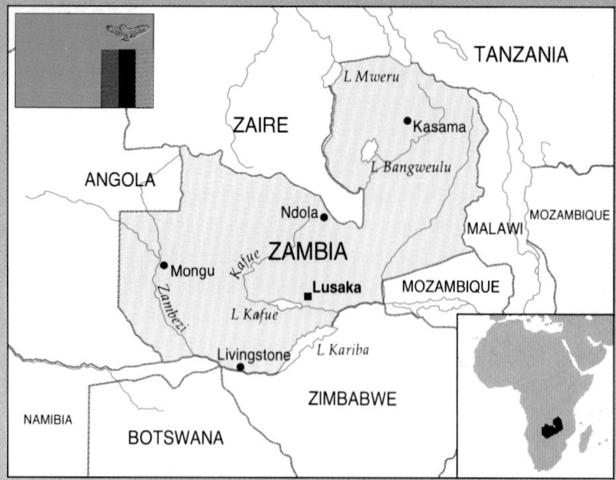

CHRONOLOGY

1889	Rhodes' British South African Company administers territory north of Zambezi river
1924	Territory named Northern Rhodesia and handed over to British colonial office
1953	Central African Federation (CAF) formed, consisting of two Rhodesias and Nyasaland
1963	CAF is dissolved
1964	Independence proclaimed under Kenneth Kaunda and the United National Independence Party (UNIP)
1972	Declaration of one-party state
1989	Demonstrations and calls for multiparty system
1990	Attempted coup follows food riots

ESSENTIAL STATISTICS

Capital Lusaka

Population (1989) 8,148,000 (UN estimate)

Area 752,614 sq km

Population per sq km (1989) 10.8

GNP per capita (1987) US$240

CONSTITUTIONAL DATA

Constitution Republic with one legislative house (National Assembly)

Date of independence 24 October 1964

Major international organizations UN; Commonwealth of Nations; OAU; ACP; SADCC

Monetary unit 1 Zambian kwacha (K) = 100 ngwee

Official language English

Major religion Christian

Head of government since independence (President) Kenneth Kaunda (1964–)

The British in Southern Africa in the late 19th century moved north of the Zambezi river at the instigation of the mining entrepreneur Cecil Rhodes, who wanted to annex the copper deposits of Katanga. In 1889 Rhodes' British South African Company was empowered to make treaties with local chiefs and administer the territory. In 1924 the territory that was to be known as Northern Rhodesia was handed over to the British Colonial Office. The colony became important to the British for copper production. Investment was almost exclusively used to develop mining and to support the white settler community that grew up around it. The Company shareholders also prospered and British revenue coffers swelled, but little was ploughed back into the colony and high social costs for the African population resulted.

Migrant labor grew up on the Copperbelt on the border with the Belgian Congo (now Zaire), where conditions were abysmal, with strict job segregation and African wages falling between 1930 and 1940. From 1945 onward, there were strikes on the railways and mines. Trade unionism came to be a focus for African nationalism in Zambia, allowing resistance politics to extend beyond the African educated elite and giving nationalism a mass base. African nationalism was a protest against not just British colonialism but, more specifically, the Central African Federation (CAF) which was formed in 1953, encompassing the two Rhodesias and Nyasaland and giving the settlers greater autonomy from Whitehall. Ultimately this Federation lost the support of Northern Rhodesian settlers as they saw it operating in favor of their counterparts in Southern Rhodesia. Their loss of heart, combined with mounting African resistance north of the Zambezi, led to the Federation being dissolved in 1963 and a new constitution introduced to phase in independence.

The United National Independence Party (UNIP), led by the young missionary-educated schoolteacher, Kenneth Kaunda, was elected to power in 1964. Although under the Federation UNIP had been banned, Kaunda and others imprisoned, and thousands injured in clashes with the colonial authorities, the first transition to independence was more peaceful and democratic than in many countries. Kaunda gained respect as a founding father and elder statesman of independent Africa, not least for his role in the Organization of African Unity (OAU), the nonaligned movement and for his peace initiatives in the Southern African region. His support for the liberation struggles of Zimbabwe, Namibia and South Africa also earned him esteem on the continent. At home, he was seen as a unifying figure who transcended ethnic rivalries, while his philosophy of humanism (akin to the African Socialism of Tanzanian president Julius Nyerere) emphasized equality.

However, his popularity steadily declined as Zambia's democratic face became increasingly pockmarked with the declaration of a one-party state (in 1972), the stifling of opposition and suppression of popular protest. The years 1989 and 1990 were punctuated by student demonstrations, worker action and street protests, with mounting calls for a multiparty system. The announcement of a massive increase in the price of the maize meal staple led to food riots and a coup attempt in June 1990. In April the embattled Kaunda released a record, written, sung and performed by himself. It was called "Tiende Pamodzi" meaning "We shall Walk Together". Its popularity was not resounding, and a multiparty system was accepted.

Economic and social trends

The colonial economy was geared almost entirely towards the export of copper, which had been discovered shortly before World War I, and developed from the mid-1920s. In the years following independence, the economy continued to grow as world copper prices remained buoyant. Investment was made in infrastructure and social services and the mining section nationalized.

Agricultural failure and decline led to massive migration to the towns. Around half of Zambia's population now lives in towns (the majority of which are in the Copperbelt), with vast disparity between urban and rural incomes. Zambia's own social problems were exacerbated by refugees fleeing from fighting in South Africa, Namibia, Angola and Mozambique.

When the economic crisis deepened in the mid-1980s after the collapse of copper prices Zambia borrowed from the IMF and the World Bank and was soon paying well over half its foreign exchange earnings in servicing its debt. This led Kaunda to try to break with the IMF and World Bank in 1987 and introduce an internal economic recovery program despite the resulting socio-political difficulties.

There are 60 different ethnic groups in Zambia, all of them Bantu. The English language, as the discourse of government, education and commerce, has assisted as a language of national unity, although many people desire to promote Zambian language and culture in the schools.

▼ The Victoria Falls on the Zambezi river between Zambia and Zimbabwe.

ZIMBABWE

REPUBLIC OF ZIMBABWE

CHRONOLOGY

1923	Southern Rhodesia becomes a self-governing colony
1962	Formation of Rhodesian Front
1965	Unilateral Declaration of Independence under Ian Smith
1966	Start of second *chimurenga* (liberation war) waged by Zimbabwe African National Union (ZANU) and Zimbabwe African Patriotic Union (ZAPU)
1976	Patriotic Front formed
1979	Internal settlement introduces transitional government
1980	Zimbabwe gains independence following Lancaster House agreement
1987	The reservation of white seats in Senate and House of Assembly abolished
1988	Constitutional change from prime ministership and ceremonial presidency to executive presidency
1990	Lancaster House constitutional restrictions lapse

ESSENTIAL STATISTICS

Capital Harare

Population (1989) 9,122,000

Area 390,759 sq km

Population per sq km (1989) 23.3

GNP per capita (1987) US$590

CONSTITUTIONAL DATA

Constitution Unitary single-party republic with two legislative houses (Senate; House of Assembly)

Date of independence 18 April 1980

Major international organizations UN; Commonwealth of Nations; OAU; ACP; SADCC

Monetary unit 1 Zimbabwe Dollar (Z$) = 100 cents

Official language English

Major religion Christian

Heads of government since independence Robert Mugabe (Prime minister 1980–87; President 1987–)

Zimbabwe is a landlocked country on the plateau between the Zambezi and Limpopo rivers. Lured by gold, from 1889 Cecil Rhodes' British South Africa Company administered the territory by royal charter and put down uprisings known as the first *chimurenga* (liberation war) by the Shona and Ndebele people. In 1923 it became a self-governing colony known as Southern Rhodesia, in which the settlers constructed a racially stratified and segregated society. Through land pressure, taxation, restrictions on movement and the growth of a cash economy, the main economic role of Africans was confirmed as migrant workers on white-owned farms and mines.

During World War II Britain showed preference for larger economic and administrative units in Africa to attract new levels of investment. In line with this, the Central African Federation (CAF) was formed in 1953, between Northern and Southern Rhodesia and Nyasaland. Under the leadership of prime minister Roy Welensky, the CAF operated in favor of Southern Rhodesia, diversifying the economy and attracting immigrants, but it was dissolved in 1963 in the context of growing African nationalism. The strength of Rhodesia's settler community could be gauged from its resistance to the "winds of change" that swept the continent at this time. In 1962 the Rhodesian Front (RF) was formed to oppose the constitution favored by Britain, leading to majority rule. Under Ian Smith in 1965, Rhodesia announced a Unilateral Declaration of Independence (UDI). The British responded meekly and even the sanctions campaign was ineffective.

African nationalists resisted by launching the second *chimurenga* in 1966. The two parties, the Zimbabwean African National Union (ZANU) and Zimbabwean African People's Union (ZAPU) formed an alliance as the Patriotic Front to wage a guerrilla war from exile. This was contained by the RF until worn down by ZANU's cooperation with FRELIMO in Mozambique and pressure for a resolution from South Africa. In 1978 an internal settlement was attempted, chiefly through the person of Bishop Abel Muzorewa, prime minister of the short-lived "Zimbabwe-Rhodesia". It was doomed to failure without the exiled leadership and fresh negotiations in 1979 were conducted on the constitution in Lancaster House, London. In April 1980 Zimbabwe's turbulent transition to independence ended with ZANU's sweeping electoral victory under Robert Mugabe, which took both the British and Rhodesian whites by surprise.

A spirit of reconciliation characterized relations with whites, many of whom stayed on. However, this was not always extended to ZAPU and other black political groupings which Mugabe tried to marginalize. Tied for ten years by the Lancaster House constitution, which ensured white political representation, private property rights and a multiparty system, his ambitions for a one-party state were contained. With the lapse of these conditions in 1990, they could be realized although there was resistance. ZAPU was persuaded in 1987 to merge with ZANU under the name of ZANU-PF but many of its supporters suspected Mugabe of trying to neutralize opponents by giving them positions, and the move to an executive presidency in 1988 increased his own personal power. Despite growing discontent and a corruption scandal, ZANU won the 1990 elections which Mugabe used as a referendum for the one-party state. Nevertheless, widespread opposition remained, not least within his own party.

Economic and social trends

Independence found a highly unequal economy in which whites, who were three percent of the population, earned 60 percent of total wages and commanded two-thirds of national income. Nevertheless, Zimbabwe had a balance of payments surplus, a low foreign debt and did not rely on a single crop for export. Originally dominated by agriculture (especially tobacco) and mining (gold, chrome, asbestos, copper), the economy diversified considerably in the postwar era and after 1965 when growth was stimulated by the need to substitute imports under the poorly-imposed economic sanctions. A range of minerals provides over thirty percent of export revenue while manufacturing accounts for about a quarter of GDP.

After 1980 growth was stimulated by the lifting of sanctions but thereafter slowed down under the impact of drought and commodity price fluctuations. Moreover, disequilibrium in the southern African region generally had an adverse effect on the economy, and discouraged foreign investment. Zimbabwe deployed troops along the "Beira Corridor" to guard its route to the sea through war-torn Mozambique. However, continued disruption of transport meant trade was directed through South Africa, creating unwelcome reliance on a hostile neighbor. Joining the Front Line States' Southern African Development Coordination Conference (SADCC) had not by 1990 reduced Zimbabwe's dependence on Pretoria.

ZANU-PF's objective at home was redistribution with growth. Despite socialist rhetoric this was largely pursued within the liberal framework laid down at Lancaster House. White-owned farms had produced 90 percent of marketed maize and virtually all the export crops, notably cotton, tobacco, coffee, tea and sugar. Their importance for employment, export earnings and food security ensured continuing favor after independence, although more attention was paid to the peasant sector. The Communal Areas (the old Tribal Trust lands or reserves) received greater assistance by way of extension services, credit and marketing, although even within the Communal Areas, benefits were not evenly distributed, with households headed by women as the most frequent losers. Resettlement schemes were designed to accommodate the dispossessed and those uprooted by the war, either on the basis of family farms or through cooperatives.

Zimbabwe comprises two main groups, both of whom are Bantu: those who speak Shona, originally settled agriculturalists, and the Ndebele, more recently arrived pastoralists. Despite different precolonial histories, they shared similar colonial experiences and acted in concert in both the first and second *chimurenga*. Distinctions surfaced again after independence, with simmering ethnic tensions, although Westernization and industrialization had as much impact on Zimbabwean culture as did ethnic diversity.

▲ **Supporters of Mugabe in 1987.**

SOUTH AFRICA

REPUBLIC OF SOUTH AFRICA

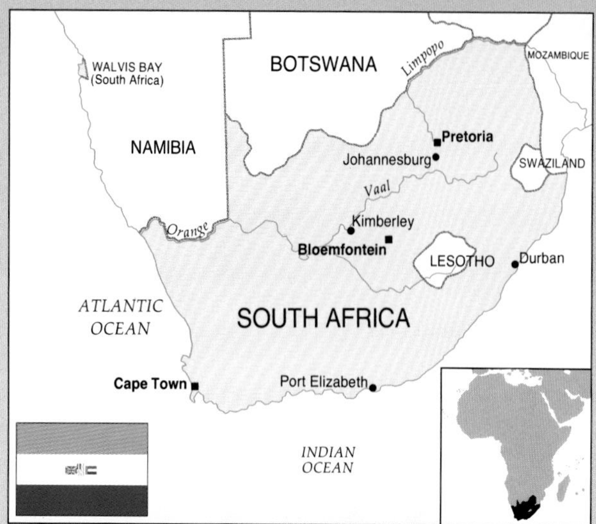

CHRONOLOGY

1910	Colonies combine to form Union of South Africa
1912	Formation of the South African Natives National Congress, later the African National Party
1960	ANC, PAC and other antiapartheid organizations banned
1961	South Africa becomes a republic
1969	Steve Biko founds Black People's Convention (BPC)
1976	Riots by schoolchildren start in Soweto township
1982	Conservative Party founded in opposition to constitutional reforms
1990	Nelson Mandela released from prison and ANC legalized
1991	Last apartheid laws removed

ESSENTIAL STATISTICS

Capitals Pretoria; Bloemfontein; Cape Town

Population (1989) 30,224,000

Area 1,123,226 sq km

Population per sq km (1989) 26.9

GNP per capita (1987) US$1,890

CONSTITUTIONAL DATA

Constitution Multiparty republic with three legislative houses (House of Representatives; House of Assembly; House of Delegates)

Date of independence June 1934

Major international organizations UN

Monetary unit 1 rand (R) = 100 cents

Official languages Afrikaans; English

Major religion Christian

Heads of government since independence (Prime minister) J. Hertzog (1924–39); Gen. J. Smuts (1939–48); D. F. Malan (1948–54); J. G. Strijdom (1954–58); H. Verwoerd (1958–66); B. J. Vorster (1966–78); P. W. Botha (1978–84); (President) P. W. Botha (1984–89); F. W. de Klerk (1989–)

South Africa was originally populated by the Khoisan (Hottentots and Bushmen) and Bantu-speaking blacks, and it was colonized by the Dutch in the 17th century, followed by the British. They in turn introduced Malay slaves and Indian indentured laborers. In a country obsessed with racial classification, to the "non-white" population was added the category "colored", mixed-race people whose presence contradicted two centuries of a political concept dedicated to racial segregation.

Battles over land between black chiefdoms and white settlers punctuated the 18th and 19th centuries. During this period too, relations soured between the Boers or Afrikaners (descendants of Dutch settlers) and the English, culminating in the South African (Anglo-Boer) War of 1899-1902, a fight for the prize of Transvaal gold. Defeated by British military strength and its scorched-earth policy in the countryside, the Boers ultimately triumphed however, with General Louis Botha becoming the first prime minister.

During both World Wars Afrikaner dissenters resisted South Africa's support for Britain as part of the Commonwealth. In order to tackle the economic crisis brought on by the Great Depression, divisions were buried under the "Fusion government", formed in 1934 with the United party linking Barry Hertzog's National party and Jan Smuts' South African party. However, when Smuts took South Africa into World War II, Hertzog resigned in protest and hostilities simmered until South Africa became a republic in 1961.

Boer and Briton were united on one issue, to exclude "non-whites" (blacks, Asians and mixed-race people) from the political process. This, together with the Land Act of 1913 which reserved for the black majority just 13 percent of the land, prompted the formation of the South African Natives National Congress, the forerunner of the African National Congress (ANC). In the 1920s it was the Industrial and Commercial Workers Union that commanded popular support. However, within the decade it had gone into decline, achieving little except a demonstration of the potential of black mass mobilization. The small but consistently effective South African Communist party (SACP) was active among white miners until the early 1920s. Thereafter it turned its energies toward the black working class and later an alliance with the ANC.

After World War II the business community, represented by the United Party, saw advantages in cultivating a settled black urban workforce which could acquire skills and become consumers for South Africa's growing manufacturing output, albeit in segregated cities. This policy was opposed by the Nationalist party (NP) who represented Afrikaner farmers and workers, neither of whom favored black urbanization. The former wanted to prevent a drain of rural labor to the cities, while the latter wanted to stem the tide of black competition for urban jobs. The NP won the 1948 election and initiated over forty years of apartheid (separateness) rule.

Legislation was promulgated, registering the population according to race, confirming residential segregation, preventing mixed marriages and introducing separate amenities. Distribution of resources for segregated education, health and other social services was heavily weighted in favor of the white minority. In the 1950s, liberal whites found a political home either in the short-lived multiracial Liberal party or later the Progressive party, which opposed apartheid from within the whites-only parliament, for many years with a single MP, Helen Suzman.

Opposition to apartheid swelled during the 1950s, with strikes, marches and campaigns. At this time the ANC confirmed itself as an organization of the black majority and worked together with progressive Indians, coloreds and whites and formed a strategic alliance with the now outlawed and underground SACP. Not all stood for nonracialism, however, and in 1959 the all-black Pan-Africanist Congress (PAC) broke away. During the 1950s, government repression escalated until all political opposition was outlawed. Many of the leaders were arrested and sentenced to life imprisonment on the notorious Robben Island. The ANC and PAC went underground within the country and launched the armed struggle from exile.

The 1960s and 1970s were decades of consolidation both for the government and its opponents. The NP extended apartheid under their leader Hendrik Verwoerd, notably by turning the reserved lands into 10 self-governing "homelands" or bantustans. Renewed opposition came in the 1970s in the form of "black consciousness" under the charismatic leadership of Stephen Biko. In 1976 a spontaneous uprising against "gutter education" was started by Soweto's schoolchildren. The police responded brutally. Crowds were fired on, people were detained without trial and many died in detention. After widespread strikes in 1973, there was a revival in working-class action and a powerful trade union federation emerged.

In the 1980s the government tried to reform apartheid. A new constitution introduced a three-chambered parliament which incorporated Indians and coloreds as junior partners to whites but left out blacks. Even these limited reforms were too much for Afrikaner hardliners who formed the Conservative party (CP) in 1982 and swelled the ranks of Eugene Terreblanche's neo-Nazi Afrikaner Weerstandsbeweging (Afrikaner Resistance Movement). On the other side, community struggles, boycotts, stayaways and strikes rocked the townships, while the ANC's military wing *Umkonto we Sizwe* stepped up its campaign of armed propaganda. In 1983 the United Democratic Front (UDF), which enjoyed a close relationship with the ANC, brought together antiapartheid groups and orchestrated a successful boycott of the elections for the new parliament. Black consciousness found expression in the Azanian People's Organization with links to the PAC. In 1984

▼ British troops fight the Boers in 1902.

South Africa's "Homelands"

Apartheid is about separation, both racial and territorial. Under the 1913 and 1936 Land Acts, South Africa's white parliament confined the African majority to 13 percent of the land. These reserves were the foundation for 10 "homelands" or bantustans created from 1958 onward and designed to provide an arena for black political aspirations. Like incomplete jigsaw puzzles, they were fragmented into numerous territorial enclaves. Their citizens were often assigned cultures to which they did not adhere and allocated "homelands" they had never seen. Thousands, dubbed "the surplus people," were forcibly removed from white-owned areas and resettled in distant and desolate camps.

To justify the policy it was argued that different ethnic groups had the right to develop their own culture. The political intention was to reduce the threat of a united African nationalism by creating or rekindling several ethnic nationalisms. The economic rationale was to continue the system of cheap labor. African migrants who worked in "white" South Africa were paid low wages on the grounds that family agriculture in the bantustans supplemented their income. The bantustans also bore the cost of supporting the nonproductive population and were aptly labeled "dumping grounds" for redundant workers and their dependents.

Destined for full political "independence", six held out against it while Transkei, Bophuthatswana, Venda and Ciskei acquiesced. The leadership was often corrupt and unstable, with the Ciskei witnessing repeated coup attempts. After the unbanning of the African National Congress (ANC) in 1990, some bantustans, such as Transkei (the birthplace of Nelson Mandela) and KwaNgwane, indicated a willingness to cooperate with the ANC and be reintegrated into a unified postapartheid South Africa. Others were reluctant. Lucas Mangope of the bantustan of Bophuthatswana remained wedded to his personal fiefdom despite popular opposition. Mangosuthu Gatsha Buthelezi, Chief Minister of the KwaZulu bantustan, welcomed unity but rejected a negotiated settlement dominated by the ANC. In the 1970s and 1980s he created a regional power base by refusing "independence" but still manipulated the bantustan system and demanded an autonomous place at the negotiating table. From the 1980s supporters of his populist and ethnic Zulu Inkatha Freedom party engaged in pitched battles with ANC supporters.

▼ A shack in a mining district.

an enlarged and more politically active trade union federation was formed, the Congress of South African Trade Unions (COSATU). The church also played an important role in the struggle, with the veteran campaigner Archbishop Desmond Tutu being awarded the Nobel Peace prize in 1985.

A state of emergency from 1985 clamped down on resistance from the UDF, COSATU and organizations of women, youth and community activists. This gave police and troops unbridled powers of repression and thousands were killed or detained without trial. Violence became endemic in the townships, notably in the troubled Natal region where the Inkatha Freedom Party, loyal to the KwaZulu bantustan leader Chief Mangosuthu Gatsha Buthelezi, attacked supporters of the ANC in a bid for local political supremacy. Their ambitions became national with the violence spreading beyond Natal to the Transvaal. This seriously undermined the peace process set in motion by president F.W. de Klerk. In 1990 he unbanned the ANC, PAC and SACP and released many political prisoners from Robben Island including Nelson Mandela. Walter Sisulu and Govan Mbeki. He allowed the return of exiles including the ANC president Oliver Tambo and he took steps toward dismantling some apartheid legislation. The international community responded in 1991 by withdrawing sanctions on South Africa's participation in international sporting events, and the United States lifted economic sanctions as well, though others waited to see liberation in action.

Social and economical trends

South Africa is a fertile land, rich in mineral wealth. Oil is the only vital mineral not found in economic quantities but an advanced oil-from-coal program produces 40 percent of South Africa's fuel requirements. The discovery of diamonds in the 1860s, and gold 20 years later, laid the basis for the country's modern development. Today mining has given way to manufacture as the leading sector but it remains of strategic importance, with South Africa supplying 70 percent of the world's gold. However, South Africa tended to rely on foreign investment and technology to maintain its rate of growth. The world recession of the 1970s and international sanctions campaign of the 1980s induced an economic crisis by 1990, though its roots lay deep in the structures of South Africa society.

The commercialization of agriculture and the mineral revolution in the 19th century unleashed an insatiable demand for unskilled labor. This brought Chinese and Indian indentured laborers to South Africa while confiscation of land, increased taxes and the lure of urban life created a system of migratory labor for thousands of black men. They worked down mines and in factories, living in single-sex hostels and returning to the countryside at the end of their annual contracts.

Skilled labor was initially provided by white artisans and apprentices attracted from abroad. They used union organization to protect their skills from the competition of cheaper black labor. This division was entrenched in law with the "color bar" that prevented blacks from doing skilled work. A policy of import substitution during World War II meant black employment in manufacturing rose dramatically and urbanization increased. Nevertheless, blacks were treated as "temporary sojourners" in the towns. Passes were issued, first to men and later to women, which admitted blacks for limited periods and for work purposes only. Otherwise they were to live out their lives in the bantustans, with forced removals sending millions of urban dwellers without passes to desolate resettlement camps. Some, through long service, were granted the right to permanent urban residence but were not allowed to own property or businesses.

Throughout the 1960s, a consistently high growth rate barely touched black real incomes, leading to industrial action and political unrest in the 1970s. The cost of apartheid began to affect whites as well. Changes in technology in the mines and capital-intensive industry meant a shift in demand from unskilled workers to skilled operators. This revealed an acute skills shortage due to poor education for blacks and resulted in low productivity. In the 1980s economic recession turned to catastrophe. The rate of growth plummeted, inflation spiraled and unemployment soared, fueling discontent. Investor confidence was shaken which boosted the international campaign of economic sanctions, orchestrated by the ANC. Loss of overseas markets for South African goods could not be compensated for by the domestic market because of widespread poverty. Without political change, solutions were impossible to find. The apartheid chickens had finally come home to roost.

Cultural trends

In South Africa the word "culture" is politically emotive because the concept has been used to justify apartheid – preserving cultural purity and integrity – and manipulated to rationalize the creation of the bantustans.

The question of culture proved a vexed one for anti-apartheid activists because in overcoming the bogus traditionalism and ethnicity of apartheid's social engineering, they did not want to undermine the cultural values of a heterogeneous population. Thus the right of all people "to use their own languages, and to develop their own folk culture and customs" is enshrined in the Freedom Charter. Despite divisions, a truly South African culture emerged. Migrant labor was central to developing a unified working-class culture. Workers took their urban experiences to the countryside. Traditional forms and references from the rural areas found their way into worker songs, poetry and theater. South African townships were noted for their jazz and *shebeen* (speakeasy) life. While rugby football remained the almost exclusive preserve of whites, soccer, along with boxing, gained mass popular appeal among black South Africans, who demonstrated considerable proficiency and eagerly awaited international competition in the postapartheid era.

▲ **ANC leader Nelson Mandela released from prison in 1990.**

LESOTHO

KINGDOM OF LESOTHO

CHRONOLOGY

1884	Basutoland (now Lesotho) declared a British High Commission Territory
1952	Basuto Congress Party (BCP) formed under Ntsu Mokhele
1958	Basuto National Party (BNP) formed under Chief Leabua Jonathan
1966	Lesotho becomes independent under Chief Leabua Jonathan
1970	Election annulled after BCP victory and state of emergency declared
1974	King Moshoeshoe II's executive power curtailed and Lesotho becomes constitutional monarchy
1982	Start of South African raids against ANC supporters and bases in Lesotho
1985	Elections cancelled by Chief Jonathan
1986	Coup establishes military junta
1988	State of emergency declared
1990	BCP calls for Lesotho to join a "Free South Africa"

ESSENTIAL STATISTICS

Capital Maseru

Population (1989) 1,715,000

Area 30,355 sq km

Population per sq km (1989) 56.5

GNP per capita (1989) US$360

CONSTITUTIONAL DATA

Constitution Monarchy assisted by a Military Council

Date of independence 4 October 1966

Major international organizations UN; Commonwealth of Nations; OAU; ACP; SADCC

Monetary unit 1 loti (m) = 100 lisentc

Official languages Sotho; English

Major religion Roman Catholic

Heads of government since independence (Prime minister) L. Jonathan (1966–86); Maj.-Gen. J. Lekhanya (1986–)

The mountainous kingdom of Lesotho is one of the world's least developed countries. Producing little itself, its chief export is labor to South African mines. It is surrounded by and is dependent on South Africa, which makes it vulnerable to economic blockades and political leverage.

In 1868, threatened by the expansion of the Afrikaner Trekboers, Moshoeshoe I of the Basuto appealed to the British for protection, and this heralded almost a century of British rule. In 1966 Lesotho became an independent constitutional monarchy under chief Leabua Jonathan of the Basuto National Party (BNP), aided by Britain and South Africa.

Jonathan proved to be intolerant of opposition and in the early 1970s the democratic process was suspended. Resistance to the BNP fermented in Lesotho, particularly from the more radical Basuto Congress party (BCP). Moreover, Jonathan's flirtation with South Africa was shortlived and he invited Pretoria's wrath by condemning apartheid in international forums and hosting the South African liberation movements. In 1986, Jonathan was ousted by a South-African-supported coup which introduced a military council under Major-General Justin Lekhanya. However, despite almost universal opposition to Jonathan, disruption and divisions persisted in Lesotho's domestic politics. In 1990 Lekhanya deposed King Moshoeshoe II in favor of his son King Letsie III. There were no signs of a return to civilian rule.

Economic and social trends

In the late 1980s over ninety percent of Lesotho's population lived in rural villages, although only 13 percent of the land is arable. In addition to the subsistence crops of maize, sorghum and wheat, there is diamond, wool and mohair production.

Lesotho is inexorably bound up with its giant neighbor. It is party to a monetary and customs union with South Africa. Above all, the economy was dominated by migrancy, with miner's remittances accounting for over half of Lesotho's GNP. In the late 1980s, it was estimated that nine out of every ten Basuto in paid employment worked in South Africa. These were predominantly male, as women were not legally permitted to work across the border. The living standard of a rural household often depended on the number of members in paid employment. The better-off used their earnings to invest in resources such as livestock, seed, implements and to secure the favorable disposition of the local chief, who had the power to allocate the right to land use. However, agricultural decline meant that the fate of even these households was precarious.

In 1990 Lesotho's economic future was uncertain except for an inevitable dependence on aid. Industrial development was in its infancy. Some optimism surrounded the ambitious and costly Highland Water Scheme which envisaged selling water to South Africa, with self-sufficiency in hydroelectricity for Lesotho as a byproduct. Tourism might also be developed.

Cultural trends

By African standards, the Basuto nation shows remarkable cultural and linguistic cohesion. It was forged by Moshoeshoe I out of tribal remnants fleeing to the mountains in the wake of the "wars of calamity" in the early 19th century when the Zulu state was being built up. This history, local political tradition and the nature of Lesotho's terrain all went towards ensuring a tightly knit social organization.

NAMIBIA

REPUBLIC OF NAMIBIA

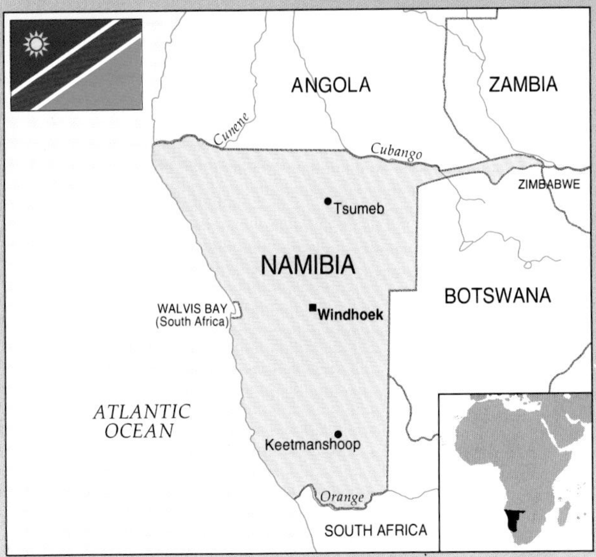

CHRONOLOGY

1920	South Africa administers the territory, previously a German protectorate, under a League of Nations mandate
1960	South West African People's Organization (SWAPO) formed
1968	The name Namibia officially adopted by the United Nations
1971	The International Court of Justice declares South Africa's occupation of Namibia illegal
1973	United Nations recognizes SWAPO as the authentic representatives of the Namibian people
1988	Geneva protocol signed between South Africa, Angola and Cuba to implement a peaceful settlement
1989	Ceasefire and Namibia's first free elections for a Constituent Assembly
1990	Namibia gains full independence in March

ESSENTIAL STATISTICS

Capital Windhoek

Population (1989) 1,270,000

Area 823,144 sq km

Population per sq km (1989) 1.5

GNP per capita (1988) US$1,300

CONSTITUTIONAL DATA

Constitution Republic with two legislative houses (National Assembly; National Council)

Date of independence 20 March 1990

Major international organizations UN affiliations; SADCC

Monetary unit 1 South African rand (R) = 100 cents

Official languages Afrikaans; English

Major religion Lutheran

Head of government since independence President Sam Nujoma (1990–)

Namibia (formerly South West Africa) was long known as Africa's last colony, with independence only coming in 1990. Its colonial status dates back to 1884 when Germany declared it a protectorate; a period of coercive colonial rule ensued. Germany lost the territory following its defeat in World War I, when its colonies were put under the control of victorious neighbors as League of Nations mandates; Namibia went to South Africa.

Throughout the 1920s South Africa used Namibia as a source of cheap land for Afrikaner farmers. They poured in, displacing the African population from two-thirds of the land and confining them to reserves. These formed labor pools for the white farms and the mines. There was resistance which gradually became more organized and in 1960 the South West African People's Organization (SWAPO) was formed. SWAPO rallied international support against South Africa's occupation.

South Africa spurned all efforts to bring about a settlement and refused to implement the United Nations Resolution 435, which set out conditions for a transition to independence. Instead it attempted a series of internal settlements. Initially Namibia was governed as part of South Africa, complete with its apartheid laws and "bantustans" or ethnic homelands. By the late 1970s, however, South Africa was seeking to construct a successor state, its vehicles being the ethnically-based and white-dominated Democratic Turnhalle Alliance (DTA) and the bantustan administrations. The DTA failed to establish popular credibility and in 1983 Namibia returned to direct South African rule. South Africa continued to foster internal parties opposed to SWAPO and to resist negotiations. South African troops and police created and controlled an army in Namibia, a police force and an array of paramilitary counterinsurgency units, the most well-known being Koevoet, which was notorious for its violence and atrocities.

South Africa finally agreed to a ceasefire in April 1989 and a transition to independence supervised by the UN Transitional Assistance Group (UNTAG). In November 1989, Namibians went to the polls to elect a Constitutent Assembly to pave the way to full independence. SWAPO won a convincing majority, obtaining 57.3 percent of the vote and 41 out of 72 seats.

Economic and social trends

Namibia was settled by Bantu-speaking people, notably the Ovambo and Okavango who practiced mixed farming in the north, and by the Nama, Herero and Damaras who were pastoralists further south. It is also the home of over 20,000 Khoisan (bushmen) and approximately 80,000 whites of German and Dutch origin. It is arid, and agricultural activity is mainly confined to cattle ranching and sheep farming. Namibia imports most of what it consumes and exports most of what it produces. Mining, principally of diamonds, uranium and base metals, contributes about a quarter of GDP. The economy remains dominated by three multinational companies, the most important being the Anglo-American Corporation of South Africa. Agriculture and fisheries provide about one-tenth of GDP.

In Namibia's guerrilla camps, villages and townships, traditional songs and dance were adapted and sung. But ethnic and historical divisions and the ravages of the war ensured that the fight for a truly national culture, striving for unity but recognizing diversity, would be a long and difficult one.

ANGOLA

PEOPLE'S REPUBLIC OF ANGOLA

CHRONOLOGY

1891	Portugal confirmed as colonial power in Angola
1926	Salazar comes to power in Portugal and colonial rule harshens
1956	Movimento Popular de Libertaçào de Angola (MPLA) founded under Agostinho Neto
1961	Armed struggle begins
1966	Uniào Nacional para a Independência Total de Angola (UNITA) founded under Jonas Savimbi
1975	Angola achieves full independence after military coup – Cuban troops arrive
1981	South African military incursions into Angola increase
1988	Namibian independence and withdrawal of Cuban troops agreed between Angola and South Africa
1989	Namibian elections
1990	First talks between UNITA and MPLA in Portugal

ESSENTIAL STATISTICS

Capital Luanda

Population (1989) 9,739,000

Area 1,246,700 sq km

Population per sq km (1989) 7.8

GNP per capita (1984) US$830

CONSTITUTIONAL DATA

Constitution People's Republic with one legislative house (People's Assembly)

Date of independence 11 November 1975

Major international organizations UN; OAU; ACP; SADCC

Monetary unit 1 kwanza (Kw) = 100 lwei

Official language Portuguese

Major religion Christian

Heads of government since independence (President) A. Neto (1975–79); J. dos Santos (1979–)

Angola is rich in mineral resources and strategically located on Africa's Atlantic seaboard. Although Portuguese trading posts had been established on the coast since the 16th century, the land was not colonized by Portugal until the late 19th century. A brutal form of colonialism persisted throughout the rule in Lisbon of the rightwing dictator Dr Salazar and his successor Marcelo Caetano, from 1926 until 1974 when Caetano was dislodged by a bloodless military coup.

The commencement of hostilities in Angola in 1961 began 13 years of guerrilla warfare against Portuguese rule in southern Africa. In Angola no single nationalist group achieved an unassailable position throughout the war. It was the Movimento Popular de Libertaçào de Angola (MPLA, later MPLA-PT) under Agostinho Neto which took power at independence in 1975. The most heterogeneous of the three parties, it included many intellectuals, had a strong urban base, and espoused an explicitly socialist program. Divisions within it were contained by Neto's successor, Jósé Eduardo dos Santos, but almost immediately it was challenged by the other liberation movements and had to rely on the assistance of Cuban troops to retain its hold. The longevity of the Frente Nacional de Libertaçào de Angola (FNLA) did not match that of the Uniào Nacional para a Independência Total de Angola (UNITA) which was engaged in a civil war with the MPLA-PT until mid-1991.

UNITA had an ethnic constituency, regional power bases and American and South African backing. The United States was uncomfortable with oil-rich Angola's allegiance to Moscow, and South Africa opposed its hospitality to South African and Namibian guerrillas. Consequently, the support of these two powers for Jonas Savimbi's UNITA was substantial, and South Africa made blatant military excursions into Angolan territory. The war cost Angola dear and, as a pawn on the chessboard of international power relations, its fate was orchestrated from beyond its own borders.

Economic and social trends

Angola has abundant arable land and mineral resources, notably oil. Under the Portuguese, cash crops included coffee, cotton, sisal and sugar. However, in the last years of colonialism there was investment in manufacturing and infrastructure, and at independence Angola was a relatively high-output economy. From the mid-1970s the economy suffered greatly on all accounts. The exodus of Portuguese settlers left the country bereft of vital human resources, while the war with UNITA and South Africa hampered productive capacity and made agricultural production difficult. 1990 witnessed a severe famine unfolding in the south. In the towns the local currency was worthless and the shops contained little to buy. The crisis led the MPLA-PT to reassess its exclusive allegiance to the Eastern bloc and to turn to the International Monetary Fund and World Bank for loans.

Angola is a secular state in which all religions are respected. The Portuguese brought Christianity to Angola, although in 1990 much of the country still followed traditional beliefs. Likewise, whilst Portuguese is used by the educated elite and is the official language, it did not supplant indigenous African languages. A sparse population, ethnic diversity, poor urban-rural links and postindependence hostilities meant that the constitutional commitment to development was persistently thwarted. Angola remained a deeply divided country.

CONGO

PEOPLE'S REPUBLIC OF THE CONGO

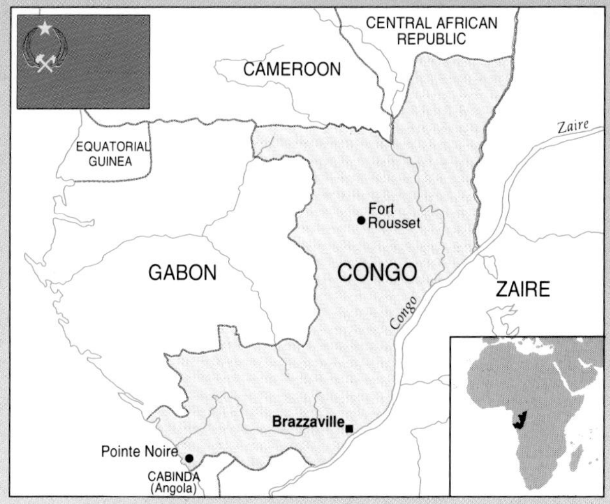

CHRONOLOGY

1908	Establishment of administrative federation of French Equatorial Africa
1960	French colony of Moyen-Congo granted independence as Republic of Congo
1963	First President overthrown in popular revolt
1969	Establishment of Marxist-Leninist regime under President Ngouabi
1977	Ngouabi assassinated
1979	Sassou-Nguesso takes over as president

ESSENTIAL STATISTICS

Capital Brazzaville

Population (1989) 2,245,000

Area 342,000 sq km

Population per sq km (1989) 6.6

GNP per capita (1987) US$880

CONSTITUTIONAL DATA

Constitution People's republic with one legislative body (People's National Assembly)

Date in independence 15 August 1960

Major international organizations UN; OAU; ACP

Monetary unit 1 CFA franc (CFAF) = 100 centimes

Official language French

Major religion Roman Catholic

Heads of government since independence (President) Foulbert Youlou (1960–63); Alphonse Massemba-Débat (1963–68); Maj. Alfred Raoul (1968–69); Capt. Marien Ngouabi(1969–77); Col. Joachim Yhombi-Opango (1977–79); Col. Denis Sassou-Nguessou (1979–)

In the 1880s and 1890s the French annexed a large part of Equatorial Africa, including the territory that now forms the People's Republic of the Congo. French Equatorial Africa was made into an administrative federation in 1908, with the Moyen-Congo, whose borders became those of the independent republic, as one of the four component parts. Brazzaville was capital both of the colony and of the federation.

It proved impossible to build up an elaborate infrastructure over so large and thinly populated a territory. To encourage development, privately owned concessionary companies were allowed a free hand. This led to gross exploitation of the native inhabitants, who were forced to hand over exportable products at minimal prices. Seventeen thousand laborers died in the building of the railroad from Pointe Noire to Brazzaville. In the 1920s the French found themselves faced with widespread resistance centered around local religious leaders. Not till 1931 was the colony effectively pacified.

Colonial reforms after World War II encouraged local political activity, but with only a tiny educated class to draw on nationalist movements never obtained the mass support they acquired in other colonial territories. Independence was granted in August 1960.

Already before 1960 it was evident that, given the ethnic differences within the country, the Congo's politics would in large part take the form of a struggle between north and south. Three-quarters of the population lived in the south but northerners compensated for their numerical inferiority by their preponderance in the armed forces.

The first president, the Abbé Foulbert Youlou, was overthrown in a popular revolt in August 1963. His successor, Alphonse Massemba-Débat, also a southerner, swung sharply to the left, establishing the *Mouvement National de la Révolution* (MNR). Before long the new president was in trouble with the party's radicals, who forced his resignation in September 1968. A northern army officer, Captain Marien Ngouabi, then took over the presidency and organized a Marxist-Leninist vanguard party, *Parti Congolais du Travail* (PCT) which gradually replaced the MNR. In January 1970 the country was renamed the People's Republic of the Congo.

Ngouabi held on to power until his assassination in March 1977. Colonel Joachim Yhombi-Opango then became head of state but lost the support of the radicals and resigned in March 1979. Another northern army officer, Colonel Denis Sassou-Nguesso, took over the presidency, survived several attempted coups and provided the country with its most stable regime. While maintaining the revolutionary rhetoric of his predecessors, President Sassou-Nguessou established cordial relations with the West and with conservative African governments. His economic policy allowed scope for private enterprise and he showed himself remarkably successful in attracting aid from both East and West.

Timber was the country's main export until the 1970s. By the mid-1980s petroleum and natural gas were providing 80 percent of export earnings. In the late 1980s about half the total population was classified as urban: Brazzaville had a population of 600,000. With an income per head of US$990 the Congo was rated among the middle-income developing countries. But external debt was close on US$4 billion by the end of the 1980s, making the Congo on a *per capita* basis the most heavily indebted country in Africa.

ZAIRE

REPUBLIC OF ZAIRE

CHRONOLOGY

1908	Belgium takes over Congo region from King Leopold II
1957	Belgium introduces political reform resulting in ethnic rivalries
1960	Belgian Congo granted independence. Army mutiny and Katangan secession leads to UN intervention
1961	Former prime minister, Patrice Lumumba, is murdered
1964	Revolt in eastern Congo
1965	General Mobutu becomes president following coup
1971	Country's name changed to Zaire
1977	Invasion of Shaba by the *Front National pour la Libération du Congo* (FNLC)

ESSENTIAL STATISTICS

Capital Kinshasa

Population (1989) 33,336,235

Area 2,345,095 sq km

Population per sq km (1989) 14.2

GNP per capita (1987) US$160

CONSTITUTIONAL DATA

Constitution Single-party republic with one legislative house (Legislative Council)

Date of independence 30 June 1960

Major international organizations UN; OAU; ACP

Monetary unit 1 zaïre (Z) = 100 makuta

Official language French

Major religion Roman Catholic

Heads of government since independence (President) J. Kasavubu (1960–65); Gen. J. D. Mobutu (1965–)

The vast extent of territory covered by modern Zaire presented a complex mosaic of different ethnic groups in the 19th century, most of them speakers of Bantu languages, organized in petty chiefdoms or independent village communities but with two substantial kingdoms formed by the Lunda in the south. In the 1880s and 1890s the entire area came under the control of the Congo Free State, an extraordinary political phenomenon, being at once the brain-child, the creation and the exclusive property of one man, Leopold II, King of the Belgians, acting in a private capacity. The Congo Free State was essentially a commercial venture designed to profit its founder. This could be done only through the ruthless exploitation of the native population. Reports of the atrocities associated with the regime aroused a storm of protest in Belgium and other countries, forcing the King to hand the territory over to the Belgian government in 1908.

The Belgians had no tradition of colonial rule. Pragmatically they set about the reform and development of their unwillingly acquired colony. With an economy firmly based on plantation agriculture and the mining of copper and other minerals, with a well-developed transport system and widespread provision of primary and technical education, the Belgian Congo was presented as a model colony. Belgian rule was, however, more autocratic than that of the British and the French: no higher education or admission into senior ranks of the administration was available for the Congolese, no freedom of the press and no representative institutions. But the colony could not be insulated from the nationalist fervor that developed elsewhere in Africa after World War II, and in 1957 the Belgians introduced political reforms. Nationalist passions and ethnic rivalries suddenly erupted. By late 1959 the colony was becoming ungovernable. The first colony-wide elections were held in May 1960; a month later the Belgian Congo became independent.

In the first weeks of independence the Congolese army mutinied against its white officers. Many Belgians fled the country and a secessionist movement, led by Moise Tshombe, seized power in the richest province, Katanga, in the southeast of the country. In desperation the prime minister, Patrice Lumumba – the most charismatic of the country's politicians – appealed to the United Nations. There followed a violent and confusing period. Early in 1961 Lumumba, who had been removed from office in September, was captured and murdered by his opponents in Katanga, but UN intervention eventually broke the force of Katangese secessionism. In 1964 the central government found itself faced with an equally serious revolt mounted by Lumumba's followers in the eastern provinces: it was put down by a force of hastily recruited white mercenaries. In November 1965 General Mobutu, who had played an active part in Congolese politics since 1960, seized power in a bloodless coup.

Mobutu went on to build up a highly personal system of rule, which by 1990 had given the country 25 years of political stability. In 1966 he established the *Mouvement Populaire de la Révolution* (MPR) as the country's only political party; the president's own standing was heightened by an officially sponsored cult of his personality. In essence Mobutuism was dependent on patronage, bribery and surveillance skillfully used to retain the loyalty, or at least the quiescence, of his subordinates. At the same time the president used his unchecked power to amass a huge personal fortune. In 1971

Mobutu changed the country's name from Congo to Zaire (the new name being an alternative native name for the country's major river) and commanded other changes of custom designed to stress the "authenticity" of his regime.

Mobutu showed himself equally adroit in his dealings with Western powers, laying stress on his country's strategic importance in negotiations with the United States and reminding aid-givers and investors in Belgium, France and other Western countries of Zaire's huge economic potential. This cultivation of foreign backers paid off in the late 1970s, when Mobutu was faced with a serious threat, an invasion of the country's richest province Shaba (formerly Katanga) launched from Angola in 1977. The country was invaded again in 1978 by a force made up mostly of old Katangese soldiers under the name of the *Front National pour la Libération du Congo* (FNLC). Military support was promptly provided by Morocco, France and Belgium, enabling Mobutu to defeat the invaders.

Muted opposition to the regime continued: in the 1980s it was most openly expressed by a small group of members of the legislature who founded the *Union pour la Démocratie et le Progrès social* (UDPS). But the president retained his skill at outwitting all opponents with blandishments and repression.

Economic and social trends

With its massive extent of underutilized agricultural land, its immense forests, its network of rivers providing both arteries of transport and sources of hydroelectric power and its rich store of minerals, Zaire is potentially the richest country in Africa. Belgian rule saw the establishment of the most elaborate economic infrastructure to be found anywhere in tropical Africa, with the colonial state playing a large part in the development of the mining industry, plantation agriculture and an efficient system of communications.

The years after independence were marked – except for a brief upsurge between 1968 and 1973 – by economic stagnation and decline. In part this was a result of the destruction resulting from internal conflicts between 1960 and 1965. The "Zairianization" program launched in 1973, by which hundreds of foreign-owned businesses, commercial, industrial and agricultural, were expropriated and put under local ownership, had a disastrous effect on productive forces. Other adverse factors were beyond Zaire's control, particularly the fluctuations and fall in the price of export commodities. This was especially serious for the mining industry, the source of 80 percent of the country's export earnings: a rise in the price of industrial diamonds in the 1980s was more than counterbalanced by falls in copper and cobalt prices. Inevitably, then, Zaire plunged deeply into debt. By 1988 external debt stood at US$7 billion and debt servicing required almost half the country's export earnings, even though the government adhered strictly to most of the austerity measures prescribed by the International Monetary Fund (IMF), and although an attempt to reschedule Zaire's debt to Belgium went through despite Belgian outcry against Mobutu's corruption, upon which he threatened to turn to France as major trading partner. For ordinary Zaireans this meant a steady decline in living standards.

▼ **Fighting in Katanga in the early 1960s.**

NIGERIA

FEDERAL REPUBLIC OF NIGERIA

CHRONOLOGY

1900	Nigeria created as a British dependency
1960	Nigeria attains independence
1963	Republican constitution adopted with Nnamdi Azikiwe as nonexecutive president
1966	Colonel Gowon heads new government following Maj.-Gen. Aguigi-Ironsi's coup
1979	President Obasanjo hands over power to elected civilian president Shehu Shagari
1983	Fourth military coup led by Muhammed Buhari
1988	Armed Forces Ruling Council (AFRC) initiate controlled return to civilian rule

ESSENTIAL STATISTICS

Capital Abuja

Population (1989) 115,973,000

Area 923,768 sq km

Population per sq km (1989) 125.5

GNP per capita (1987) US$370

CONSTITUTIONAL DATA

Constitution Federal Republic temporarily governed under emergency powers by Armed Forces Ruling Council (AFRC)

Date of independence 1 October 1960

Major international organizations UN; Commonwealth of Nations; OAU; ACP; ECOWAS; OPEC

Monetary unit 1 Nigerian naira (N) = 100 kobo

Official language English

Major religion Muslim

Heads of government since independence (Prime minister) Sir A. T. Balewa (1960–66); (Military ruler) Maj.-Gen. J. Aguigi-Ironsi (Jan–Jul 1966); Lt.-Col. Y. Gowon (1966–75); Gen. M. Ramat Mohammed (1975–76); (Head of state) Gen. O. Obasañjo (1976–79); (President) A. Shehu Shagari (1979–83); (Military ruler) Gen. M. Buhari (1984–85); (President) Maj.-Gen. I. Babangida (1985–)

Nigeria is the most populous state in Africa. Its government, after independence, alternated between multiparty democracy and an uneasy imposition of a military rule.

Nigeria was occupied by Britain between 1861 and 1903. In 1914, north and south were amalgamated under Lord Lugard as Governor-General, but there was no union of peoples, and no all-Nigerian representative institutions. In 1939, the south was divided into the eastern region, whose predominant tribe were the Ibos, and the western region, whose predominant tribe were the Yoruba. Missionary education, permitted in the south but not the north, produced a situation of great educational imbalance.

Dr Nnamdi Azikiwe, an Ibo, formed the National Council of Nigeria and the Cameroons (NCNC) in 1944 to fight for independence, and Obafemi Awolowo, a Yoruba lawyer, started the Action Group to fight the 1951 election. The north had 56 percent of the population, but reacted in a defensive manner, forming a conservative Northern People's Congress (NPC) under Ahmadu Bello, the Sardauna of Sokoto, and Abubakar

The Ibo people

The energy of the Ibos in professional life and all aspects of modernization, together with their intense tribal loyalty, were at once their strength and the cause of the suspicions which they provoked among other Nigerian peoples. Before colonial rule, the Ibos had a long tradition of smallscale political organization based on competition for status rather than on inherited chieftainship.

Under colonial rule the Ibos, strengthened by missionary education and church adherence, leaped into the process of modernization; families and clans combined to build schools and to send pupils abroad to acquire qualifications. Ibos left their own overcrowded areas for jobs in other regions, in the middle echelons of government and for private trade. They prospered, but remained culturally and often residentially separate. At home the Ibos rejected indirect rule by imposed "warrant chiefs" and proceeded rapidly to elected local government. The super-efficient Ibo tribal unions were suspected by other tribes of carving up the job market, and making secret political plans.

The Ibo people felt alienated after independence, since the policy of northernization excluded them from jobs in that region, and they were seen as the junior partners in the NPC/NCNC ruling coalition. The murderous military coup of January 1966 was seen as an Ibo plot. The pressure of Ibo "young Turks," hungry for job opportunities, induced General Ironsi to introduce the disastrous unification decree, in May 1966, and the Ibos suffered in the subsequent massacres. Emeka Ojukwu, the Governor of the East, was propelled by the overconfident Ibo advisors into Biafran secession. For two and a half years the splendid cooperative efforts of the Ibo people at every level sustained Biafra against overwhelming odds. With the collapse of Biafra, the Ibos changed course and achieved a remarkably quick reintegration into Nigerian economic and political life.

Tafawa Balewa, a reforming conservative. Sir John Macpherson, the Governor-General, responded to the pressure by a series of new constitutions, moving toward self-government. On 1 October 1960, Nigeria attained independence as a federation where the federal prime minister, Abubakar Tafawa Balewa, was overshadowed by the Sardauna, the northern regional premier. Dr Azikiwe became the titular president of Nigeria and Chief Awolowo became leader of the opposition and soon entered radical politics and was imprisoned for treason. The federal government declared an emergency and took over the government of the western region for six months before installing Samuel Akintola, a client of a Sardauna, or regional premier. The subsequent regional elections were rigged, and resulted in large-scale violence in the west. Finally, in January 1966, seven army majors, predominantly Ibo, attempted a coup and murdered Abubakar, the Sardauna and Akintola, and four of the five northern senior officers. The coup-makers were arrested but the frightened remnants of the federal cabinet handed power to Major-General Aguiyi Ironsi, himself an Ibo. Ironsi failed to heal the wounds of this one-sided coup and caused northern resentment by a unification decree. In July he was himself murdered, with 200 other Ibo soldiers, in a northern counter-coup, which brought to power Colonel Yakubu Gowon, a northern Christian from a minority tribe. Gowon did not establish his rule in eastern Nigeria. Further massacres of Ibos occurred in the north and eight months later the east declared independence, under its regional governor Emeka Ojukwu, as the independent "Republic of Biafra".

After two and a half years of bitter civil war, Nigerian unity was reestablished by federal victory and made secure by Gowon's policy of healing the nation's wounds through the "three Rs" – reconciliation, reconstruction and rehabilitation. The 12-state structure created by Gowon provided a secure base for a more center-leaning federal system.

In July 1975, however, Gowon's government, which had become increasingly characterized by corruption, was overthrown in a bloodless coup. Mohammed Murtala, who succeeded, instituted a vigorous policy of corrective government. He also appointed 49 elders to draw up a draft constitution. In February 1976, Murtala was murdered in an unsuccessful coup and his successor, General Obasanjo, who had been his Chief of Staff, continued the Murtala policies and handed over power to an elected civilian president, Shehu Shagari, on 1 October 1979. The Nigerian constitution was most unsuccessful in restraining corruption and managing the economy. This incompetence, coupled with a rigged election in 1983 and a disastrous downturn in the oil price, resulted in a further coup on 31 December 1983. General Buhari, who took power in the coup, instituted a punitive military policy detaining several hundred of the former political leaders and imprisoning journalists, but this harshness resulted in his removal in a further coup in 1985 by General Ibrahim Babangida, the Chief of Staff. Babangida involved many civilians in his policy of amending the constitution, forming two political parties and preparing Nigeria for a third period of civilian rule. Two new states were created, making 21 in all. Local government elections were held in 1990, state elections planned for 1991, and in 1992 Babangida was due to hand over power to an elected president.

Economic and social trends

During the colonial period, Nigeria's economy was based on the export of groundnuts and cotton from the north, cocoa from the west and palm oil from the east, but with the coming on stream of substantial oil production at a favorable world price after 1973, cash-crop production collapsed, and Nigeria had also to import large quantities of food. After the civil war, oil income was distributed almost equally by the federal government to all states in Nigeria. Federal expenditure also increased dramatically, and created great improvements in infrastructure. Unfortunately, massive corruption and extravagance occurred and when the oil price was halved after 1981, an acute cash crisis forced the Nigerian government to cut down on imports and government expenditure. By 1990 Nigeria acquired unpayable debts of US$32 billion, 119 percent of GNP, and in spite of very severe structural adjustment terms enforced by the IMF, debt continued to increase. Meanwhile, the resulting unemployment, currency depreciation and business slump created widespread discontent.

◀ **1980s Lagos was one of the world's most crowded cities.**

GHANA

REPUBLIC OF GHANA

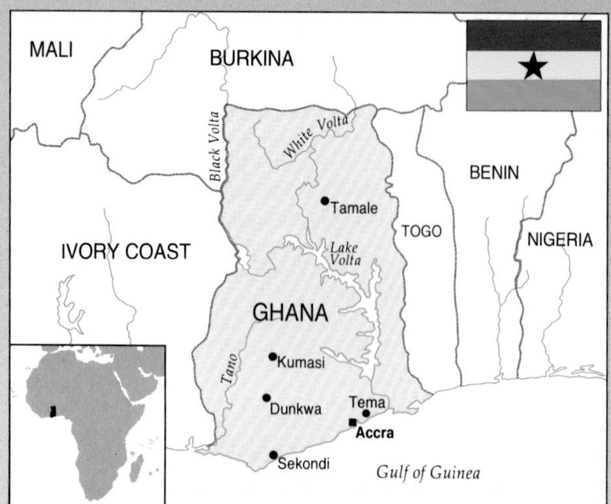

CHRONOLOGY

1879	British Gold Coast Colony created
1957	Gold Coast achieves independence as Ghana
1966	Military coup overthrows government establishing National Liberation Council
1969	Return to civilian rule following national elections
1972	Armed forces and police seize power from civilian regime
1979	Supreme Military Council (SMC) toppled in coup
1981	Provisional National Defence Council (PNDC) dismisses government and suspends constitution

ESSENTIAL STATISTICS

Capital Accra

Population (1989) 14,566,000

Area 238,533 sq km

Population per sq km (1988) 61.1

GNP per capita (1987) US$390

CONSTITUTIONAL DATA

Constitution Military regime

Date of independence 6 March 1957

Major international organizations UN; Commonwealth of Nations; OAU; ACP; ECOWAS; CEAO

Monetary unit 1 cedi (C) = 100 pesewas

Official language English

Major religion Christian

Heads of government (Prime minister) K. Nkrumah (1957–60); (President) K. Nkrumah (1960–66); (Military ruler) Gen. J. Ankrah (1966–69); Lt.-Gen A. Afrita (1969–70); (Prime minister) K. Busia (1969–72); (Military ruler) I. Acheampong (1972–78); F. Akuffo (1978–79); (President) H. Limann (1979–81); (Chairman of Provisional National Defense Council) Flt.-Lt. J. Rawlings (1981–)

The Colony of the Gold Coast and Asante (later the Gold Coast Colony) was ruled by the British until it became independent in 1957 as the Republic of Ghana. British rule in the Gold Coast was based on the cash-crop economy (cocoa) and exploitation of natural resources (gold, timber). Transport was structured to foster economic development. The British also encouraged a degree of education and limited political participation. After World War II a new generation of nationalists came to the fore. These demanded speedy decolonization, and they were opposed to chiefs and critical of earlier forms of constitutional protest. The most prominent such nationalist was the charismatic Kwame Nkrumah. He wrested the nationalist initiative from his opponents, and his Convention Peoples Party (CPP) led Ghana to independence as the first British African colony to achieve that status, in 1957.

Nkrumah's CPP ruled Ghana's first republic from 1957 to 1966. In 1961 a referendum allowed Nkrumah to declare Ghana a one-party state. Nkrumah's international reputation as a spokesman for Africa continued in parallel with increasing authoritarianism at home. Corruption and ill-judged CPP economic plans, based on a nebulous African socialism, eroded the fiscal strength of Ghana. In 1966 Nkrumah was overthrown by an army coup that enjoyed a good deal of popular support. In 1969 elections were held, and the second republic was inaugurated under the leadership of Nkrumah's old rival Kofi Busia and his Progress Party. The Busia regime enjoyed much initial goodwill, but its effectiveness was severely affected by the fall of cocoa prices and by a general economic decline resulting from mismanagement as well as misfortune.

From 1972 until 1979 Ghana was under the rule of the military, headed by General I. K. Acheampong (1972–78) and, following his removal for corrupt incompetence, by General F. Akuffo (1978–79). In this period the economy declined catastrophically amid widespread corruption and profiteering. The military dragged its feet on any return to civilian rule, and added alienation of the political nation to its economic errors. In 1979 the army began to implement a return to civilian rule in response to universal protest at its inability to resolve Ghana's problems. In the course of this process a revolt took place in June 1979. This was led by the youthful Flight-Lieutenant J. Rawlings, and involved junior military personnel. Rawlings moved publicly against corruption, executing both Acheampong and Akuffo. Charismatic in the manner of Nkrumah, Rawlings promised to restore Ghana's dignity and economic wellbeing together with political democracy.

In 1979, therefore, elections took place for Ghana's Third Republic. The new head of state was Dr H. Limann, but his regime singularly failed to reverse the catastrophic economic decline of the 1970s. In 1981 another popular coup took place, again headed by Rawlings, and again involving junior military personnel. This time Rawlings' regime institutionalized itself as the Provisional National Defence Council (PNDC), and ruled Ghana into the 1990s. The PNDC's priority was to reinvigorate the economy, recognizing this as the key to Ghana's political future. The PNDC courted anti-Western powers, but at the same time struck economic agreements with the World Bank and IMF. Rawlings remained in power, and Ghana painfully began to confront the great economic and political problems that beset it.

IVORY COAST

REPUBLIC OF CÔTE D'IVOIRE

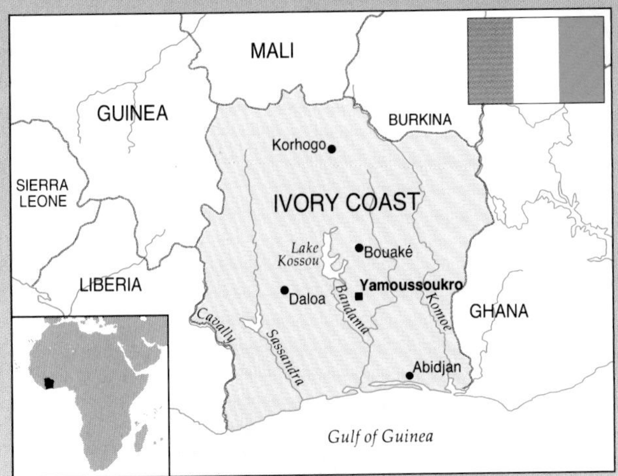

CHRONOLOGY

1900	French colony
1944	Syndicate Agricole Africaine (SAA) founded to voice interests of African planters. Led by Felix Houphouët-Boigny
1946	Formation of French Fourth Republic allows for African political representation
1948	Period of intense confrontation between PDCI and French colonial government
1951	Compromise reached between PDCI and French
1960	Ivory Coast declares independence
1986	Côte d'Ivoire declared to be sole international form of name of country
1989	One-party state ended and new opposition groups given legal recognition

ESSENTIAL STATISTICS

Capital Yamoussoukro

Population (1989) 12, 135,000

Area 320,763 sq km

Population per sq km (1989) 37.8

GNP per capita (1987) US$750

CONSTITUTIONAL DATA

Constitution Single-party republic with one legislative house (National Assembly)

Date of independence 7 August 1960

Major international organizations UN; OAU; ACP; ECOWAS

Monetary unit 1 CFA franc (CFAF) = 100 centimes

Official language French

Major religion Christian

Head of government since independence (President) Félix Houphouët-Boigny (1960–)

The Ivory Coast is the richest of the eight states of the former French West African Federation and in consequence preferred to enjoy its wealth alone rather than to share it in a federal union. Under its ruler since independence, Félix Houphouët-Boigny, it also remained remarkably stable.

The Ivory Coast was brought under firm military control by the French governor, Gabriel Angoulavant, at the beginning of the 20th century. It had a larger number of French settlers than any other state in West Africa and the early history of the RDA in the French West African Federation and of the PDCI, its territorial branch in the Ivory Coast, was characterized by conflict with the settlers and by the demand for the abolition of forced labor. Houphouët-Boigny, a Baoule medical assistant and large-scale farmer, was elected to the French parliament and achieved great acclaim in Africa when he sponsored a bill abolishing forced labor. He went on to become a minister in several French governments as well as being the acknowledged political leader in the Ivory Coast.

In 1950, when the Communists left the French Government and an unsympathetic governor, Laurent Péchou, was posted to the Ivory Coast, Houphouët-Boigny and the PDCI came into sharp conflict with the colonial administration but after three years both sides saw that this confrontation was unnecessary. The Governor was posted away and Houphouët-Boigny, the President of the RDA throughout West Africa, broke his links with the French Communist party. Houphouët became the acknowledged intermediary for the French government in its dealings with the Ivory Coast, and the PDCI made reconciliation with French planters.

The Ivory Coast enjoyed a remarkable rate of economic growth based on cocoa, coffee and "agribusiness". This contentment made it natural for the Ivory Coast to vote for the French community as opposed to independence in the 1958 referendum. In 1959, however, when Houphouët saw that his rival Senegal and the Mali Federation had been promised independence by de Gaulle, he also demanded and received it in 1960. Despite considerable ethnic diversity and the rivalry of the two largest ethnic groups, the Baoule and the Agni, the Ivory Coast (now officially named Côte d'Ivoire) achieved a remarkable sense of national unity.

In 1963 an alleged coup plot by radical members of the PDCI and youth leaders was uncovered and punished, but after this Houphouët was able to exercise more or less unchallenged authority for some 25 years. His status as a distinguished French statesman was reinforced by his skill in incorporating rather than suppressing opposition. Nonetheless, the Ivory Coast was a de facto one-party state until 1989.

In the late 1980s and early 1990s the failing political skills of Houphouët and the failure to designate a successor to the president interrupted the good order of the state. The PDCI had to accept multiparty politics but still retained power.

The Côte d'Ivoire replaced Ghana as the world's major cocoa producer, but the disastrous fall in world cocoa prices so aroused the anger of Houphouët that he withdrew from the Cocoa Producers' Alliance and lost one hundred million dollars. By 1991 the country was in financial crisis and the total overseas debt had risen to $15 billion (182 percent of GNP), far more than could ever be repaid. Côte d'Ivoire has developed its own national pride and sophisticated political culture based on prosperity and pragmatism.

LIBERIA

REPUBLIC OF LIBERIA

CHRONOLOGY

1900	Liberia ruled as effective one-party state
1926	Firestone Agreement
1927	League of Nations inquiry into forced labor agreements
1944	President William V.S. Tubman assumes office with promise of a "new deal" for the indigenous population
1979	"Rice Riots" – troops called in from Guinea
1980	Military coup headed by Master Sergeant Samuel K. Doe
1981	Attempted coup led by General Thomas Wen Syson
1985	Attempted coup led by General Thomas Quiwonkpa
1990	Taylor's National Patriotic Forces of Liberia overrun country

ESSENTIAL STATISTICS

Capital Monrovia

Population (1989) 2,508,000

Area 99,067 sq km

Population per sq km (1989) 25.3

GNP per capita (1987) US$440

CONSTITUTIONAL DATA

Constitution Multiparty republic with two legislative houses (Senate; House of Representatives)

Major international organizations UN; OAU; ACP; ECOWAS; CEAO

Monetary unit 1 Liberian dollar (L$) = 100 cents

Official language English

Major religion Christian

Heads of government since 1900 (President) Garretson Wilmot Gibson (1900–04); Arthur Barclay (1904–12); Daniel Edward Howard (1912–20); Charles Dunbar Burgess King (1920–30); Edwin Barclay (1930–44); William Vacanarat Shadrach Tubman (1944–71); William R. Tolbert (1971–80); Samuel Kanan Doe (Chairman 1980–85; President 1985–90); A. Sawyer (1990–)

Liberia is the only country in Africa never to have experienced colonial rule and has, in consequence, known neither the modernizing effects of a colonial state, nor the unity that comes from forming an anticolonial nationalist movement. It was further inhibited by the cleavage between the small group of returned black Americans – so-called Americo-Liberians who were totally dominant until the coup of 1980 – and the mass of the African hinterland people of Liberia.

Liberia has been an independent state since 1847, but in the early 20th century it had to defend that independence against international creditors and the attempt of the League of Nations to impose an "advisor" with executive powers.

President Edwin Barclay began the policy of unification of Americo-Liberians and the hinterland. This policy was continued and developed under the distinguished leadership of President William Tubman from 1944 to 1971.

The relationship with the hinterland people was still one of paternalistic dominance but it was somewhat softened by the custom of adopting hinterland young people and incorporating them into the privileged Americo-Liberian society. Political power rested with the True Whig Party, which bound the elite together. It was by far the oldest political party in Africa. The electoral system was grotesquely tilted in favor of the Americo-Liberians who composed only about five percent of the population, but occupied most of the government offices and owned large estates in the hinterland. The Firestone Rubber Company also possessed very large rubber estates. Considerable revenue has also been derived from iron ore.

On the death of Tubman, in 1971, William Tolbert, another Americo-Liberian, succeeded as president and pursued a policy of alternate liberalization and authoritarian action to maintain the dominance of the regime against constitutional protest led by the Progressive Alliance of Liberia, whose leaders were Amos Sawyer, a distinguished academic, and Baccus Matthews. In 1979, a jump in the price of rice caused riots in which 78 people were killed and 548 injured. In 1980, noncommissioned officers of the army, all from hinterland tribes, led by Master Sergeant Samuel Doe, seized the opportunity to murder President Tolbert and to execute twelve other high officials in public. The coup put an end to overt Americo-Liberian dominance, but several members of this group still played an important part in Liberian politics. Doe's government was characterized by extreme harshness and by quarrels with his former coup-making colleagues. The economy soon deteriorated through mismanagement and corruption. Doe survived a number of coup attempts and rigged the election in 1985. In December 1989, a successful guerrilla opposition was established by Charles Taylor and spread from Nimba County to control most of Liberia outside Monrovia. Prince Johnson's rival insurgent force, the Independent National Patriotic Front of Liberia, took control of most of Monrovia.

The economic community of West African states sent a 7,000-strong "ECOMOG" peacekeeping force under Major General Joshua Dogonyaro, a Nigerian soldier. President Doe, unwisely, came out of his guarded headquarters to visit the ECOMOG commander and was promptly ambushed and murdered by troops of Prince Johnson. Meanwhile, the meeting of exiles chose Amos Sawyer as interim president and head of government. A ceasefire was established and representatives of all three factions met to try to establish peace.

SENEGAL

REPUBLIC OF SENEGAL

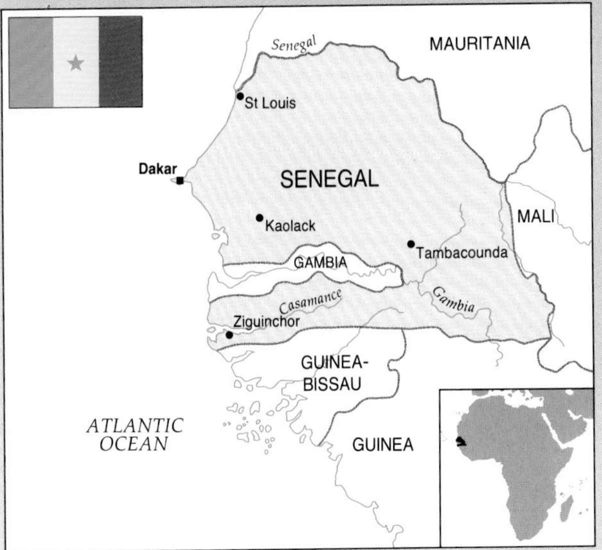

CHRONOLOGY

1902	Senegal becomes part of French West Africa
1959	Senegal and French Sudan join to form shortlived Federation of Mali
1960	Senegal becomes independent
1963	Opposition parties banned
1968	Widespread unrest
1974	Dia and other detainees released and a measure of political pluralism permitted
1976	Three-party system introduced
1981	Senegalese forces intervene in The Gambia
1989	Border clashes between Senegal and its northern neighbor Mauritania

ESSENTIAL STATISTICS

Capital Dakar

Population (1989) 7,400,000

Area 196,722 sq km

Population per sq km (1989) 37.6

GNP per capita (1987) US$510

CONSTITUTIONAL DATA

Constitution Multiparty republic with one legislative house (National Assembly)

Date of independence 20 August 1960

Major international organizations UN; OAU; ACP; ECOWAS; IDB

Monetary unit 1 CFA franc (CFAF)= 100 centimes

Official language French

Major religion Sunnite Muslim

Heads of government since independence (President) Léopold Sédar Senghor (1960–1979); Abdou Diouf (1980–)

Senegal, as the earliest French colony in Africa and the center of the West African Federation in colonial times, combined an abiding influence of the three great Muslim *tariqas* (brotherhoods), the Mouride, the Tijania and the Quadariya, with an intense devotion to French language and culture among the elite. The four communes of those granted French citizenship, Dakar, Gore'e, Rufrique and Saint-Louis, enjoyed direct representation in the French parliament, which was denied to the rest of Senegal until 1946. In 1917, Blaise Diagne, a black Senegalais, was elected to the French parliament and exercised great power. Senegalese soldiers served with distinction in the French army, especially in France, during World War I.

After World War II, Senegalese aspirations found their expression at first in the *Bloc Africaine*, led by Lamine Gueye, based on the support of the four communes, but in 1948, Lamine Gueye's disciple, Léopold Sédar Senghor, broke away, formed the BDS and appealed to the hinterland majority, easily winning the 1951 election to the French parliament. Senghor was also supported by the other major tribes, the Wolof, who constitute a third of Senegal, the Tukulor, the Bambara, the Mandingo and the Fulani (Peule). Senghor learnt to bargain in the sophisticated world of French politics in Paris, and unlike other African leaders, he allied himself to the old Socialist party (SFIO) rather than to the Communists. He is a considerable poet and the philosopher of the Negritude movement but retained an immense pride in the French language.

In 1958, Senegal voted to accept President de Gaulle's proposal of membership of the French community rather than independence, but persuaded de Gaulle, on his visit to Dakar, to grant independence in 1960. In 1962 Mamadou Dia, the prime minister, who was more radical than Senghor, was accused of illegality and imprisoned for life. Under the subsequent constitution, power was concentrated in the office of president and when a prime minister was finally appointed, he was clearly only a lieutenant. The ruling party achieved a virtual monopoly of power by merging with some opposition groups and by marginalizing others from government patronage. In 1976, Senghor formally instituted a three-party system by allowing opposition on right and left, but the ruling BDS continued to win the vast majority of the votes.

In 1979, Senghor retired as president but continued to be regarded with respect, and was succeeded in January 1980 by Abdou Diouf, who continued the policy of liberalization allowing for a full multiparty system and achieved his own political stature. Nonetheless, there were serious riots in the university and elsewhere, led by radical groups.

The excellent relations between Senegal and France until the late 1980s assured Senegal of an enhanced price for its crop of groundnuts in the European market. Subsequently Senegal's income declined, causing some political discontent. Senegal's foreign policy was based on a black interpretation of pan-Africanism and resulted in indifferent relations with its northern neighbor, Mauritania, which was disturbed by conflict between its black and white inhabitants. Relations with The Gambia were dominated by Senegal's desire to unite the two countries. But Senegambia has failed to become a political reality, although it existed loosely from 1982 to 1989. Dakar remained a center of a highly sophisticated African and Francophone culture, although some of the new elites were critical of what they regarded as overclose ties to France.

BENIN

(People's Republic of Benin) **Capital** Porto Novo **Population** (1989) 4,593,000 **Area** 112,622 sq km **GNP per capita** (1987) US $300 **International organizations** UN, OAU, ACP, ECOWAS, ICO, NAM, IDB, CEAO

Benin, previously Dahomey, gained independence from France in 1960, but the various tribes did not lie easy as one nation, and by 1972 there had been six coups. Major Mathieu Kerekou announced a Marxist-Leninist regime in 1974. In 1975 the state was renamed Benin, after an ancient African kingdom, and in 1976 the PRPB (Revolutionary party) was formed. In 1977 a National Revolutionary Assembly was set up with directly elected members (representing social classes), who would elect a president. In 1979 Kerekou (the sole candidate) was reelected. In 1986, the country's foreign loan debt was rescheduled. Kerekou was again reelected in 1989, continuing to resist the introduction of a multiparty system. In early 1990 the National Conference of Active Forces of the Nation took control and set up an interim government; in 1991 multiparty elections were conducted peacefully. This was the first time on the African mainland that a head of state had been removed democratically.

BOTSWANA

(Republic of Botswana) **Capital** Gaborone **Population** (1988) 1,211,000 **Area** 47,000 sq km **GNP per capita** (1986) US$840 **International organizations** UN, Commonwealth, OAU, ACP, SADCC.

Botswana (then Bechuanaland) began the 20th century as a British protectorate, administered by a resident commissioner with an advisory council of chiefs until 1965, when the Bechuanaland Democratic party won the first general election and led the country to independence in 1966 under the presidency of Seretse Khama. Botswana was among the founding members of the Southern African Development Coordination Conference, created to counter South Africa's dominance in the area. A haven for Rhodesian refugees and African National Congress (ANC) guerrillas, Botswana, because of the latter, was vulnerable to South African military incursions. This led in 1977 to the creation of a permanent border patrol. In 1980 Khama died and fellow-Democrat Dr. Quett Masire became president; he was reelected in 1984 and 1989.

BURKINA FASO

(Burkina Faso) **Capital** Ouagadougou **Population** (1988) 8,530,000 **Area** 274,200 sq km **GNP per capita** (1986) US$150 **International organizations** UN, OAU, ACP, ECOWAS, IDB, CEAO.

Burkina was first colonized by the French as part of Upper Senegal-Niger (so called 1904–20) in the last decade of the 19th century. The dominant Gourma tribe accepted French protectorate status in 1897. In 1919 the French made the country a separate colony, Upper Volta. In 1932 this was partitioned between Ivory Coast, Niger and French Sudan, but reformed in 1947 to become a French overseas territory. In 1960 independence was proclaimed. The military have overshadowed the political life of Burkina Faso (so called from 1984), staging five coups by 1983, when Captain Thomas Sankara installed a radical socialist regime; he was murdered in 1987 after proposing one-party rule. In 1990 President Compaoré's Popular Front installed a new constitution providing for a multiparty system and direct presidential election.

BURUNDI

(Republic of Burundi) **Capital** Bujumbura **Population** (1988) 5,131,000 **Area** 27,834 sq km **GNP per capita** (1986) US$240. **International organizations** UN, OAU, ACP.

Burundi has the densest population in Africa, consisting of 85 percent Hutu and less than 15 percent Tutsi ethnic groups. Burundi was part of German East Africa from 1890, then from 1919 administered, as part of Rwanda-Urundi, by Belgium. Independence came in 1962. In 1965 the elected Tutsi premier was assassinated in an attempted Hutu coup, answered by brutally repressive measures. In 1966 the king was deposed and Burundi was proclaimed a republic. In 1972 after a Hutu uprising, virtually the entire educated Hutu population was eliminated in Tutsi reprisals. In the late 1980s the country was still largely dominated by the Tutsi. In 1987 the Tutsi military leadership ousted President Bagaza, and Major Pierre Buyoya became president. "Black apartheid" continued and in 1988 a number of massacres, took place on both sides. Tension lessened during 1989, and 43,000 Hutu refugees returned from Rwanda.

CAMEROON

(Republic of Cameroon) **Capital** Yaoundé **Population** (1988) 11,206,000 **Area** 465,458 sq km **GNP per capita** (1986) US$910 **International organizations** UN, OAU, ACP, IDB.

After World War I Britain and France divided Cameroon, previously a German colony, between them. The British sector was integrated into Nigeria, remaining so until 1961, when after a plebiscite its southern part was reunified with French Cameroon. The French sector was reabsorbed into French Equatorial Africa, from which Germany had taken it in 1911, and joined the Free French Movement in 1940; for a time the town of Douala was the headquarters of General de Gaulle. In 1960, it became the Republic of Cameroon, further enlarged the following year by the absorption of southern British Cameroon. The country became a federal republic until 1972 and the formation of the United Republic of Cameroon. The first president, Ahmadou Ahijdo, resigned in 1982 and appointed Paul Biya as his successor. Biya resisted multiparty rule until 1990 after violent demonstrations and in an economic crisis, when he passed the necessary legislation.

CAPE VERDE

(Republic of Cape Verde) **Capital** Praia **Population** (1988) 359,000 **Area**: 4,033 sq km **GNP per capita** (1986) US$460 **International organizations** UN, OAU, ACP, ECOWAS.

In 1951 Cape Verde became a Portuguese overseas province, with all citizens granted Portuguese citizenship in 1961. In 1975 it became an independent republic as a one-party state under president Aristides Pereira, secretary-general of the African Party for the Independence of Guinea and Cape Verde (PAIGC) since 1973. In 1980 Cape Verde withdrew from PAIGC and in 1981 formed the African Party for the Independence of Cape Verde (PAICV); but in 1988 relations between Cape Verde and Guinea were restored. In 1990 direct presidential election and a multiparty system were introduced; in 1991 Carlos Viega (Democratic Movement) was elected president.

CENTRAL AFRICAN REPUBLIC

(Central African Republic) **Capital** Bangui **Population** (1988) 2,843,000 **Area** 622,436 sq km **GNP per capita** (1986) US$290 **International organizations** UN, OAU, ACP.

Conflicting French and British interests in this area were resolved in 1899. The French began a systematic campaign against the slave-trading sultans, and commercial exploitation of the colony. In 1946 it became a French overseas territory, dominated until independence in 1960 by Barthélemy Boganda's Movement for Black African Social Evolution (MESAN). Boganda was succeeded in 1959 by David Dacko, who was overthrown in 1965 by Jean-Bédel Bokassa, who made himself president in 1972, transferred all power to his cabinet and in 1976 proclaimed himself "emperor", as Bokassa I. His extravagance and bloody repression aggravated the country's economic problems and brought condemnation from the international community. In 1979 he was removed by the French government, who returned Dacko to power, though he was ousted in 1981 by Gen. André Kolingba. In 1986 Kolingba was granted a six-year mandate as head of state by referendum. In 1991 he reluctantly promised a multiparty system, to be established speedily.

CHAD

(Republic of Chad) **Capital** N'Djamena **Population** (1988) 5,395,000 **Area** 1,284,000 sq km **GNP per capita** (1986) US$160 **Internal organizations** UN, OAU, ACP, IDB.

Chad became a part of French Equatorial Africa in 1908, and in 1946 a French overseas territory. Independence came in 1960. In 1963 a state of emergency was proclaimed due to a suspected Muslim conspiracy, initiating a period of one-party (PPT) government. In the mid-1960s, a civil war was incited by two antigovernment and anti-French groups, in an attempt to align Chad more with neighboring North African Arab states. French forces intervened from 1968; the war continued throughout the 1970s. In 1973 Libya occupied the Aozou strip in north Chad, and Libyan troops intervened in 1980 in the civil war but were withdrawn in 1981, whereupon the Forces Armées du Nord reoccupied all the major towns in eastern Chad. In 1982 two governments were formed, one under Hissène Habré (backed by France), the other under President Goukouni Queddei (backed by Libya). In 1987 Goukouni acknowledged Habré as official head of state; in late 1990 Habré was overthrown by the rebel MPS, whose leader, Colonel Idriss Deby, as president promised liberalization and free elections; but in 1991 he delayed the process.

COMOROS

(Federal Island Republic of the Comoros) **Capital** Moroni **Population** (1989) 448,000 **Area** 1795 sq km (without Mayotte) **GNP per capita** (1988) US$440 **International organizations** UN,OAU, ACP, IDB, PTA, IOC

In 1843 the French took Mayotte, and the other Comoro Islands in 1886, placing them under Madagascar's rule until its independence in 1960, when they attained internal autonomy. In 1975 all except Mayotte became independent after a referendum. In 1978, after a coup by European mercenaries on behalf of the exiled ex-president Abdallah, the Comoros was declared a Federal Islamic Republic; the continued presence of the mercenaries brought about expulsion from the OAU. Abdallah was reelected president in 1984 and in 1989 he amended the constitution to allow himself to serve a third term. After unrest, he was killed within the month; mercenary leader Bob Denard took over and disarmed the military. In December, after French military pressure, Denard left and in 1990 the interim president Djohar was properly elected at the first multiparty elections, only to be ousted in 1991 by Ibrahim Haled, head of the supreme court.

DJIBOUTI

(Republic of Djibouti) **Capital** Djibouti **Population** (1988) 484,000 **Area** 23,200 sq km **GNP per capita** (1984) US$740 **International organizations** UN, LAS, OAU, ACP, IDB.

The Côte Française des Somaliland Protectorate was defined in 1888, with Djibouti its capital. In 1940 the colony aligned itself with the Vichy regime and was blockaded by the Allies until 1942 when it declared itself for the Free French Movement. After World War II, it federated with Ethiopia to maintain its economy. In 1958 French Somaliland voted to become a French overseas territory. A dispute arose between the Somalis, who wanted complete independence, and the Afars, who wanted to maintain the tie with France. In 1977 independence was proclaimed; but the dispute continued; and the economy was strained by an influx of Ethiopian refugees. In 1981, in the first direct presidential election, Hassan Gouled was reelected, and again in 1987.

EQUATORIAL GUINEA

(Republic of Equatorial Guinea) **Capital** Malabo **Population** (1988) 335,000 **Area** 28,051 sq km **GNP per capita** (1987) US$330 **International organizations.** UN, OAU, ACP.

Equatorial Guinea was called Formosa ("beautiful") by the 15th-century Portuguese. In 1778 they ceded Spain two islands and rights on the mainland coast for the slave trade. In 1960, Spanish Guinea was reorganized into two overseas provinces of Spain. Equatorial Guinea (as it was now called) won independence in 1968 under president Francisco Macia Nguema, who was executed during a military takeover in 1979. A new constitution was promulgated in 1982, and the new president Obiang Nguema repaired relations with Spain and the USA, while reducing ties with the Soviet Union.

GABON

(Republic of Gabon) **Capital** Libreville **Population** (1989)1,245,000 **Area** 267,667 sq km **GNP per capita** (1986) US$3,020 **International organizations** UN, OAU, ACP, IDB, OPEC.

Gabon was attached to the French Congo from 1897 to 1910, then to French Equatorial Africa. In 1946 it became a French overseas territory, finally becoming independent in 1960. The dominant figure in Gabonese politics until 1967 was Leon M'ba, elected president in 1961, whose pro-French policies were hated. In 1968 Gabon became a one-party state. On M'ba's death, Albert Bernard (Omar) Bongo was made president, and sustained the close relations with France as well as the economic policies that had made Gabon prosperous. An opposition group, the Movement for National Realignment (MORENA) was set up in the early 1980s, accusing Bongo of corruption and demanding a multiparty democracy. In 1985 the party leadership formed a government-in-exile in Paris. Reelected in 1986, Bongo legalized multiparty government, after widespread unrest, in 1990. The opposition party leader was found poisoned shortly after; and Bongo won the subsequent elections.

GAMBIA

(Republic of the Gambia) **Capital** Banjul **Population** (1988) 811,000 **Area** 10,689 sq km **GNP per capita** (1986) US$230 **International organizations** UN, Commonwealth, OAU, ACP, ECOWAS, IDB.

A British colony from 1888, Gambia became a staging post on air routes to the Middle and Far East during World War II. In 1965 the territory became independent, and a republic in 1970. The first president, Sir Dawda Jawara, survived an attempted coup to be reelected in 1982. In the same year the confederation of Senegal and Gambia was approved, forming Senegambia. A parliament was created, and economic and military resources were merged, though each state preserved its independence. In 1989 the confederation was dissolved.

GUINEA

(Republic of Guinea) **Capital** Conakry **Population** (1989) 6,705,000 **Area** 245,857 sq km **GNP per capita** (1986) US$240 **International organizations** UN, OAU, ECOWAS, ACP, IDB, ICO, NAM

In 1958 Guinea gained its independence from France under Sekou Touré, who severed relations with France and conducted an increasingly repressive rule. Touré held frequent purges of suspected pro-French conspirators; but Guinea and France were reconciled in the late 1970s. After Touré's death in 1984, a military coup dissolved the National Assembly and the new president, Colonel L. Conté, began to dismantle the central planning system. President Conté received World Bank and IMF aid after agreeing in late 1985 to devolve power regionally and bring civilians into government. Growing unrest culminated in 1989 in the start of a transfer to civilian rule; and the new constitution of 1990 prescribed a 5-year transition period to a two-party system.

GUINEA-BISSAU

(Republic of Guinea-Bissau) **Capital** Bissau **Population** (1989) 953,000 **Area** 36,125 sq km **GNP per capita** (1987) US$160 **International organizations** UN, OAU, ACP, ECOWAS, NAM.

Until 1879 Guinea-Bissau was administered by Portugal with Cape Verde. Portugal settled border disputes with France by 1905, and had imposed its rule, often violently, by 1915. 1956 saw the formation of the African Independence party (PAIGC) which had overall control by 1972. After considerable conflict, independence was declared in 1973 and accepted by Portugal in 1974. In 1977 regional councils were elected and a National People's Assembly (NPA) chosen from them. Attempts at unity with Cape Verde were made until 1980 when Commander Vieira took over with a coup and dissolved the NPA. In 1988 a cooperation agreement was signed with Cape Verde, and in 1989, a treaty with Egypt. In January 1991 PAIGC approved the introduction of a multiparty system.

MADAGASCAR

(Democratic Republic of Madagascar) **Capital** Antananarivo **Population** (1989) 11,602,000 **Area** 587,041 sq km **GNP per capita** (1987) US$200 **International organizations** UN, OAU, ACP, PTA, IOC.

Madagascar became a French protectorate in 1896. The populace is chiefly Indonesian in descent and spirit, and can be seen therefore as a bridge between Africa and Asia, and the government embraces this role. The French improved the infrastructure and generally turned things to their own advantage. In 1920 Madagasacar became a French overseas territory, with some Malagasays acquiring French citizenship in 1946. After 11 years of unrest it became autonomous in 1958. Independence came in 1960, and the new President Tsiranana introduced "grassroots socialism". In 1972 demonstrations forced Tsiranana to hand over power to the military. The new president, Maj.-Gen. Ramanantsoa, reduced economic dependency on France, establishing aid links with other countries in both Western and Eastern blocs. In 1975 Ramanantsoa resigned in favor of the radical Colonel Ratsimandrava, who was assassinated within a week. The next president, Didier Ratsiraka, set up a single-party system under the revolutionary party AREMA. Ratsiraka declared a socialist revolution; he cultivated relations with Arab and Communist countries and set out to restructure the domestic economy, leaving the market free. Despite two attempted coups, Ratsiraka was returned in 1982 for a second term as president, and again in 1989. He dismissed opposition government members, but in 1990 admitted a multiparty state; a coup ousted him in 1991.

MALI

(Republic of Mali) **Capital** Bamako **Population** (1989) 7,911,000 **Area**: 1,240,192 sq km **GNP per capita** (1988) US$230 **International organizations** UN, OAU, ACP, ECOWAS, IDB, ICO, NAM, CEAO.

From 1898 until 1960, Mali was part of French Sudan. In 1959 the name Mali was suggested for the federation of Senegal, the Sudanese Republic, Dahomey and Upper Volta. The two latter withdrew on the point of federation, and in August 1960 Senegal seceded. Under president Modibu Keita, all ties with France were cut and the republic of Mali declared. In 1965 Keita dissolved the National Assembly and purged possible opponents. He was overthrown by a military coup in 1968. The new president, Lieutenant Traoré, promised to restore civilian rule after overhauling the economy. A new constitution establishing a one-party state with interim military rule for five years was approved by referendum in 1974. Traoré was reelected in 1985 after manipulating the constitution. He tightened the economic reins at home, and turmoil ensued in 1988. In early 1991 rioting was answered with further repression. By the end of March, Traoré was overthrown. The new leader, Lieutenant Touré, promised multiparty government, but the violence continued.

MAURITANIA

(Islamic Republic of Mauritania) **Capital** Nouakchott **Population** (1989) 1,946,000 **Area** 1,030,700 sq km **GNP per capita** (1988) US$480 **International organizations** UN, OAU, ACP, LAS, ECOWAS, IDB, AMU, ICO, NAM, CEAO, UAM.

In 1903 Mauritania became a French protectorate and in 1920 part of French West Africa. Economic pressures forced the largely nomadic population to begin to settle in the 1960s, and the urban population grew. In 1964 President Daddah imposed a one-party state. In 1975–76, Spain ceded Western Sahara to Mauritania and Morocco, and fighting ensued between them, and also the Western Saharan guerrilla force, the Polisario. This struggle practically bankrupted the country, and in 1978 Daddah was deposed, his party banned, and a military leadership installed, which in 1979 suspended the constitution. They renounced Western Sahara and made peace with the Polisario. In 1984 Colonel Taya took power with a bloodless coup, introduced a reform program, and restored diplomatic relations with Morocco. Relations with Senegal, already poor, were suspended as a result of a bloody and prolonged dispute over border grazing rights. In 1989 Mauritania was a co-founder of the Arab Maghreb. In 1990 the FLAM (African front) accused the government of genocidal racism against blacks.

MAURITIUS

(Mauritius) **Capital** Port Louis Population (1989) 1,061,000 **Area** 2,040 sq km **GNP per capita** (1988) US$1,810 **International organizations** UN, Commonwealth, OAU, Nonaligned movement, ACP, NAM, PTA, IOC.

Mauritius gained independence from Britain as a member of the Commonwealth in 1968, a multiparty democracy as set out in the constitution of 1969 by the Labour (MLP) and Social Democrat (PMSD) coalition government under P.M.S.Ramgoolam (MLP leader). In 1973 the coalition collapsed, but the MLP formed a new ruling coalition with the CAM (Muslims). In 1976 Ramgoolam formed another coalition government, including the PMSD; but in 1982 the leftwing MMM, in alliance with the Socialists (PSM) won all 60 contested seats. The two ruling parties soon fell out and PSM prime minister A. Jugnauth expelled the MMM, which became the official opposition party, and set up the Militant Socialists (MSM). In 1986 under pressure from MSM and MMM Jugnauth set up a commission to investigate allegations of drug dealing on the part of government members. In 1990 Jugnauth did not get enough support to make Mauritius a republic, and a new coalition was formed.

NIGER

(Republic of Niger) **Capital** Niamey **Population** (1989) 7,523,000 **Area** 1,189,000 sq km **GNP per capita** (1987) US$280 **International organizations** UN, OAU, ACP, ECOWAS, IDB, CEAO

In 1890 Niger was divided between the British and the French, who subdued the militant Tuareg nomads. In 1922 it became part of French West Africa, and in 1946 a French Overseas Territory. Independence, in 1960, did not entail economic emancipation from France. In 1974 President Diori was ousted, and Colonel Kountché became head of state, banning all political groups and suspending the constitution. He made efforts to effect economic recovery after the drought of 1968–74; he reduced Niger's dependence on France, and cultivated Arab connections; but with another drought in 1984–85, Niger's dependency on foreign aid was increased. In 1987 the prime minister, Colonel Ali Seybou, became head of state on Kountché's death. The new constitution of 1989 allowed a single ruling party, MNSD, and a directly elected president and National Assembly. Seybou, the only candidate, was reelected. 1990 saw a recurrence of the student demonstrations which had forced the constitutional changes of 1989, this time against austerity measures. Three ministers were dismissed for provoking the uproar, and for ordering the deaths of several students. Later that year President Seybou announced the introduction of a multiparty democracy.

RWANDA

(Republic of Rwanda) **Capital** Kigali **Population** (1988) 6,709,000 **Area** 26,338 sq km **GNP per capita** (1986) US$290 **International organizations** UN, OAU, ACP.

Tutsi people from Ethiopia subdued the Hutu and founded the kingdoms Rwanda and Urundi, which fell into Belgian hands during World War I. The territory was in 1925 linked to the Belgian Congo by an administrative union. Within Rwanda-Urundi each native kingdom remained separate. In 1959 a tribal war was followed by the expulsion of many Tutsi. Hutu parties won the 1960 communal elections in both kingdoms. In 1961 an internal coup took place and Rwanda declared itself a republic. Elections in 1961 were won by the Hutu in Rwanda and by the Tutsi in Urundi. Both kingdoms rejected UN proposals to federate. Complete independence was declared in 1962 under president Kayibanda (a Hutu). In 1963 Tutsi refugees invaded the country in an attempt to restore the monarchy, and thousands were killed in Hutu reprisals. A resurgence of anti-Tutsi feelings and violence led to a military coup in 1973. A new party (MNRD) was formed under a new constitution adopted in 1978, and centralized administration introduced. Civilian rule was reintroduced in 1981; MNRD leader Juvenal Habyarimana was the first elected president. In 1990 he announced a move to multiparty rule.

SAO TOME AND PRINCIPE

(Democratic Republic of São Tomé and Príncipe) **Capital** São Tomé **Population** (1989) 118,000 **Area** 963 sq km **GNP per capita** (1987) US$280 **International organizations** UN, OAU, ACP.

By 1900 the cocoa trade flourished in these islands under harsh Portuguese rule until autonomy was granted in 1973. São Tomé and Príncipe was granted independence in 1975, with President da Costa head of a one-party state. Droughts and price drops throughout the 1980s brought about a sharp decline in the cocoa-based economy. The government declared nonaligned status in 1984, opening the door to Western aid. In 1986 expatriate opposition groups demanded democratic elections. In São Tomé and Príncipe, the FRNSTP (National Resistance Front) sought South Africa's help against the government, which introduced direct presidential election, albeit with only one candidate. Two people died in 1988 in a major coup attempt by FRNSTP dissidents. In 1990 a multiparty system was introduced. In January 1991 the government resigned before release of the results of an election in which the Democratic Convergence Party beat them.

SEYCHELLES

(Republic of Seychelles) **Capital** Victoria **Population** (1989) 67,100 **Area** 455 sq km **GNP per capita** (1988) US$3590 **International organizations** UN, Commonwealth, OAU, ACP.

In 1903 the Seychelles became a separate British colony. Full voting rights were granted in 1967, and the Democrats (SDR) won the 1970 and 1974 elections. Independence followed in 1976, and with it the restoration of several islands which had been part of the British Indian Ocean Territory. The Democratic element of the coalition government was ousted by a coup in 1977, and President Mancham replaced by the socialist prime minister, France René, who set up a series of five-year plans, to reform Seychellois society by the 1980s. A new constitution of 1979 established a one-party state, under the new Progressive Front (SPPF, formerly SPUP); René, the sole candidate, was reelected then, in 1984 and, despite repeated coup attempts, in 1989.

SIERRA LEONE

(Republic of Sierra Leone) **Capital** Freetown **Population** (1989) 3,957,000 **Area** 71,740 sq km **GNP per capita** (1988) US$240 **International organizations** UN, OAU, ECOWAS, Commonwealth, ACP, ADB, IDB.

Sierra Leone, named in the 15th century by a Portuguese sailor after the hills around Freetown (= lion mountain), became a British protectorate in 1896. In 1961 it gained independence under prime minister Sir Milton Margai, with an opposition party (APC) already in place. In 1967 a military government took power until 1968, when the lower ranks mutinied and restored parliament, with a coalition government headed by Siaka Stevens. In 1971 Stevens declared a republic, with himself as president. From 1981 there was constant violent unrest; and in 1985 Stevens resigned and Major General Joseph Momoh took over. Inflation continued to soar, an economic reform program failed, and from 1987 to 1989 Momoh declared an economic emergency. The IMF withdrew its support in 1988 because of debt repayment arrears. In 1990 public pressure for a multiparty system mounted and a commission was set up to review the constitution.

SOMALIA

(Somali Democratic Republic) **Capital** Mogadishu **Population** (1988) 6,334,000 **Area** 637,000 sq km **GNP per capita** (1986) US$280 **International organizations** UN, LAS, OAU, ACP, IDB.

Somalia entered the 20th century divided into British (northern) Somaliland and Italian (southern and northeastern) Somaliland. British rule was challenged by Sayyid Mohammad, who had united Muslims against the colonizers and held power in the interior until his death in 1920. Settlement was encouraged in Italian Somaliland, however, and in 1936 the colony was incorporated into the Italian East Africa empire by Mussolini. In 1940 the Italians invaded British Somaliland, though the area was retaken by the British a year later, and they administered both British and Italian Somaliland until 1950, when Italian Somaliland was made a UN trust territory administered by Italy. In 1960 both areas were united to form the Independent Republic of Somalia, and in 1967 Abdirashid Ali Shermarke became president, with Mohammad Haji Ibrahim Egal as prime minister. The former was assassinated in 1969, and the latter was toppled in a coup by Maj. Gen. Mohammad Siyad Barre, who set up a Supreme Revolutionary Council of military officers to form the Somali Democratic Republic. He established ties with the Soviet Union, but they were broken when the latter supported Ethiopia against Somalia's attempt in 1977 to annex its Ogaden region (whose inhabitants were almost all ethnic Somalis); and the USA became Somalia's ally. In 1988 Barre reached accord with Ethiopia's Lt-Col Mengistu over Ogaden; but civil war erupted; and after much bloodshed and repression Barre promised multiparty elections in 1990. After months of internecine fighting, Barre was ousted and Ali Mahdi Mohammed sworn in in January 1991. He was not universally accepted, and four months later northeast Somalia seceded to form the "Somaliland Republic", with multiparty elections scheduled for 1993.

SWAZILAND

(Swaziland) **Capital** (administrative) Mbabane; (royal and legislative) Lobamba **Population** (1988) 716,000 **Area** 17,364 sq km **GNP per capita**(1986) US$600 **Intetnational organizations** UN, Commonwealth, OAU, ACP, SADCC.

Swaziland was administered by Britain from the end of the South African (Boer) war. In 1967 the country became a protected state with the title of the Kingdom of Swaziland, and in 1968 became fully independent. In 1973 King Sobhuza II restored the old system of government, with power concentrated in the royal capital and villages administered by local government, the *tinkundlu*. Sobhuza tried to stay friendly with both neighbors, Marxist Mozambique and apartheid-ruled South Africa, though he opposed both regimes. Swaziland enjoyed stability and some economic growth under Sobhuza. After his death in 1982 a power struggle ensued between the queen mother and her supporters, and those of the teenage heir Prince Makhosetive; the Prince was crowned King Mswati II in 1986. In 1990, following worker/student unrest, the king undertook to set up an *indaba* (popular parliament), as a forum for complaints.

TOGO

(Republic of Togo) **Capital** Lomé **Population** (1989622,000 **Area** 56,785 sq km **GNP per capita** (1988) US$370 **International organizations** UN, OAU, ECOWAS, ACP, NAM

In 1884 German protection was foisted on Togo's native chiefs. After World War I Britain and France shared its administration,and by the end of World War II French Togoland was part of French West Africa. In 1956 British Togoland seceded to Ghana and French Togoland gained autonomy, and independence in 1960. Togo resisted Ghana's overtures for integration; and a trade embargo followed. In 1963 President Olympio was assassinated and a provisional civilian government was endorsed in elections. In 1966 relations with Ghana improved post-Nkrumah. In 1967 General Eyadema took power in a military coup. He voided the constitution and banned political parties. In 1969 he introduced a single-party system under the People's Assembly (RPT), and in 1980 he produced a new constitution. Terrorism mushroomed, and the economy was damaged by low cocoa and coffee prices; but Eyadema was reelected in 1986. In 1990 after prodemocracy rioting Eyadema announced yet another constitution, with the transition to a multiparty system to be completed in 1991. The people did not trust him, and in 1991 rioting flared again.

▲ **The great mosque at Djenne, Mali.**

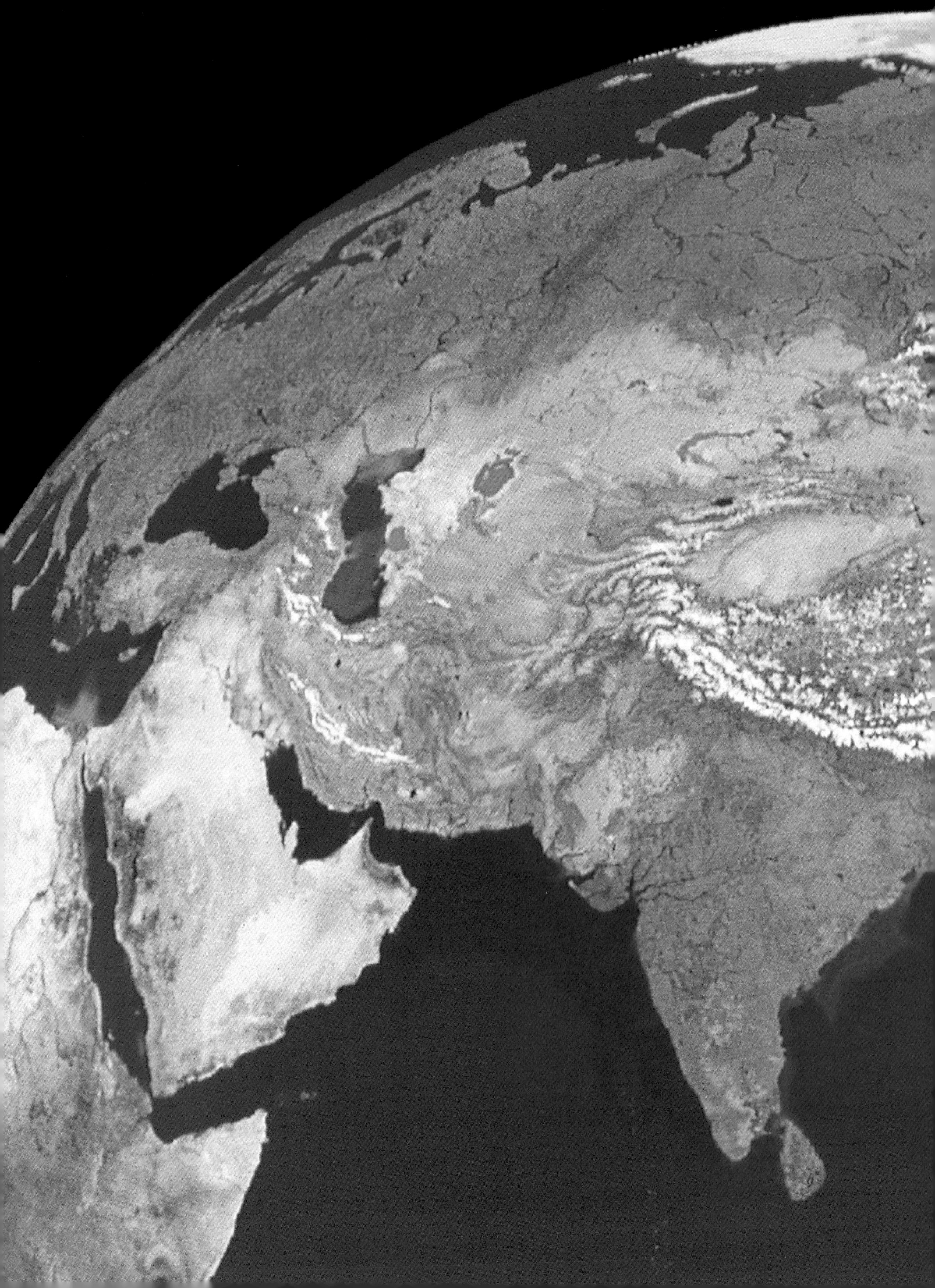

WEST AND SOUTH ASIA

INDEX OF COUNTRIES

TURKEY

REPUBLIC OF TURKEY

The Republic of Turkey, proclaimed on 29 October 1923, was heir to the Ottoman Empire which finally collapsed in World War I. But Greek troops landing at Izmir in 1919 provoked a nationalist movement in Turkey led by Kemal Atatürk.

Atatürk imposed a secular revolution on Turkey, transferring the capital from Istanbul to Ankara. The caliphate and Muslim religion were excluded from public life. A new, European-derived legal code replaced Islamic and Ottoman law. From 1934 women were allowed to vote and stand for election. The Gregorian calendar was introduced. In 1928 the Latin alphabet was inculcated throughout the country. The constitution provided for a single legislative chamber, which elected the president, who appointed the prime minister. In 1924 Atatürk organized his supporters into the Republican People's party (RPP). The assembly had real powers, but the government was a benevolent dictatorship.

The territorial settlement of 1923 had left Turkey in control of the Dardanelles, which were to be open to ships of all nations unless Turkey was at war. In World War II Turkey remained neutral until February 1945, when, by declaring war on Germany, Turkey won the right to be a founder-member of the United Nations.

Competitive party politics were introduced in postwar Turkey by the formation of the Democrat party (DP) to fight the 1946 elections. A new, fairer electoral law in 1950 gave the DP 55 percent of the votes and it formed a government under Adnan Menderes. In 1947 Turkey was offered economic and military support under the Truman doctrine for the containment of Communism. For a few years increased agricultural production and low inflation stimulated industrial production.

CHRONOLOGY

1919	War of independence
1923	Declaration of independence
1924	Benevolent dictatorship under Kemal Atatürk
1946	Formation of the Democrat Party (DP) under Mendris leads to competitive politics
1960	Military coup dissolves DP party; Mendris executed
1961	Return to civilian rule
1974	Turkish troops invade Cyprus following military coup
1980	Political violence leads to military rule
1983	Return to civilian rule – Motherland party wins election
1987	Turkey applies to join European Community

ESSENTIAL STATISTICS

Capital Ankara

Population (1989) 55, 541,000

Area 779,452 sq km

Population per sq km (1989) 71.3

GNP per capita (1987) US$1,200

CONSTITUTIONAL DATA

Constitution Multiparty republic with one legislative house (Turkish Grand National Assembly)

Date of independence 29 October 1923

Major international organizations UN; EEC; IDB; NATO

Monetary unit 1 Turkish lira (LT) = 100 kurush

Official language Turkish

Major religion Sunnite Muslim

Heads of government since independence (President) M. Kemal Atatürk (1923–38); Gen. I. Inönü (1938–50); M. Bayar (1950–60); Gen. C. Gürsel (1961–66); Gen. C. Sunay (1966–73); Adm. F. Korutürk (1973–80); (Military ruler) Gen. K. Evren (1980–82); (President) Gen. K. Evren (1982–89); T. Özal (1989–)

In the Atatürk tradition the army did not itself govern, but ensured what it saw as orderly government. After the 1960 coup the DP was dissolved and Menderes tried and executed. Then civilian government was restored under a new constitution. A new party, the Justice party (JP), succeeded the DP. In 1965 it won a majority and formed a government under Suleyman Demirel. The economy grew, agricultural products were marketed into Europe, the industrial sector expanded. Foreign aid continued, since Turkey, which belonged to both NATO and CENTO, was a key member of the Western alliance. But population growth and urbanization were rapid and created an urban poor; industrialization and a high rate of inflation produced a militant trade union movement. The military establishment was a target for the left; Turkey's Western commitment a target for the Islamic right.

In 1971 the high command intervened again, dictating the formation of a "strong and credible government". Tension between the military and civilians made the experiment unworkable and in 1973 fresh elections were held. Leftwing and Islamic parties were forbidden and the two major parties, the RPP and the JP, increasingly resembled West European parties. The RPP, led from 1972 by Bulent Ecevit, favored a corporatist social democracy; the JP clearly favored the private sector and received peasant support. The balance was held by the National Salvation party, which stood for the re-Islamization of Turkish life. It joined an Ecevit government in 1974. It was this government which responded to a coup by Greek army officers in Cyprus, where there was a substantial Turkish population, by invading the island and establishing the Federated Government of North Cyprus.

Armenians

The Armenians are a people with an enduring sense of nationalism who in modern times have only enjoyed statehood for a brief moment of time. The sense of nationhood derives from history, language and religion. The Armenians trace their history to the collapse of the ancient kingdom of Urartu in the 7th century BC. The Armenian language is an independent language, with its own script. There is a substantial classical literature deriving both from Russian and Turkish centers of Armenian culture. The history of the Armenian church begins with the conversion of King Tiridates in AD 300.

When the 20th century opened, the historic territory of Armenia was ruled by czarist Russia and Ottoman Turkey. Armenians experienced a historical renaissance but its effect was to alarm the decaying Russian and Turkish governments. The Russian government confiscated Armenian church property in 1903. In 1895 the Turkish government massacred thousands of Armenians and in 1915 deported the entire population, causing the death of hundreds of thousands.

At the end of World War I the Allies sponsored the creation of an Armenian state, although the United States refused the offer of a mandate. The state was recognized by the Treaty of Sèvres (1920) but was destroyed by Russian and Turkish armies. Armenia, with its capital at Erevan, became part of the Soviet Union, from 1936 as the Armenian Soviet Socialist Republic. Fifty years later Armenian nationalism reasserted itself and in 1990 Armenia declared its independence.

Ecevit and Demirel governments alternated but the electoral system gave neither a clear majority. The economy, dependent on imported oil, faltered. Political violence returned, with 3,500 deaths in 1980, when the military intervened again.

The return to civilian rule in 1983, under a new constitution, promised a new period of stability. Once again the old parties and politicians were banned, to be restored in 1987. More important, a new political party called the Motherland party emerged under the leadership of Turgut Ozal. Having formed a government in 1983, he won an increased majority in 1987.

The new government abandoned the Atatürk tradition, initiating a major program of privatization of basic industries and developing water resources in east Turkey, building the massive Atatürk dam. In 1987 Turkey made formal application to join the European Community (EC).

The political order remained fragile. In 1989 Ozal was elected president – the first civilian to become head of state. But his Motherland party had lost its electoral popularity and was divided by his election. The bid to join Europe would take time and require major adjustments if it came about. At the same time the economy was profoundly affected by events in the east. The liberal democracy of 1983 had not resolved the issue of the Kurdish nationality, which became increasingly urgent as the Turks sought to suppress Kurdish culture and institutions. Meanwhile, popular attachment to Islam, never removed by the Atatürk revolution, was regenerated in Turkey as in the whole Muslim world.

◀ **German battleships in Istanbul in 1914.**

SYRIA

SYRIAN ARAB REPUBLIC

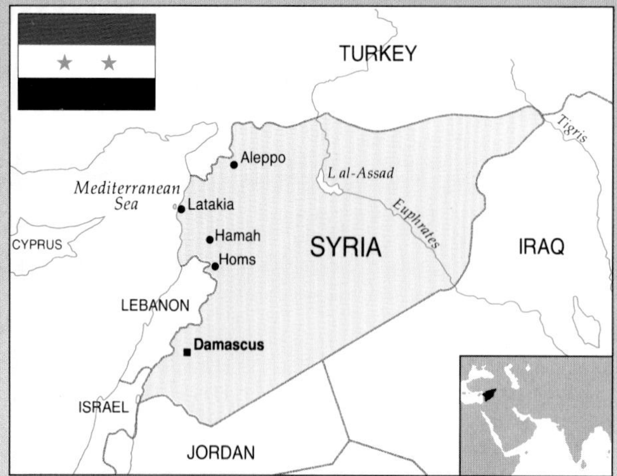

CHRONOLOGY

1920	League of Nations mandate given to France
1936	Franco–Syrian Treaty negotiated
1946	Independence achieved under President Quwatli of the National party
1949	First military coup in the Arab world
1958	Syria merges with Egypt to form the United Arab Republic (UAR)
1961	Syrian army officers break up UAR
1970	Coup led by general Hafiz al-Asad

ESSENTIAL STATISTICS

Capital Damascus

Population (1989) 11,719,000

Area 185,180 sq km

Population per sq km (1989) 63.3

GNP per capita (1987) US$1,820

CONSTITUTIONAL DATA

Constitution Unitary multiparty republic with one legislative house (People's Republic)

Date of independence 17 April 1946

Major international organizations UN; LAS; IDB

Monetary unit 1 Syrian Pound (LS) = 100 piastres

Official language Arabic

Major religion Muslim

Heads of government since independence (President) S. Shukri al Quwatli (1943–49); Col. H. Zaim (Jun–Aug 1949); (Military ruler) Col. A. es-Shishakli (1949–53); (President) A. es-Shishakli (1953–54); H. al-Atassi (1954–55); S. Shukri al Quwatli (1955–58); Col. G. Nasser (Egypt) (1958–61); N. el-Kudsi (1961–63); Maj.-Gen. L. Atassi (Mar–Apr 1963); Gen. A. el-Hafez (1963–66); N. Atassi (1966–70); A. Khatib (1970–71); Lt.-Gen. H. al-Asad (1971–)

Syria has no natural frontiers and historically it was at the crossing of routes from the Mediterranean to Mesopotamia and from Turkey to Arabia. It is the birthplace of Arab nationalism but its frontiers were drawn by the Allied Powers at the end of World War I.

Arab nationalism first found expression in Syria in opposition to the Ottoman Empire and grew in strength during World War I. Sharif Husain of Mecca was persuaded to lead the Arab revolt to Ottoman rule, by a British promise to "recognize and uphold the independence of the Arabs". The promise excluded the area of French interests, which Britain formally recognized in the Sykes-Picot agreement with France (1916). In March 1920 a Syrian Congress offered the crown of Syria and Palestine to Husain's son Faisal, who accepted. But the French army then occupied Damascus; the League of Nations allotted a mandate over Syria and Lebanon to France and the frontiers of the new states were then drawn in agreements with Britain and Turkey.

When France fell to the Germans in 1940, French authorities in Syria accepted the Vichy government. But effective authority came to be exercised by the British who occupied Syria and Lebanon, with the support of the Free French, to safeguard their military position in the Middle East. In 1941 General Catroux, on behalf of the Free French leader Charles de Gaulle, issued a declaration of Syrian independence, although it was limited by the necessities of war. It then became clear that de Gaulle's government wanted to link independence to a treaty safeguarding French interests, and would not transfer control of the army until this was achieved. But the British would not allow the French to suppress the nationalist movement and, by April 1946, all foreign troops were withdrawn.

Independence was thus achieved under President Quwatli leading a National party. By then their nationalism was being overtaken by more populist movements, of which the most important was the Ba'ath party founded by Michel Aflaq and Salah al-Din Bitah. The party's ideals were a mixture of European Marxism with Arab nationalism. It eventually took power in both Syria and Iraq, but not before it had fallen prey to the intense factionalism of Arab politics.

Before the Ba'ath took power Syria passed through a period of turbulence. The shortcomings revealed by the 1948 war with Israel led to the first military coup in the Arab world in March 1949; but the longest period of stable government was the dictatorship of Shishakli (1950–54). Syria became a prize pursued in competition between Jordan, Iraq and Egypt. The pressure was intensified when the British tried to link Jordan into their anti-Communist Baghdad Pact with Iraq and, when, after the Suez crisis between Egypt and Britain in 1956, there was pressure from the United States to adhere to the Eisenhower Doctrine, also aiming to resist the spread of Communist influence. A natural response for Syria (which had an important Communist party) was to tip toward the Soviet Union. As protection against these competing pressures, the Ba'athists promoted a union with Egypt and the United Arab Republic – made up of Egypt and Syria – was formed in 1958, but the union was run from Cairo and Syrian political parties were suppressed. In 1961 the union was broken by Syrian army officers. This was the prelude to three successive Ba'athist coups. Those of 1963 and 1966 had an important part in the origin of the 1967 Arab–Israeli war as the Syrian government

supported the liberationist movements of the Palestinians. The Syrians did not join the war until Jordan and Egypt had been defeated; against strong resistance the Israelis stormed and captured the Golan Heights, a strategically border area between the two countries.

The 1970 coup was led by General Hafiz al-Asad against the civilian government which had risked Syrian security by sending tanks to support the Palestinians in Jordan as they were crushed by King Hussein. The Asad regime remained in power for more than 20 years. Its core was a group of men who, like Asad, were Alawis from the Latakia region of Syria. Asad broadened his support by bringing Nasserists, Communists and independents into the government, but retained a highly personal system of authority with a tightly run security service. President Asad's greatest success was in defending Syria's international position. He negotiated the disengagement of Israeli and Syrian forces on the Golan sought to achieve "strategic parity" with Israel. In part this could be found (until the late 1980s) in Soviet support. In addition it called for a common negotiating position by Arab states. He was opposed therefore to the Camp David agreement of 1979 between Israel and Egypt, and resisted negotiations by Jordan and the Palestine Liberation Organization (PLO) to negotiate separately with Israel. He supported Iran in its war with Iraq while retaining Saudi support for his resistance to Israel, and continued to oppose Iraq in the 1990 Gulf crisis.

Economic and social trends

Traditionally Syria, part of the Fertile Crescent which stretches from the mouth of the Euphrates to Lebanon, was a major source of agricultural produce. Since World War II the economic potential of the land has not been fully realized and planning has failed to produce an efficient industry. Economic policy was socialist and state-directed, and periodic encouragement given to the private sector remained limited.

Oil is an important resource and Syria is a member of the Organization of Arab Petroleum Exporting Countries. A new find by Pecten-Shell revived declining production levels and in the 1980s refining capacity was expanded for the domestic use of oil. The economy supports a massive defense expenditure, consuming two-thirds of the budget. President Asad was adroit in securing aid not only from the Soviet Union but from Saudi Arabia and Iran.

Much was done to keep up a decent standard for the swiftly increasing population, which benefited from better housing, electricity and consumer durables. Industry and agriculture provided for many domestic needs, however inefficient their management. Crises arising from changes in the oil price and bad rainfall were somehow overcome. A vigorous black market and smuggling across the Lebanese border mitigated the constraints of a state-directed economy.

▲ **Daily life under the gaze of General Asad.**

LEBANON

REPUBLIC OF LEBANON

CHRONOLOGY

1916	Sykes-Picot agreement recognizes French interest in Lebanon
1920	League of Nations mandate given to France
1936	French Popular Front negotiate end of direct French rule
1943	Independence is asserted
1958	President Chamoun requests US intervention to restore government authority
1975	Outbreak of civil war
1978	Israel establish "security zone" in south Lebanon
1985	Israel withdraw troops
1989	Arab peacemaking mission fails

ESSENTIAL STATISTICS

Capital Beirut

Population (1989) 2,897,000

Area 10,230 sq km

Population per sq km (1989) 283.2

GNP per capita (1984) US$690

CONSTITUTIONAL DATA

Constitution Multiparty republic with one legislative house (National Assembly)

Date of independence 26 November 1941

Major international organizations UN; IDB; LAS

Monetary unit Lebanese pound (LL) = 100 piastres

Official language Arabic

Major religion Shiite Muslim (by unofficial 1984 census)

Heads of government (President) Bishara al-Khuri (1944–52); Camille Chamoun (1952–58); Gen. Fuad Shihab (1958–64); Charles Hélou (1964–70); Suleiman Franjieh (1970–76); Elias Sarkis (1976–82); Bashir Gemayel (Aug–Sep 1982); Amin Gemayel (1982–88); René Moawad (5–22 Nov 1989); Elias Hrawi (1989–)

The history of Lebanon owes much to its geography. The Lebanon Mountains with their deep valleys have provided a haven for sectarian communities which have coexisted while retaining their separateness, for the most part peaceably until 1975. At the same time the Mountains are physically continuous with neighboring Syria, while the coastal plain facing the Mediterranean connects Lebanon by land to the Arab world and by sea to Europe.

Like all the surrounding region, this area began the 20th century under Ottoman rule. The defeat of Turkey in World War I meant that the area was put in the hands of the victorious Allies. The state in its present borders came into existence under French mandate in 1920. It achieved independence in November 1943 and full sovereignty in 1946. But the major communities forming Lebanese society had been there for at least a thousand years. They are Maronite Christians, descendants of Christianized Arabs who emigrated from north Syria in the 6th century; Sunnite Muslims; Shi'ite Muslims who first sought refuge in Lebanon in the 7th century and Druze, a dissident Shi'ite sect who also looked for safety in the mountains after their formation in Cairo in 1017. Minority Christian sects include Greek Orthodox and Greek Catholic, and Armenians who fled the Turkish massacres of the late 1910s.

From the 18th century France acted to protect Catholics in Lebanon. Britain supported the Druze and Russia the Orthodox as counterweights, but the last major intervention in 1860–61 left France dominant. French interests were recognized by the British in the Sykes-Picot agreement (1916) and Britain and France combined to thwart the establishment of a Sharifian kingdom in Damascus. France was then given a mandate over Syria and "greater Lebanon", which now included the coastal cities, the Akkar plain, the Beqaa valley and Jabil Amil.

The structure of Lebanese government was formed under the mandate. The 1926 constitution established a republic with a single legislative assembly elected from districts that cut across sectarian boundaries. The assembly elected a president who appointed the prime minister. In 1936 the French Popular Front government negotiated the end of direct French rule in Lebanon. The treaty was never ratified in Paris, but it provoked anxiety in Lebanon amongst both Christians and Muslims as to which would dominate. The Maronite Kataeb (Phalange) party was formed with a paramilitary wing. But anxieties were allayed when the Maronite president Edde appointed a Sunni prime minister.

In 1942 this form of power-sharing was extended in an unofficial National Pact, which was agreed between the Maronite president Bishara al-Khuri and his Sunni prime minister Riyadh al-Sulh. Henceforth the speaker would be Shi'ite; the commander of the army would be Maronite, his chief of staff Druze; top posts in the administration would be shared proportionately between the sects; seats in the assembly would be divided six to five to the advantage of Christians.

Independence was asserted by the Lebanese parliament in 1943. It was resisted by the French who tried to regain their special military and economic position in 1945 until they were forced to concede by the British. Their withdrawal was completed in 1946. Lebanon was a founder member of the Arab League and joined the United Nations.

Lebanon at its best then developed as a harmonious and prosperous country. The Western-style constitution provided the framework within which power was held by local bosses (*za'im*) and politics conducted on a patron-client basis within each sect. Lebanese were well-known for their energy and acumen as traders and their readiness to establish businesses abroad. Business networks were linked to family and community ties. Arabs were ready to spend and invest in Lebanon. From 1959 the central bank disciplined the banking system and pursued a conservative policy as an entrepôt bank. The large degree of personal freedom and freedom of expression made Beirut an attractive center for Arab and Western interests, writers and artists.

But there was an underlying tension between Christian and Muslim communities which was sharpened by international politics. From 1952 President Chamoun took a very pro-Western position while Arab nationalism, rising to a crescendo with the Suez affair in 1956, pulled Muslims toward the Arab world. His acceptance of the Eisenhower Doctrine of 1957, resisting Communist influence in the Middle East, and his attempt to be reelected provoked a civil war in Lebanon, leading to an American landing in Beirut in 1958. The shock of American intervention (although the Americans fired no shots) brought a reconciliation and General Fuad Shihab, a descendant of the old princely family, was elected president.

An important period of reform followed but it could not keep pace with destructive social and political change. The population grew and although no census was taken it was agreed that by 1968 Muslims outnumbered Christians, though the distribution of political positions was unchanged. Urbanization accompanied population growth, creating an explosive urban poor. The 1967 Arab–Israeli war sent a shock wave through Lebanon, in spite of its minimal active involvement. In 1970 Palestinians, driven from Jordan, established themselves in Lebanon, joined by second-generation refugees who had grown up in camps there.

The most important political change was in the Shi'ite community, which came to outnumber the Sunnis. Their traditional economy in the south was undermined by the creation of Israel, and their homeland became the target of Israeli attacks against Palestinians. From being a docile minority they became a highly active political force.

These were the components of a civil war which was sparked off by a Christian attack on a Palestinian bus in 1975. Through the next 15 years the war became more complex, anarchic and ruthless. Syria intervened in 1976 to contain the war and Syrian troops were still in Lebanon in 1990. President Asad's objective was to prevent any other power challenging Syria's dominant influence in Lebanon and to that end he attempted, without success, to engineer a reformed sectarian balance. Israel intervened first in 1978, establishing a "security zone" in south Lebanon. In 1982 Israel invaded again with the more ambitious objective of establishing a Maronite proxy in Beirut and exercising indirect control over Lebanon, but it failed. The president whose election the Israelis supported, Bashir Gemayel, was assassinated (he was succeeded by his brother, Amin). The United States mediated an agreement between Israel and Lebanon covering Israeli withdrawal (17 May 1983), but Israeli forces were then forced to withdraw by Syrian-supported terrorism and guerrilla war.

No dominant force emerged. Instead rival Christian militias fought each other. The Sunnis had no effective militia and disappeared from the field. The Iranian revolution of 1979 split the Shi'ites: the Hezbollah, supported by Iran, opposed the Amal. Palestinian factions fought each other. Only the tight-knit social system of the Druze kept a single army together. The result was that almost any alliance of convenience was possible to avoid defeat. The outstanding success of the peacekeeping group was to negotiate a powersharing agreement between parliamentarians, outside Lebanon, in Taif and organize the election of a president, René Moawad, and even find a successor, Elias Hrawi, when he was assassinated. But he was opposed by one of the Christian militias and the civil war continued, though a treaty of "brotherhood" signed with Syria in 1991 promised to reduce the bloodshed.

▼ **A young Druze militiaman in 1980s Beirut.**

ISRAEL

STATE OF ISRAEL

CHRONOLOGY

1897	First Zionist Congress takes place in Basle
1917	Steps taken to establish home for Jews in Palestine
1937	British government propose partition of Palestine
1947	United Nations accept partition plan leading to War of Independence
1949	De facto borders of Israel drawn up
1964	Palestinian Liberation Organization (PLO) founded
1967	Six-Day War against Egypt, Jordan, Syria
1973	Yom Kippur Arab-Israeli war
1979	Camp David Agreement
1988	PLO recognize existence of Israel
1990	Israel attacked by Iraq in Gulf War

ESSENTIAL STATISTICS

Capital Jerusalem

Population (1989) 4,563,000

Area 20,700 sq km

Population per sq km (1989) 220.4

GNP per capita (1987) US$4,370

CONSTITUTIONAL DATA

Constitution Multiparty republic with one legislative house (Knesset)

Date of independence 14 May 1948

Major international organizations UN; I-ADB

Monetary unit 1 New (Israeli) sheqel (NIS) = 100 agorot

Official languages Hebrew; Arabic

Major religion Jewish

Heads of government since independence (Prime minister) D. Ben-Gurion (1948–53); M. Sharett (1953–55); D. Ben-Gurion (1955–63); L. Eshkol (1963–69); G. Meir (1969–74); Gen. Y. Rabin (1974–77); M. Begin (1977–83); Y. Shamir (1984); S. Peres (1984–86); Y. Shamir (1986–)

The state of Israel was proclaimed on 14 May 1948. It was immediately recognized by the United States of America and the Soviet Union and admitted to membership of the United Nations in May 1949. Its origins lie in the Zionist movement and in the British government's Balfour Declaration of 1917.

The first Zionist Congress was held in Basle in August 1897. Inspired by Theodor Herzl's pamphlet *Der Judenstaat,* it set up a Zionist organization committed to the establishment of a Jewish homeland. World War I gave the Zionists the first opportunity to establish a claim on Palestine, as the Allied Powers made agreements for the dismemberment of the Ottoman Empire which included Palestine. In a letter from the British Foreign Secretary to Lord Rothschild (2 November 1917) the British government undertook to use their best endeavors to establish "in Palestine a national home for the Jewish people" without prejudice to "the civil and religious rights of existing non-Jewish communities in Palestine".

This commitment was incorporated in the British mandate for Palestine in 1920. But the British government, with the approval of the League of Nations, restricted Jewish immigration to the area west of the Jordan. The Zionist Organization was recognized as the Jewish Agency responsible for cooperating with the British government; by 1929 the Jewish Agency, incorporating Zionists and non-Zionists, became a proto-government of a Jewish state yet to be established.

Jewish immigration to Palestine, slow in the 1920s, increased with the world slump and then Hitler's persecutions of the Jews. Between 1932 and 1935 the Jewish population doubled from 185,000 to 375,000; but the Arab population had grown from 600,000 in 1920 to 960,000 in 1935. In 1937 the British government proposed the partition of Palestine, establishing a Jewish and an Arab state, but then abandoned the plan in the face of an Arab revolt. Instead, in 1939, it placed a limit on Jewish immigration, to a total over five years of 75,000.

After World War II the efforts to limit immigration appeared increasingly impracticable and inhuman. The British government sought American help through an Anglo-American Commission but rejected its first recommendation for the immediate immigration of 100,000 Jews. In April 1947 it referred the Palestine question to the United Nations, which in November accepted a partition plan including the establishment of a Jewish state. The British then withdrew. On the last day of the mandate the Jewish leader David Ben-Gurion broadcast a declaration of independence.

The armies of Lebanon, Syria, Iraq and Jordan immediately made an uncoordinated military attack on the new state. Jewish units conducted a successful defense. They were welded together by Ben-Gurion into the Israeli Defense Forces and in 1949 completed the war of independence with a successful offensive against Egyptian forces in the south. The United Nations then mediated armistice agreements between Israel and its Arab neighbors (but not Iraq) on the island of Rhodes. This established the *de facto* borders of the country until 1967. Israeli territory was 21 percent greater than under the UN partition plan and included west Jerusalem (east Jerusalem and the remainder of Arab Palestine becoming part of Jordan). The armistice lines were arbitrarily drawn and inevitably created border clashes; Gaza was controlled by Egypt under nominal UN administration; Arab refugees numbered 720,000 or 70 percent of the Arab population of Palestine.

The prestate organization of the Jewish Agency was adapted to become a provisional government. A "transition law", generally known as the "small constitution", established a parliamentary form of government. No other constitution was enacted; constitutional practice was developed by convention and legislation. There is a single assembly (the Knesset) elected by proportional representation in a single constituency on the basis of party lists. The president is chosen by the Assembly and appoints the prime minister by asking the leader of the largest party to form a government. There is a dual court system which leaves matters of personal status to Jewish, Muslim and Christian religious tribunals.

A multiplicity of parties competed in elections to the Knesset but the dominance of the moderate socialist Mapai party gave Israel stability for nearly 30 years. This was reinforced by the strength of the Jewish Labor Organization, the Histadrut, which Mapai dominated. From the prestate days it had developed purposes far beyond those of a trade union, with the aim of establishing a Jewish workers' society in Palestine. It embraced a variety of economic, medical and social security enterprises and published its own newspaper.

National security remained a dominant issue. In 1956 the Anglo-French response to the nationalization of the Suez Canal gave Israel the opportunity for a preemptive strike against Egypt. Militarily this first Sinai campaign, coordinated with Britain and France, was brilliantly successful, but American pressure forced Israel to withdraw from all occupied territory. In 1967 a crisis provoked by the Egyptian president Nasser, closing the Gulf of Aqaba, led to the Six-Day War, fought against Egypt, Jordan and Syria. This resulted in Israeli occupation of the West Bank of Jordan, the Golan Heights (a strategically important border area in the northeast), Sinai and the Gaza strip on the Mediterranean coast.

These two wars brought great prestige to the Israeli Defense Forces (IDF), their flexibility, leadership from the front and use of airpower. Defense became the single most important industry. Intellectual talent was developed in major universities in Jerusalem, Tel Aviv and Haifa. Political and strategic studies and scientific research devoted to defense technology, including nuclear weapons, flourished.

The fourth Arab-Israeli war, that of Yom Kippur in October 1973, began with a successful limited attack by Egypt and Syria. Israeli forces counterattacked successfully as the United States and the Soviet Union remained in the background. The United States then intervened actively and negotiated disengagement agreements between Israel and Egypt and Israel and Syria. Egypt transferred its alliance from the Soviet Union to the United States. In 1977 the Egyptian president Sadat affirmed

▲ **Jewish girls on a 1940s kibbutz.**

his readiness to negotiate; in March 1979 United States president Carter mediated the Camp David agreements which included a peace treaty between Israel and Egypt and abortive provisions for autonomy for the occupied territories.

The Yom Kippur war had helped change the political order in Israel. The Mapai and the old leadership were discredited and sharp divisions appeared in society. In 1977 Israel's perennial opposition came to power. The Likud coalition led by Menachem Begin replaced Labor as the largest party. Its leadership derived from the hardline tradition of the old Zionist leader Vladimir Jabotinsky and the terrorist campaigns under the mandate; its following came from second-generation oriental Jews. The 1981 elections brought a new Likud government.

After 1977 the Labor party did not regain its dominance. Politics were characterized by a rough parity between Likud and Labor, such that governments were either a "grand coalition" between the two (1984–88, 1988–90) or a Likud coalition supported by small religious and rightwing nationalist parties (1977–81, 1981–84, 1990–).

A divided society

Israel developed a distinctive economic system, characterized by a lack of natural resources and local capital. This contrasted with a high level of skills among its European immigrants and a plentiful supply of unskilled labor amongst the immigrant population of oriental Jews. The public sector was exceptionally large, employing 60 percent of Israel's wage-earners by 1960. But the public sector itself was distinctive since it included the Histadrut with its industrial and banking conglomerates. A comprehensive system of social welfare developed, although the expense was borne on a negotiated voluntary basis between employers and employees rather than by the state. Capital came from abroad. In the 1950s net receipts from United States government sources were between US$40 million and US$60 million. Gifts from Jews outside Israel and the sale of bonds brought an estimated US$4 billion between 1948 and 1973. Reparations were paid by the German government to the Israeli government and to Israeli individuals. In the 1950s world Jewry covered 59 percent of Israel's balance of payments deficit, the United States 12 percent and Germany 29 percent. The economy sustained one of the highest growth rates in the world at around ten percent and also sustained an exceptionally large defense budget. There was a steadily improving standard of living.

Underlying electoral and parliamentary politics, Israeli society remained deeply divided. The left of the political spectrum favored some withdrawal from the occupied territories in exchange for peace, while the right wanted to retain and settle them. Secular rationalist Ashkenazi Jews in the old Zionist tradition lived in a different culture from the Orthodox who, although a minority, held a balance in parliament. Oriental Jews were culturally separate from Ashkenazi and continued to support the Likud against the old Labor establishment.

In spite of the peace treaty with Egypt, Israel remained throughout the 1980s a country under stress. The economy was shaken by the oil price increases of 1973 and 1979 and went through a period of hyperinflation. Defense spending increased as a result of the wars and the sophistication of technology; financial support from the United States, increasingly in grants rather than loans, only mitigated the problem.

In 1987 the resistance of the Arabs in the occupied territories sharpened. Young people and children took the lead in unarmed attacks on Israeli soldiers in a new "uprising" (*intifada*). Spurred into a change of policy, the Palestine Liberation Organization (PLO) declared the establishment of a Palestinian state and its acceptance of Israel, and so was able to open a dialog with the United States. This brought fresh pressure on the Israeli government to enter a "peace process". The Labor party, ready to enter an Israeli-Palestinian dialog and to exchange land for peace, broke the coalition. But Yitzhak Shamir then formed a rightwing government with the support of the religious parties, who remained intent on retaining the occupied territories. At the same time the new policies of liberalization in the Soviet Union allowed the emigration of Soviet Jews to Israel. Their arrival put pressure on resources and increased the tension and animosity from the Arab world.

The Palestinians

Palestinians are those Arabs who live in or derive from the area of Palestine (Arabic *filastin*, from the Roman Palestina), of which Israel now occupies the major part. Palestinians have a deep sense of national identity derived from longstanding traditions. Palestinian nationalism grew in response to the pressure of Zionism and Israel. It tended to fragment, at first because of family, religious and local loyalties and then over ideology and tactics. The splits underlying organizational divisions were between political conservatives, whose only radical program was the liberation of Palestine, and leftwing Marxist groups; between those who continued to believe that armed struggle alone could liberate Palestine and those who saw diplomatic activity as more likely to ensure success; between those whose focus was on Palestine and those who saw Palestine as a cause which only Arabs together could win.

The first stage of Palestinian nationalism exploded in the Arab revolt of 1936-38 which defeated British plans to partition Palestine but, followed by World War II, left Palestinians defenseless when the state of Israel was created. Recovery came in the 1960s: Fateh (an acronym in reverse from Harakat Tahrir Filastin – the Palestine Liberation Movement), the largest organization, was formed in 1959; the Palestine Liberation Organization (PLO) was established with Egyptian sponsorship in 1964. The 1967 Arab–Israeli "6-day" war was a turning point: the West Bank and Gaza were occupied by Israel, a second wave of refugees created and Egypt discredited by an annihilating defeat.

The PLO freed itself from Egyptian tutelage. Fateh was the largest single part of the organization and Yasser Arafat was elected chairman of the PLO; but neither controlled the other constituent organizations, notably the Popular Front for the Liberation of Palestine and its splinters the PFLP General Command and the Democratic Front for the Liberation of Palestine.

Palestinians have no territorial base outside the occupied territories, the PLO being expelled from Jordan in 1970 and reduced to one of the number of warring factions in Lebanon. The PLO is governed by its National Council. Its constitution is the National Charter, which expresses the maximalist position of the recovery of the whole of Palestine, an objective which by 1990 few Palestinians thought realistic. Palestinians thus found themselves divided between Israel (600,000), the West Bank (800,000), Gaza (500,000) and possibly two million in Arab states. Like many deprived of a homeland, they sought status and social mobility through education. The most important change in PLO policy was prompted by the *intifada* (uprising) in the occupied territories, starting in December 1987. The PLO then declared a Palestinian state and accepted the existence of Israel, renouncing terrorism as a tactic, and entered a dialog with the United States.

JORDAN

HASHEMITE KINGDOM OF JORDAN

CHRONOLOGY

1923	British mandate gives Jordan nominal independence
1946	Full independence achieved
1949	Jordan takes control of West Bank and part of Jerusalem
1951	King Abdullah is assassinated
1970	King Hussein expels Palestinian Liberation Organization (PLO) to Lebanon
1989	Riots spread through Jordan
1991	Jordan assists Iraq in Gulf War

ESSENTIAL STATISTICS

Capital Amman

Population (1989) 3,059,000

Area 88,947 sq km

Population per sq km (1989) 34.4

GNP per capita (1987) US$1,540

CONSTITUTIONAL DATA

Constitution Constitutional monarchy with two legislative houses (Senate; House of Deputies)

Date of independence 25 May 1946

Major international organizations UN; LAS; IDB

Monetary unit 1 Jordan Dinar (JD) = 1,000 fils

Official language Arabic

Major religion Sunnite Muslim

Heads of government since independence (King) Abdullah ibn Hussein (1946–51); Talal ibn Abdullah (1951–52); Hussein ibn Talal (1952–)

The Hashemite Kingdom of Jordan originated in the settlement of the Middle East that followed World War I and the collapse of the Ottoman Empire. After the Amir Faisal had been driven out of Syria by the French, his older brother Abdullah was persuaded by the British to establish an Amirate of Transjordan in part of Palestine. It became nominally independent, under a British mandate, in May 1923.

In 1948 its British-trained army fought with great effect against the newly proclaimed state of Israel. Under the 1949 armistice which followed, Transjordan, independent since 1946, took over the part of Palestine that was not Israel – including east Jerusalem – and Abdullah became king of Jordan. Abdullah established contact with the Israelis with a view to a peace settlement but in July 1951 he was assassinated. He was succeeded briefly by his son who was obliged by illness to abdicate in favor of his own son who, at the age of 17, became King Hussein of Jordan.

The enlargement of the kingdom and the incorporation of Palestinians, who owed no natural allegiance to the Hashemite monarchy, was not accomplished easily. Nationalism was excited by efforts to bring Jordan into the Baghdad Pact (a British-led alliance against Soviet powers in the Middle East) and then by the Suez affair of 1956. In 1957 Hussein took preemptive action against an army coup and in 1958 accepted British intervention to support his throne at the time of the Iraqi revolution. But in the Six-Day War of 1967, Hussein believed that he had no choice but to fight alongside Egypt, thereby losing to the Israelis east Jerusalem and the West Bank of the Jordan.

After that time Hussein's objective was to recover the occupied territories, especially Jerusalem, always in an ambiguous relationship with the Palestine Liberation Organization (PLO) which was set up in 1964 and with Israel. In 1974 he was pushed by other Arab states into accepting that the PLO were the representatives of the Palestinian people but continued to fund local government in the occupied territories until 1988. In 1978–79 he resisted American pressure to participate in the Camp David negotiations between Israel and Egypt.

Jordan has few natural resources. Agriculture is developed in the Jordan valley, fruits and vegetables being exported to the Gulf. Phosphates are exported into a volatile world market. Economic survival has depended on subsidies, from Britain, then the United States, then the Gulf countries. The economy has been highly dependent on the oil market because of subsidies, and on remittances from Jordanians working abroad. The collapse of the oil price in the late 1980s was one of the factors which brought Jordanian politics to crisis point in 1989.

In the 1980s King Hussein gave full support to Iraq during its war with Iran. The loss of the economic benefit from the war when it ended sharpened the economic crisis. Discontent expressed itself in riots in April 1989, occurring in areas where traditional support for Hussein had been strongest. In response the king dismissed his prime minister, Zaid Rifai, who had a reputation for massive corruption but whose Syrian connection had helped balance support for Iraq. King Hussein then moved closer to Saddam Hussein of Iraq. He restored parliamentary democracy and organized fair elections. Political parties were not allowed but the strongest representation was that of Islamic candidates. For a time there was a drift toward constitutional monarchy, until the Gulf crisis of 1990 put intolerable strains on the kingdom.

SAUDI ARABIA

KINGDOM OF SAUDI ARABIA

CHRONOLOGY

1902	Abdul Aziz captures Riyadh
1916	Britain recognizes Abdul Aziz as sole ruler of Nejd and Al-Hasa
1924	Saudis capture Jeddah, Mecca and Medina
1929	World depression slashes pilgrimage revenues
1932	Proclamation of Kingdom of Saudi Arabia
1933	Socal (Standard Oil of California) granted exclusive oil rights in the Eastern Province
1938	First major oil discoveries
1943	Major US military and economic commitments
1984	Slump in Saudi oil revenues begins
1991	Gulf war costs Saudi Arabia US$48 billion

ESSENTIAL STATISTICS

Capital Riyadh

Population (1989) 13,592,000

Area 2,240,000 sq km

Population per sq km (1989) 6.1

GNP per capita (1986) US$6,930

CONSTITUTIONAL DATA

Constitution Monarchy

Date of independence 23 September 1932

Major international organizations UN; LAS; IDB; OPEC

Monetary unit 1 Saudi riyal (SR1s) = 100 halalah

Official language Arabic

Major religion Muslim

Heads of government since independence (King) Abdul Aziz ibn Abdul Rahman Al Saud (1926–53); Saud (1953–64); Feisal (1964–75); Khalid (1975–82); Fahd (1982–)

Saudi Arabia is the only state named after its ruling dynasty, the House of Saud. The "father" of the modern Saudi state was Abdul Aziz ibn Abdul Rahman Al Saud (1881–1953), also known as "Ibn Saud". In 1902, aged 21, he permanently recaptured Riyadh. In 1902–06 he consolidated his control of Nejd and in 1906–13 he won control of Al-Hasa, bordering on Kuwait and Qatar. In 1916, Britain recognised Abdul Aziz as sole ruler of Nejd and Al-Hasa, granting him a temporary subsidy of £60,000 a year and encouraging him to expand north toward Hail. They thereby won his acceptance of the pro-British Sharif Hussein as ruler of Mecca and Hijaz and his participation in the wider British-backed "Arab revolt" against the Ottoman Empire and its Rashidi allies (based in Hail). Thus strengthened, Abdul Aziz took control of Asir in 1920 and Hail in 1921, thwarting the expansionist ambitions of Sharif Hussein, although his sons Faisal and Abdullah were installed as British client-rulers of Iraq and Jordan.

Sharif Hussein proclaimed himself king of Arabia (much of which was in Saudi hands) and caliph of Islam. This was too much for the Saudis. They invaded and annexed Taif, Jeddah, Mecca and Medina in 1924–25. In 1926 Abdul Aziz was proclaimed king of Hijaz and Sultan of Nejd and its dependencies (nearly eighty percent of the Arabian peninsula), winning due recognition by Britain in 1927, and in 1932 he proclaimed the kingdom of Saudi Arabia.

In 1953 a Council of Ministers was established, to further more collective deliberation and responsibility. But it was merely advisory, as only the king had final executive and legislative authority, and since 1964 the king has also been prime minister. Since 1953 Saudi Arabia has been ruled by sons of Abdul Aziz: Kings Saud (1953–64), Feisal (1964–75), Khalid (1975–82) and Fahd (1982-). Their father bequeathed to them the world's most important oil-exporting economy, which was catapulted "from rugs to riches" between the 1940s and the 1970s.

Economic and social trends

At the turn of the century, Arabians mainly lived by herding dromedaries, sheep and goats. About 60 percent of what is now Saudi Arabia was used for grazing and less than one percent was cultivated. The principal crops were date palms, coconut palms, some coffee, qat and some coarse grains. The main towns – the capital Riyadh, the holy cities of Mecca and Medina and the ports of Jeddah, Dammam and Jubail – were local centers of trade, administration and crafts. The population was 30 percent urban in 1960 (compared with 75 percent in 1989).

In 1930 King Abdul Aziz entered an acute fiscal crisis, as pilgrimage fees from Mecca and Medina (his main revenue) slumped under the impact of the great Depression. In 1933 the crisis was alleviated by granting Socal (Standard Oil of California) exclusive oil rights in Saudi Arabia's Eastern Province for 60 years. Drilling began in 1935, the first big oil finds were made in 1938 and 65,618 tons of crude oil were produced in 1939. Socal brought other American oil companies as partners and the joint venture was renamed "Aramco" in 1944. King Abdul Aziz received useful cash advances. But, when World War II again slashed his revenues from pilgrimage fees at Mecca, he demanded much larger advances. The company demurred so the British government stepped in.

Mistrusting Britain's motives, the United States president Roosevelt finally decided in February 1943 that "the defense of Saudi Arabia is vital to the defense of the United States". In 1943–47 Saudi Arabia received around $100 million in US aid and the United States built its largest mainland Asian airbase, at Dhahran.

Saudi oil output soared from eight million tons in 1946 to 26 million tons in 1950, 188 million tons in 1970 and 493 million tons in 1980 when it was the world's second largest producer, before a slump which began in 1984, taking output to 173 million tons in 1985, before it recovered to reach 257 million tons in 1989. Saudi oil revenues rose even more dramatically: from US $39 million in 1949 to an astronomical $113 billion in 1981, aided by accession to the Organization of Oil Producing and Exporting Countries (OPEC) in 1960.

From 1965 onward, burgeoning oil revenues were used to build physical infrastructure (particularly roads, ports, dams, irrigation, power stations, desalination plants, public buildings, housing and telecommunications), to educate and train the young Saudi workforce (in the 1980s nearly sixty percent of Saudis were under 21), to build up the armed forces and purchase expensive US and British "air defense systems" manned by Western personnel (defense absorbed nine percent of GNP and over twenty percent of public expenditure by the early 1970s), to expand public services and social welfare provisions, especially schooling, to develop oil-refining, petrochemical and energy-intensive electro-metallurgical industries, and to subsidize irrigated agriculture and hydroponics.

These massive programs were mostly launched under the competent, pragmatic, technocratic stewardship of Faisal bin Abdul Aziz, first as prime minister (1958–64) and then as king (1964–75). They were coordinated by the state petroleum, mineral and petrochemical company "Petromin" and the Central Planning Office; by successive five-year development plans (starting 1970–75); and by SABIC, the 70-percent state-owned Saudi Arabian Basic Industries Corporation (established 1976). Faisal also vigorously opposed Nasserist Arab nationalism; fought off Egyptian backing for the Yemeni Arab Republic (1962–67); cemented the close relationship with the Americans following the 1956 Suez crisis and the promulgation of the "Eisenhower Doctrine" pledging US assistance against "subversion" in the Middle East; and promoted the astute US-trained lawyer Sheikh Ahmed Zaki Yamani to the post of Oil minister (1962–86). In 1972 Yamani completed negotiations for the Saudi state to purchase a 25-percent stake in Aramco. In 1980 the Saudi holding rose to 100 percent, but the Saudis retained a large American staff.

The Iraqi invasion of Kuwait in August 1990 highlighted Saudi Arabia's acute vulnerability. Its major cities and oilfields were clearly utterly vulnerable to attack, while its armed forces were small. The Iraqis were expelled from Kuwait in February 1991 by an American-led United Nations alliance, relying heavily on Saudi Arabia for military bases, but Saudi Arabia incurred costs in excess of US $48 billion.

▼ **Ibn Saud (seated) with his advisors in the 1920s.**

YEMEN

REPUBLIC OF YEMEN

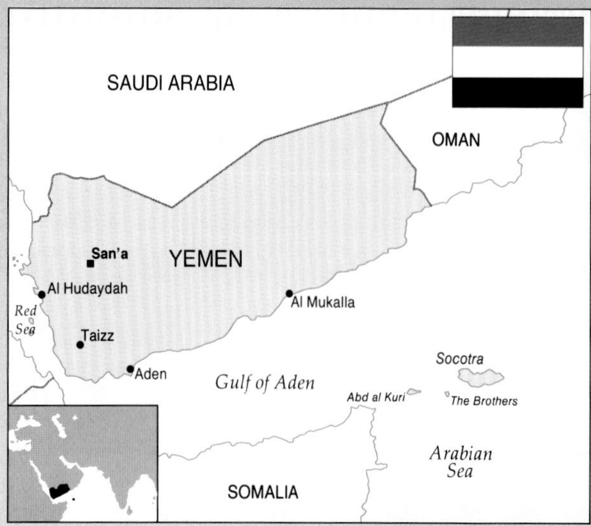

CHRONOLOGY

1839	British occupy Aden
1958	Yemen joins United Arab Republic
1962	Military coup installs Yemen Arab Republic (YAR) government in southern Yemen – civil war follows
1967	People's Democratic Republic of Yemen established [Yemen (Aden)]
1978	President of YAR assassinated
1987	YAR begin to export oil
1990	Yemen Arab Republic and People's Democratic Republic of Yemen unite

ESSENTIAL STATISTICS

Capital San'a

Population (1990) 12,000,000

Area 531,000 sq km

Population per sq km (1990) 22.5

GNP per capita (1988) US$540

CONSTITUTIONAL DATA

Constitution Multiparty republic with one legislative house (House of Representatives)

Date of unification 22 May 1990

Major international organizations UN; LAS; IDB

Monetary unit 1 Yemen Rial (YR1) = 100 fils; 1 Yemen dinar (YD) = 1,000 fils

Official language Arabic

Major religion Muslim

Head of government since unification (President) Lt. Gen. Ali Abdullah Saleh (1990–)

Yemen's mountains catch enough rainfall to grow coffee, traditionally exported through Mocha. The narrow coastal plains are inhabited by Shafi Muslims, the interior by mountain tribes of the Zaydi Shi'a sect.

Yemen's recent history begins in the 19th century with the reimposition of Ottoman rule from the north and the occupation of Aden by the British in 1839. The British had no interest in Yemen outside Aden, which served them as a fueling station on the route to India and they agreed a frontier demarcation with the Turks in 1904, which lasted until May 1990.

From the beginning of the century until 1948 Yemen was governed by the Imam Yahya, who established a working arrangement with the Turks until their collapse in World War I. His response to the modern world was to exclude it and Yemen remained a medieval kingdom. His son Ahmad, succeeding him in 1948, brought Yemen out of its isolation. He set the direction of foreign policy, away from Saudi Arabia and its ally, the United States. Russians contributed to the development of the port of Hodeida, Chinese to the road to San'a.

In September 1962 a group of army officers carried out a coup and installed the Yemen Arab Republic (YAR). It was immediately recognized by Nasser who sent troops in support. But Ahmad's successor, Imam Badr, survived the coup, the Saudis supported the royalists and the ensuing Yemen war continued until after the defeat of Egypt in the 1967 war with Israel. Foreign troops were withdrawn and the republic stabilized under General Hassan al-Amri.

Meanwhile Aden was transformed. Its importance as a British base increased as a result of the British withdrawal from India and Egypt, and because of the Abadan crisis the British-based oil company BP constructed a major refinery. It was the most important port in South Arabia, even when Hodeida was expanded. In this heavily populated conurbation a new social structure emerged, with a vigorous trade union movement and a nationalist ferment. The British faced increasingly active opposition and, soon after the Labour government came to power in Britain in 1964, the decision was taken to withdraw from Aden. To the surprise of many, the Marxist National Liberation Front, rather than the Egyptian-backed nationalist movement, seized power and set up the People's Democratic Republic of Yemen (PDRY), also known as Yemen (Aden). They took over a poor country, since the British withdrawal took away the main source of the country's income. From 1969 the government took an increasingly leftward direction.

For more than a decade the two Yemens were relatively stable, beneath a surface that was often turbulent. Union between the two was frequently discussed and in 1978 seemed very close, until the YAR president was assassinated by a bomb, probably from Aden. In 1986 a bitter civil war in Aden brought a change of government in which the strong man was seen to be Ali Salem al-Bidh, of the Yemen Socialist party.

Change came in the late 1980s with the discovery of oil. The YAR began exporting oil in 1987. At the same time Soviet exploration found oil at Shabwa in the south. Politically more significant was the promise of oil on the border of north and south. In 1988 a commitment was signed to move towards union. In the YAR President Ali Abdullah Saleh began cautious liberalization; the PDRY established diplomatic relations with the United States. In May 1990 the two countries merged, under the the Presidency of Ali Abdullah Saleh.

KUWAIT

STATE OF KUWAIT

CHRONOLOGY

1896	Shaikh Mubarak assassinates his two brothers and becomes ruler
1899	Treaty of protection signed with Great Britain
1914	Britain recognises Kuwait as an independent government under British protection
1934	Kuwait Oil Company formed
1950	Assembly government introduced
1961	Kuwait becomes independent – Iraqi government claim it as part of Iraq
1965	Foreign workers outnumber Kuwaitis
1976	Assembly dissolved
1990	Iraq invades Kuwait
1991	Allied forces liberate Kuwait

ESSENTIAL STATISTICS

Capital Kuwait

Population (1989) 2,048,000

Area 17,818 sq km

Population per sq km (1989) 114.9

GNP per capita (1987) US$14,870

CONSTITUTIONAL DATA

Constitution Constitutional monarchy with a single parliamentary house (National Assembly)

Date of independence 19 June 1961

Major international organizations UN; LAS; IDB; OPEC

Monetary unit 1 Kuwaiti diner (KD) = 1,000 fils

Official language Arabic

Major religion Sunnite Muslim

Heads of government since independence (Emir) Abdulla Al-Salem Alsabas (1950–65); Saloas Al-Salem Alsabas (1965–77); Gaber Al-Ahmad Alsabas (1977–)

Kuwait was settled early in the 18th century by families of the Utub tribe. Some decades later they agreed a division of labor amongst themselves; one group were pearl-fishers, a second merchants, the third were to organize the affairs of the community. The Al Sabah family were given this responsibility of government.

In 1896 Shaikh Mubarak became ruler. On his initiative a treaty of protection with Britain was signed in 1899 and was immediately effective in preventing the Ottoman government from belatedly asserting its authority over Kuwait. In 1914 the British formally recognized that the shaikhdom of Kuwait was "an independent government under British protection".

When Mubarak's successor Salim died in 1921 the merchants extracted a promise from the candidate most likely to succeed him, Shaikh Ahmad, that he would introduce a consultative council. Shaikh Ahmad ruled until 1950 and this was the beginning of intermittent and limited assembly government in Kuwait, which lasted until 1986. After the failure of the 1921 council, a municipal council was established in 1934 and then, in 1938, a national legislative council. This was responsible for a number of important reforms, extending the education system, reforming the corrupt customs machinery, modernizing the legal system and establishing a police force.

The Japanese development of cultured pearls destroyed the pearl-diving industry. Oil was discovered in 1932; in 1934 the Shaikh signed a concession agreement with Anglo-Iranian and Gulf Oil, who set up the Kuwait Oil Company. As a result he had an income independent of his people. Oil began to flow in 1938 but was not fully exploited until 1946. It transformed Kuwaiti society. Under Shaikh Abdullah (1950–65) oil revenues were widely distributed. The civil service was expanded and the bereft pearl-divers were given government jobs. Health services and social security were made available to everyone. The oil economy attracted foreign workers. Citizenship was largely restricted to pre-1920 residents and their descendants. By 1965 non-Kuwaitis slightly outnumbered Kuwaiti citizens.

In June 1961, Kuwait became independent, a friendship treaty with Britain replacing that of 1899. The Iraqi government claimed Kuwait as part of Iraq. The result was to strengthen the prestige and authority of the ruler, making it easier to establish constitutional government. From 1962 to 1976, through the turbulent years of two Arab-Israeli wars, an uneasy balance was maintained between the ruling family and successive national assemblies. In 1976 the ruler suddenly dissolved the assembly, which was not reconstituted until 1981. In 1986 the assembly was again prorogued during the Iran-Iraq war.

Kuwait took full control of the Kuwait Oil Company in 1975. It developed its oil industry, producing 1.7 million barrels per day at the beginning of 1990 and used its oil revenues in a distinctive way. It developed a large refining capacity and a "downstream operation", selling directly to the consumer. The Kuwait Investment Authority invested actively abroad: by 1990 investment earnings were roughly equal to oil revenues. A heritage fund was established to ensure Kuwaiti wealth for future generations. A development fund made grants to developing countries, both Arab and non-Arab, so that, with government grants in addition, Kuwait became the largest overseas donor per capita.

In August 1990 Kuwait was invaded by Iraq; the country was liberated in 1991 and greater democracy promised.

IRAQ

REPUBLIC OF IRAQ

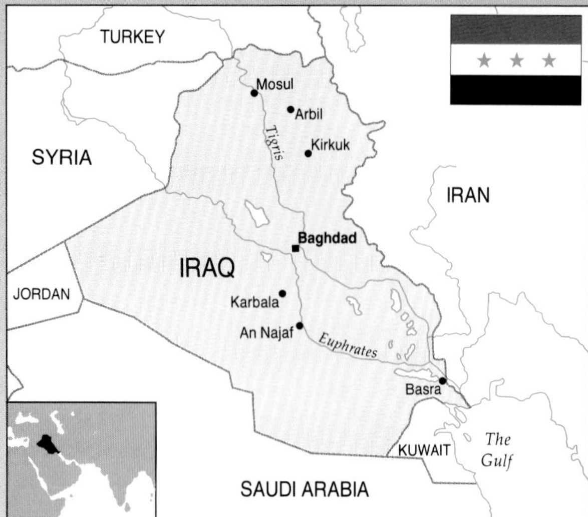

CHRONOLOGY

1920	League of Nations mandate awarded to Britain
1932	Independence declared under the Hashemite Dynasty
1958	Military coup overthrows monarchy and establishes a republic under General Qasim
1962	Kurdish minority attempt seccession
1963	General al-Salam Arif overthrows Qasim and becomes president
1968	Ba'th party revolution leads to formation of the Revolutionary Command Council
1978	Power is peacefully transferred to Saddam Hussein
1980	Iraq attempts to seize Iranian territory
1990	Iraq annexes Kuwait – UN demands Iraqi withdrawal
1991	UN-backed Gulf War breaks out – Iraq withdraws from Kuwait

ESSENTIAL STATISTICS

Capital Baghdad

Population (1989) 17,215,000

Area 435,052 sq km

Population per sq km (1989) 39.7

GNP per capita (1987) US$2,420

CONSTITUTIONAL DATA

Constitution Unitary single-party republic with one legislative house (National Assembly)

Date of independence 3 October 1932

Major international organizations UN; LAS; IDB; OPEC; OIC

Monetary unit 1 Iraqi dinar (ID) = 20 dirhams = 1,000 fils

Official language Arabic

Major religion Shiite Muslim

Heads of government since independence (King) Faisal I (1921–33); Ghazi (1933–39); Faisal II (1939–58); (President) Gen. A. al-Karim Qasim (1958–63); Field-Marshal A. al-Salam Arif (1963–66); Maj.-Gen. A. al-Rahman Arif (1966–68); Gen. A. Hassan al-Bakr (1968–79); Saddam Hussein (1979–)

Iraq used to be known as Mesopotamia, "the land between the rivers", (Tigris and Euphrates), whose illustrious history extended back to 10,000 BC. From 1831 until World War I Mesopotamia was an outpost of the Ottoman Empire made up of the *vilayets* (provinces) of Basra, Baghdad and Mosul.

The opening of the Suez Canal in 1869 encouraged increasing British presence in the Basra and Baghdad regions. In 1888, however, the Ottomans conferred Mesopotamian oil rights on the Anatolian Railway Company, owned and organized by the Deutsche Bank. These resurfaced in German negotiations during 1899–1903 to build a Berlin-to-Baghdad railway. In 1904 Calouste Gulbenkian (son of a Baku Armenian oil tycoon) presented the Ottoman government with an appetizing report on the petroleum potential of Mesopotamia and the Arabian peninsula. This encouraged British as well as German groups to compete for oil rights in Mesopotamia.

In 1911 Gulbenkian formed the Turkish Petroleum Company (TPC), to exploit Mesopotamian oil. In March 1914 the Anglo-Persian Oil Company (BP), acquired a 50-percent share of TPC, leaving the Deutsche Bank and Royal Dutch Shell with a quarter each. The concurrent agreement pledged the partners in TPC to refrain from any oil activities within the Ottoman territories other than through the TPC.

When the Ottoman empire declared war on Britain and France in 1914, British and Indian troops occupied Basra and captured Baghdad in March 1917 and Mosul in October 1918.

In March 1920 Iraq was awarded to Britain as a League of Nations "mandated territory". From July to October 1920, however, there were widespread revolts against British rule. Suppression of these revolts cost around 2,000 British lives, and in 1921 Britain installed Faisal, son of Sharif Husain of Mecca and a British "client", as king of Iraq – a country he had never visited before 1921.

There were major oil discoveries at Kirkuk in 1927. The American "oil majors" were brought into the TPC monopoly in 1928 and in 1931, faced with sharply falling grain prices and state revenues, King Faisal granted the Iraq Petroleum Company (IPC, as TPC had become in 1929) exclusive oil rights to a large part of Iraq in return for advances on future oil royalties.

From 1936 to 1941 there were seven changes of government by coup, partly reflecting growing Arab nationalist, anti-British and pro-Axis sentiments among Iraqi officers. A bold refusal to allow Britain transit rights across Iraq in early 1941 led to an Anglo-Iraqi war in May 1941 and renewed British military occupation of Iraq until 1945.

A limited legalization of political parties and trade unions in 1946 led to major strikes and unrest, which culminated in savage repression of the opposition in 1949. During the 1950s, amid burgeoning oil revenues, Iraq was increasingly administered by a British-dominated "development board", which invested in expensive infrastructure, while relatively neglecting politically risky agrarian reform, housing, education and training.

Most Iraqis became increasingly incensed by the monarchy's subservience to Britain and the 1955 Baghdad Pact, allying Britain, Iraq, Iran, Turkey and Pakistan against the USSR and "radical" Arab nationalist Egypt and Syria. In the "July Revolution" of 1958, Iraq's royal family and its collaborators were murdered by a republican Arab nationalist coup committed to closer relations with Nasser's Egypt and the Soviet Union.

In 1961, when Britain ended its protectorate in Kuwait, President Qasim declared that Kuwait was part of Iraq and threatened to annex it, but annexation was averted by a prompt return of British forces to Kuwait. Qasim retaliated by revoking part of IPC's oil concession, allowing it to keep only those oilfields already under exploitation.

In February 1963 Qasim was bloodily overthrown by Brigadier 'Ab al-Salam 'Arif, in alliance with the radical Pan-Arab Ba'ath party In November 1963, however, the brutal Ba'athists and their paramilitary National Guard were ousted by the more pragmatic military wing of the new regime.

In July 1968 another Ba'athist group, headed by Ahmad Hassan al-Bakr, seized power in Iraq. Saddam Hussein, deputy leader of the Ba'ath party, took on the tasks of emasculating opponents and "Ba'athizing" the army, while concentrating absolute power in the hands of a few Ba'ath leaders. Having seized power, the Ba'ath regime proceeded to enact radical agrarian reform, nationalize IPC in 1972 and adopt "anti-imperialist" and anti-Israeli postures. Iraq's 1972 Treaty of Friendship and Cooperation with the Soviet Union inaugurated a brief alliance with the Communist party (1972–78). In 1975 Saddam Hussein concluded an implicitly conservative agreement with the Shah of Iran. This seemingly settled frontier disputes and closed off Iranian support for the Kurds of northern Iraq, who had been in rebellion intermittently since 1961. Moreover, in 1978-79, Saddam Hussein purged and executed thousands of Communists, Pan-Arabs and pro-Soviet or pro-Syrian politicians and launched his own brutal dictatorship.

In September 1980 Iraq invaded Iran, seizing territory beyond the vital Shatt al-'Arab waterway. This launched an Iran–Iraq war which continued until 1988 and cost Iraq over 100,000 deaths and over US$200 billion in military outlays, physical destruction and lost output. In 1986–87 both sides came under growing pressures to call a ceasefire, amid growing Kurdish military support for the Iranians and international condemnation of Iraqi use of chemical weapons (both against Iranians and against their own Kurdish rebels). A ceasefire was agreed in August 1988, but Iraq's severe reconstruction and debt problems were aggravated by the decline in world oil prices. Saddam Hussein endeavored to escape his financial difficulties by annexing Kuwait in August 1990. This precipitated a damaging UN trade embargo against Iraq and UN demands for Iraqi withdrawal from Kuwait, culminating in a UN-backed "Gulf War" from 17 January to 1 March 1991, which forced Iraq to withdraw and cost it many thousands of lives, several weeks of devastating aerial bombardment and over US$200 million in lost assets and physical destruction. Enough of Saddam Hussein's forces remained intact to suppress major postwar rebellions among the Shiite Muslims in the south and the Kurds in the north, but Iraq's peoples faced very grim prospects for the rest of the 1990s.

The Kurds

In 1990 there were around 20 million Kurds, including 10–12 million in Turkey, 4 million in Iran, over 3 million in Iraq, over 700,000 in Syria, over 100,000 in the USSR and about a million working in Western Europe.

Until recently Kurds were characteristically tribal mountain-dwellers and herdsmen. Nominally subject to either Ottoman or Persian (Qajar) jurisdiction up to 1918, with the disintegration of Ottoman and Qajar rule in the wake of World War I Kurds found an opportunity to press for an independent Kurdistan. The Treaty of Sèvres of 1920 provided for "independent states" of Armenia and Kurdistan. Unfortunately, the predominantly Kurdish Mosul *vilayet* was incorporated into Iraq in the 1920s, while most of Kurdistan was incorporated into the Turkish Republic; the rest was included in Iran and Syria.

Until the late 1980s, Turkey refused to recognize the separate ethnic identity of its Kurdish minority and brutally suppressed autonomist demands in the 1920s, 1930s and early 1980s. In the late 1980s Kurds were granted greater recognition and Turkey launched a development program to boost agriculture and industrial investment in the Kurdish homeland.

Repeated attempts to create an autonomous Kurdish enclave centered on Mahabad in northwestern Iran in 1918–22, 1945–46 and 1979–83, were bloodily suppressed. In 1986–91, however, Mahabad Kurds increasingly collaborated with Iran against the Iraqi regime of Saddam Hussein and, in return, Iran gave sanctuary to large inflows of Kurdish refugees and *peshmerga* (freedom-fighters) fleeing from Iraqi "genocide" in 1988 and 1991 when the UN tried to establish "safe havens" for Kurds within Iraq.

◀ **Saddam Hussein built a huge personality cult.**

IRAN

ISLAMIC REPUBLIC OF IRAN

CHRONOLOGY

1906	First constitution grants independence
1908	Major oil find leads to formation of Anglo–Persian Oil Company (later BP)
1918	De facto British "protectorate" over Persia
1925	Reza Khan takes control after coup d'état deposing the last Shah of the Qājar Dynasty
1935	Persia changes its name to Iran
1953	Shah forced into exile until CIA overthrow Mossadeq restoring Shah as royal dictator
1963	Imprisonment of Ayatollah Khomeini
1979	Shah flees Iran, Ayatollah Khomeini returns and proclaims an Islamic Republic
1980	Iran–Iraq war breaks out
1988	UN arranges ceasefire

ESSENTIAL STATISTICS

Capital Tehran

Population (1989) 54,333,000

Area 1,648,196 sq km

Population per sq km (1989) 33.1

GNP per capita (1987) US$1,800

CONSTITUTIONAL DATA

Constitution Unitary Islamic Republic with a single legislative house (Islamic Consultative Assembly)

Date of independence 7 October 1906

Major international organizations UN; OPEC

Monetary unit 1 rial (R1s) = 100 dinars

Official language Farsì (Persian)

Major religion Shiite Muslim

Heads of government (Shah) Ahmed Mirza Shah (1909–25); Reza Shah Pahlavi (1925–41); Mohammad Reza Shah Pahlavi (1941–79); (President) A. Bani-Sadr (1980–81); M. Raja'i (Jul–Aug 1981); H. Khamenei (1981–89); H. Rafsanjani (1989–)

Iran, known until the 1920s as Persia, entered the 20th century under the ailing Qājar dynasty (1796–1921). During the 1900s growing opposition to royal absolutism, official corruption and Qājar subservience to Britain and Russia culminated in the "Constitutional Revolution" of 1905 to 1911. This established an elected *majlis* (parliament) to which ministers were to be accountable and whose assent would be needed before legislation, taxation, and foreign agreements could be ratified. The Fundamental Laws (1906 and 1907) also proclaimed equality before the law and freedoms of expression, conscience and assembly. This constitution survived until the "Islamic Constitution" of 1979.

In 1907 Britain and Russia divided Persia into "spheres of influence" (British in the south and Russian in the north, with a "neutral" central zone). In 1908 a Russian-backed coup by Mohammed Ali Shah closed the *majlis* and imprisoned or executed opposition leaders and from 1909 to 1917 Persia's northernmost regions came under Russian control. But a rebellion by the Bakhtiari tribe forced the Shah to abdicate; the constitution and the *majlis* were temporarily reinstated under a mild Oxford-educated regent. But when the Persian government employed an American to reorganize and enforce the collection of taxes in 1911, a threatened Russian military intervention forced the regent to dismiss him and suppress the *majlis*.

British interest in southern Persia was intensified by the discovery of oil in 1908 at Masjed-Suleiman – the first major oil find in the Middle East. This led to the formation of the Anglo–Persian Oil Company (later BP) in 1909.

During World War I, increasing German and Ottoman influences in western Persia resulted in further erosion of central authority and growing British military "protection" of southern and central Persia. In 1918–20 the victorious British tried to consolidate their "protectorate", embodied in the (unratified) Anglo–Persian Treaty of 1919, while the new Soviet regime renounced most of Russia's former imperialist policies toward Persia. In 1920, however, escalating costs led to British troop withdrawals. The ensuing power vacuum encouraged Reza Khan to seize power in 1921. He was crowned Reza Shah in April 1926. In 1925–28 he disarmed and forcibly settled nomadic tribesmen, enacted new secular legal codes, outlawed socialism and trade unions, expanded the army, introduced compulsory military service and established a bureaucracy centered in Tehran. The *majlis* launched a secular state education system in 1926, but under Reza Shah less than four percent of public expenditure went to education. Reza Shah decreed Western-style dress for men in 1929 and for women in 1936. In 1931 he established a state monopoly of foreign trade to restrict imports and encourage import substitution. In 1933, furthermore, he unilaterally cancelled the Anglo–Persian (BP) oil concession and, after much wrangling, accepted a new flat-rate royalty on oil exported.

In August 1941, following Hitler's invasion of the Soviet Union, Iran came under joint Anglo–Soviet military occupation and Reza Shah was deposed on account of his increasingly close links with Nazi Germany. The Allies installed Reza Shah's pliant son on the throne, legalized trade unions and a fast-growing Communist party, relaxed censorship, encouraged separatist movements, disrupted Iranian production and distribution, precipitated a sharp increase in retail prices and brought famine to some regions.

From 1945 to 1947 Iran was convulsed by widespread labor unrest, major separatist movements and deep nationalist and religious outrage at the scale of British, Soviet and American violations of Iran's sovereignty. British and Soviet troops were finally withdrawn in 1946, but their interference in Iranian affairs continued. In 1947–49 royal control was partially restored. There was renewed economic crisis in 1948–50, as inflation revived, Western imports swamped Iranian markets and transport came close to collapse. This and bitter wrangling over oil royalties in 1949–50 rallied support for an "anti-imperialist" National Front coalition led by the charismatic Mohammad Mossadeq, who in 1951 became prime minister and nationalized the oil industry. This provoked a crippling Western boycott of Iranian oil exports in 1951–53. But the Shah's attempts to dismiss Mossadeq backfired, triggering antiroyalist protests and pushing the Shah into temporary exile. In August 1953, however, the American CIA organized a coup which overthrew Mossadeq, restored the Shah to power as a royal dictator and transferred control of the oil industry to a Western consortium.

From 1954 to 1977 Iran experienced an accelerating oil boom. Burgeoning oil revenues underpinned the Shah's royal dictatorship and financed increasingly ambitious public expenditure programs, a massive expansion of Iran's military and internal security services and major expansions of urban schooling and higher education.

In the early 1960s the Shah surmounted a crisis induced by a temporary economic downturn and mounting dissatisfaction over blatant electoral fraud, by launching his so-called "White Revolution": land redistribution; privatization of state enterprises; female suffrage; nationalization of forests; a national literacy campaign; and profit-sharing. The Shah thus won some support from the secular white-collar classes. Starting in 1962, however, Ayatollah Ruhollah Musavi Khomeini strongly denounced the Shah's corruption, westernization, secularization and subservience to the United States. Khomeini was exiled to Iraq from 1965 to 1978. From there he was able to fulminate against the Shah more freely and gradually won a mass following in Iran.

The oil boom nevertheless proceeded until 1974–77, when Iran's oil revenues peaked at around $20 billion per annum. Then the economy became seriously overheated by ambitious expenditure and construction programs, causing recession in 1977–78. The expulsion of Ayatollah Khomeini from Iraq to Paris in 1978 allowed him to communicate more directly with his Muslim followers in Iran. The Shah declared martial law in September 1978 and was persuaded to leave Iran in January 1979. After the triumphal return of Khomeini in February 1979, Iran was at first governed by a moderate Muslim coalition, but in November 1979 Muslim militants professing allegiance to Khomeini occupied the American embassy in Tehran and held embassy staff "hostage" for over a year, resulting in a trade embargo against Iran and the freezing of US$11 billion of Iran's overseas assets. During 1980–82 power shifted from the presidency and the government to Khomeini, the Islamic Republican party, local "*Komitehs*" (committees), *pasdaran* (revolutionary guards) and The Party of God (*Hezbollahis*).

Iran's self-destructive infighting diminished significantly in the course of the long-drawn out Iran–Iraq war, which was precipitated by an Iraqi invasion of Iran in September 1980. This war, which ended in stalemate in August 1988, cost Iran nearly a million dead, over 600,000 permanently disabled and over US$300 billion in physical destruction, lost output and lost oil revenues. After the war, in 1989–91, Iran suffered high unemployment; acute shortages of food, housing, schools and medical supplies; inflation rates in excess of 50 percent per annum; and rampant and pervasive corruption. In 1989–90 Iran vainly sought US$27 billion in fresh foreign investment and reconstruction loans. Relations with the West were made easier after the death of Ayatollah Khomeini in June 1989, by the election of the "pragmatist" Hojatoleslam Ali Akbar Hashemi Rafsanjani to a considerably strengthened secular presidency in July 1989 and by Iran's dignified neutrality during the 1991 Gulf War between Iraq and the United Nations allies.

▲ **A pro-Khomeini demonstration in Tehran, 1985.**

AFGHANISTAN

REPUBLIC OF AFGHANISTAN

CHRONOLOGY

1919	Proclamation of independence
1963	Zahir Shah attempts to set up a constitutional monarchy
1965	Pro-Communist People's Democratic Party established (PDPA)
1973	Prince Mohammad Daoud leads military coup and seizes power
1978	Daoud is assassinated in military coup led by PDPA
1979	Civil war breaks out – Soviet troops arrive
1988	Geneva accords signed – Soviet troops begin to withdraw

ESSENTIAL STATISTICS

Capital Kabul

Population (1989) 14,825,000

Area 652,225 sq km

Population per sq km (1989) 22.7

GNP per capita (1985) US$230

CONSTITUTIONAL DATA

Constitution Unitary single-party people's republic with two legislative houses (Council of Representatives; Council of Elders)

Date of independence 19 August 1919

Major international organizations UN; IDB

Monetary unit 1 afghani (AF) = 100 puls

Official languages Pushtu; Dari (Persian)

Major religion Sunnite Muslim

Overseas territories none

Heads of government since independence (King): Amanollah Khan (1919–29); Habibollah (Jan–Oct 1929); Muhammad Nader Shah (1929–33); Muhammad Zahir Shah (1933–73); (President) M. Daoud (1973–78); D. Kadir (Apr–May 1978); N. Taraki (1978–79); H. Amin (Oct–Dec 1979); B. Karmal (1979–86); M. Najibullah (1986–)

The modern state of Afghanistan is usually dated from the reign of Ahmad Shah Abdali (1747–72), who had established his kingdom over much of Afghanistan and northern India. His successors lost India but consolidated the boundaries of present-day Afghanistan: the western boundaries with Iran in 1872; those in the north with Russia in 1885; and those in the south and east with the British Indian empire in 1893. Because of its strategic location and a persistent British fear of southward expansion by the Russians, Afghanistan was made into a buffer state following the two Anglo–Afghan wars of 1839–42 and 1878–80. The Afghans, in return for a British subsidy, accepted British control in matters of foreign policy.

Attempts to modernize Afghanistan started with King Amanollah who came to power in 1919. But as in Iran and Turkey, the so-called modernization (in reality imitation of the Europeans) was largely superficial and confined to a few urban centers. The countryside remained virtually unaffected and the fiercely independent tribes continued to live according to their traditional codes. The half-hearted attempts at modernization not only alienated the vast majority of the population from the urban elites but, because "modernization" was not accompanied by constitutional reforms, it also alienated the very elites which the "modernization" process threw up.

The process of "modernizing" education, the economy and the armed forces continued under Zahir Shah, who came to the throne in 1933 at the age of 19. Zahir took control in 1963 and attempted to establish a constitutional monarchy with limited powers in the hands of a popularly elected assembly. But in the absence of established political parties and the lack of political tradition, the politics of Afghanistan remained turbulent and unstable during the entire period of the constitutional experiment (1963–73).

As successive governments failed to assimilate the dissenting urban elites, the dissidents organized themselves in semi-clandestine parties; the first of these was the Pro-Communist People's Democratic Party of Afghanistan (PDPA), established in 1965. The PDPA subsequently fissured into the predominantly Pushtun supported *Khalq* (the people) and the *Parcham* (the flag), backed by the Dari-speaking elites. There were also a number of leftwing parties, the most important of which was *Shola-e-Jawid* (the eternal flame); and some Islamic groups drawing support from the Madrasahs (Islamic schools) and the university.

By the early 1970s it had become clear that neither the *Parcham* nor the *Khalq*, because of their remoteness from the traditional tribal power structure and more importantly because of their failure to mobilize the rural masses, were able to secure their political objectives through the parliamentary process. They therefore joined hands with the procommunist armed forces in ousting Zahir on 17 July 1973. Prince Mohammad Daoud, who replaced Zahir, was an opportunist. He had cooperated with the PDPA to consolidate his position but once his position appeared to be secure he dismissed the *Parchami* minister and allied himself with the tribal elders. But Daoud had failed to create an independent popular support base and was assassinated in a military coup in 1978 which brought into power a PDPA government under Nur Mohammad Taraki.

The doctrinaire PDPA regime, ignorant of Afghan social and political dynamics, introduced numerous socialist reforms – a

ceiling on land ownership, the remission of rural debts and mortgages on land, restriction on marriage dowries and compulsory literacy. These measures were introduced without creating a popular consensus and alienated the traditional elite, whose authority was threatened. The situation was further aggravated by the PDPA's attempt to create a strong centralized government. This was seen as a direct challenge to the autonomous status of the tribal leaders. Moreover the PDPA's excessive dependence upon the Soviet Union split the modernized elites themselves. An influential group of urban politicians, inspired by nationalist fervor and Islamic ideals, sought to challenge the PDPA with a view to reducing Soviet influence in Afghanistan. Society was fissured three ways. First, there was a contest between the monarchists and the modernists who were seeking greater constitutional authority; second, tensions among the old tribal order, who resisted the encroachment of their traditional power by the new centralized government; and finally, increasingly fierce rivalry between competing modernized elites.

By the end of 1978 Afghanistan was in turmoil and a civil war broke out by mid-1979. The Soviet Union, which had considerable influence in the country, despatched 85,000 troops in December to install Babrak Karmal in a bid to end the civil war and salvage their own interests. The Soviet intervention, instead of easing the civil war, actually intensified it and brought other foreign powers into the conflict. Nine years later when guerrilla warfare forced the Soviet forces to leave Afghanistan, the civil war was still far from over even though the Western-backed Mujahiddin were militarily marginalized.

The withdrawal of Soviet forces did not lead to the capture of Kabul by the rebels or the overthrow of the Soviet-backed regime of Najibullah. Instead the regime strengthened its control over the urban and main population centers; and the rebels – together with four million Afghan refugees – were still in camps in Pakistan and Iran.

Economic and social trends

Afghanistan is one of the poorest countries in the world. It is barely possible to estimate the size of the population: the first census carried out in 1979 estimated it at 15.5 million but even that figure (whose accuracy is questionable) was rendered obsolete by the civil war in which several million Afghans are estimated to have been killed and many more to have fled.

Agriculture – mainly fresh and dried fruits, cotton and oil seeds – was the major contributor to the Afghan economy prior to the development of natural gas, which in 1983–84 accounted for more than half of the country's total exports. Afghanistan may have extensive deposits of coal, salt, iron ore, copper and also some gold, silver and chromium but these may not be commercially viable owing to the difficulties of access.

▲ **Mujahiddin in the mountains of northern Afghanistan.**

PAKISTAN

ISLAMIC REPUBLIC OF PAKISTAN

CHRONOLOGY

1905	Earthquake at Quetta kills 20,000
1956	Islamic Republic proclaimed
1958	General Mohammed Ayub Khan establishes military dictatorship
1967	Zulfikar Ali Bhutto forms Pakistani People's party (PPP)
1969	Ayub Khan is overthrown in mass uprising
1971	East Pakistan declares independence as Bangladesh
1977	Bhutto's party win electoral majority, leading to military coup
1979	Bhutto is executed
1988	PPP emerges as largest party under Benazir Bhutto
1990	President dismisses Prime Minister Bhutto

ESSENTIAL STATISTICS

Capital Islamabad

Population (1989) 118,820,000

Area 796,095 sq km

Population per sq km (1989) 149.3

GNP per capita (1987) US$350

CONSTITUTIONAL DATA

Constitution Multiparty federal Islamic republic with two legislative houses (Senate; National Assembly)

Date of independence 14 August 1947

Major international organizations UN; Commonwealth of Nations; IDB

Monetary unit 1 Pakistan Rupee (PRs) = 100 paisa

Official language Urdu

Major religion Muslim

Heads of government since independence (President) M. Ali Jinnah (1947–48); K. Nazimuddin (1948–51); G. Muhammed (1951–55); Major General I. Mirza (1956–58); Field Marshal M. Ayub Khan (1958–69); General A. Yahya Khan (1969–71); (Prime minister) Z. Ali Bhutto (1971–77); (President) General Zia ul-Haq (1978–88); G. I. Khan (1988–)

At the beginning of the 20th century the creation of Pakistan could not have been foreseen. Although the Muslims had ruled India for nearly 1,000 years prior to losing power to the British in the mid-18th century, they made up no more than 27 percent of India's population. No amount of constitutional innovation could give them equality with the Hindu majority. And once it became clear in the 1940s that the British were going to leave India, the leader of the Muslim League, Mohammed Ali Jinnah, sought to achieve parity with the Hindus by claiming that Muslims were a separate nation and therefore entitled to their own state. As India became gripped by sectarian violence, the British agreed to divide and quit.

Pakistan was born in haste. Its leaders themselves were mostly migrant from India, and as such they lacked a popular constituency. Not surprisingly the constitution and democratic institutions were derailed from the very outset. In October 1958 General Mohammed Ayub Khan, the army chief, established his military dictatorship.

The imposition of martial law had disastrous consequences. Although in the short term there was rapid economic development, it was achieved at the expense of creating both regional disparities and an inequitable society as the benefits of development were confined to a particular province and to a handful of families. East Pakistan, 1,500 kilometers (1,000 miles) away, was particularly neglected, even though more than half the population of Pakistan lived in that province. Ayub Khan was overthrown in 1969 and his successor General Yahya Khan was compelled to hold a general election, the first in Pakistan for 23 years. The party of the Bengalis, the Awami League, won an absolute majority but the military rulers tried to nullify the election results through armed repression. In the civil war that ensued, the Bengali freedom fighters (backed by the Indian army) defeated the Pakistan army and East Pakistan broke away as independent Bangladesh in December 1971.

Pakistan, albeit truncated, had a popularly elected government for the first time under Zulfiqar Ali Bhutto. Pakistan made rapid economic and political strides. Parliamentary democracy was established, and the armed forces subordinated to civilian authorities. However, much of Bhutto's good work was undone when he tried to destroy the opposition and increasingly resorted to repressive measures. In the 1977 parliamentary elections Bhutto won handsomely, but the opposition refused to accept the outcome. General Zia ul-Haq took over in July 1977 and thus ushered Pakistan's third, and the longest, period of martial law rule, which lasted until the dictator's death in August 1988.

Zia was determined to establish the military as the ultimate authority in Pakistan's political decision-making. All opposition to this regime was ruthlessly removed: politicians and journalists were flogged and the ousted prime minister was hanged after a most dubious trial. The constitution was changed to concentrate powers in the hands of the president; and Zia combined the offices of the president and the army chief to ensure his personal control over the army.

The death of Zia in an airplane crash in August 1988 compelled the acting president to allow a general election. As was expected Pakistan's People's party (PPP), now led by Bhutto's daughter Benazir, emerged as the largest party. But the transition of power was far from smooth. The army insisted on, and obtained, an undertaking that the popularly elected government would not interfere with matters relating to the armed forces. It was soon obvious that the relationship between the elected government and the armed forces was too tense to be sustained. In August 1990 the president, under pressure from the armed forces, dismissed yet again another elected prime minister. A pragmatic electorate, already disillusioned by Benazir, voted into office a motley group under the banner of the Islamic Democratic Alliance in October 1990. Although a popularly elected government was installed under Nawaz Sharif, it would be difficult to describe the process as the restoration of democracy. It is at best a scheme for power sharing between the military and the elected representatives.

Economic and social trends

The record of Pakistan's economic development is relatively impressive. Despite major downturns in the international economy and domestic political upheavals, in the 1970s and 1980s it grew at an annual rate of about six percent. Agriculture remained the predominant contributor to GDP but Pakistan managed to carry some structural adjustment. The main crops are wheat, rice, cotton, sugar, millet, maize and oil seeds. The share of industries in the make-up of the GDP has risen in 1947 from almost negligible to about 18 percent in 1988–89. The main mineral deposits are natural gas, oil, coal, rock salt, gypsum, limestone and chromite.

While Pakistan's economic performance has much potential, there are also acute problems: rapidly increasing population, massive current budgetary and balance of payments deficit, a demand for energy which outstrips capacity, and the distorting impact of a disproportionately high defense expenditure.

▼ **Benazir Bhutto campaigns in the 1988 elections.**

INDIA

REPUBLIC OF INDIA

CHRONOLOGY

1919	Unarmed civilians killed, increasing anti-British movement
1950	India becomes a republic under Jawaharlal Nehru
1966	Indira Gandhi installed as prime minister following death of Shastri
1973	Severe drought brings India close to starvation
1984	Assassination of Indira Gandhi – succeeded by her son Rajiv
1991	Assassination of Rajiv Gandhi

ESSENTIAL STATISTICS

Capital New Delhi

Population (1989) 835,812,000

Area 3,166,414 sq km

Population per sq km (1988) 264

GNP per capita (1987) US$300

CONSTITUTIONAL DATA

Constitution Multiparty federal republic with two legislative houses (Council of States; House of the People)

Date of independence 15 August 1947

Major international organizations UN; Commonwealth of Nations

Monetary unit 1 Indian rupee (Rs) = 100 paisa

Official languages Hindī; English

Major religion Hindu

Heads of government since independence (Governor-General) Lord Mountbatten (1947–48); C. Rajagopalachari (1948–49); (Prime minister) J. Nehru (1949–64); G. Nanda (May–Jun 1964); L. Shastri (1964–66); G. Nanda (Jan 1966); I. Gandhi (1966–77); M. Desai (1977–79); C. Singh (1979–80); I. Gandhi (1980–84); R. Gandhi (1984–89); V.P. Singh (1989–90); C. Shekhar (1990–91); P. Rao (1991–)

At the beginning of the century the British control (*Raj*) over India remained firmly entrenched. The outbreak of World War I, the impact of the Russian revolution, United States' president Wilson's proclamation of national self-determination as the basis for organizing the international system and the return of Mohandas Karamchand Gandhi to India from South Africa in 1914 radically altered the political environment in India. Gandhi declared that the British rule had no legitimacy and mobilized the masses to oust the colonial rulers. The British prevaricated, devised numerous constitutions which sought to give Indians the semblance of power while retaining the real authority in British hands. The anti-British movement was swollen by repression, notably in 1919 when 400 unarmed civilians were killed in Amritsar. The following year Gandhi began a campaign of nonviolent noncooperation (*satyagraha*). Meanwhile the Indian National Congress emerged as a well-organized political force.

In 1942, India was threatened by Japanese expansionism in Southeast Asia, and defended by a vast volunteer army in Burma (now Myanmar) in 1942–44. The danger led the British to concede the need for independence and after the war Britain was compelled to leave India. As the Muslims and the Hindus could not agree to a joint inheritance, in 1947 the British transferred power to two successor dominions, India and Pakistan.

The Indian National Congress, the major political organization, had developed enormous political and organizational experience over 60 years since its foundation in 1885 and was able to fill the gap caused by the departure of the British. India was also fortunate in having leaders like Gandhi, Jawaharlal Nehru, Vallabhai Patel and Abul Kalam Azad who combined vision with pragmatism. The most serious challenge confronting the new government was to curb the orgy of violence in which hundreds of thousands of Hindus and Muslims killed each other; and to shelter, feed and clothe millions more who had fled into India from across the Pakistani border.

The government also had to deal with the integration of 500 or so princely states, which occupied nearly two-fifths of India. These states were compelled to accede to either India or Pakistan. The integration of the states to India was achieved with ease. Only Junagadh and Hyderabad resisted, but both were easily coerced. However the state of Jammu and Kashmir proved to be intractable and the problem remains unresolved.

On 26 January 1950 the Indian constitution was adopted. India became a republic but chose to remain within the British Commonwealth of Nations. The constitution was essentially an essay in liberal democracy, drawing from the experiences of both Britain and the United States. India adopted a parliamentary government akin to the Westminster model, where the executive was responsible to the legislature and the president was a titular head. The parliament consisted of an elected lower house (*Lok Sabha*) which had the real powers and an indirectly elected *Rajya Sabha* with mainly a revisionary role. While parliament is sovereign, the supreme court has the power to decide constitutional questions and acts as the protector of fundamental rights enshrined in the constitution.

One of the most difficult problems addressed by the constitution was the relationship between the center and the states (provinces). India is one of the most heterogeneous countries in the world and it was not easy to reconcile the particular demands of the various states, regions, religions and

castes with the imperatives of the central authority. The center not only controlled defense, foreign affairs, currency and railroads but was also vested with residuary powers. Moreover the center had the power to impose presidential rule in the states in emergency and override them in certain matters. The concentration of power at the center had contributed to tensions such that in some cases the states threatened to secede. However the experience of the last four decades coupled with growing self-confidence in the resilience of the Indian Union caused some rethinking and the trend in the 1990s was toward greater devolution of power to the states.

Jawaharlal Nehru, India's first prime minister, is by common consensus regarded as the maker of modern India. He created a cohesive state out of heterogeneous nations, trained the people for democracy and proved that political literacy is not synonymous with formal schooling; constructed a model for economic development based on consensus and, in foreign policy, advocated nonalignment which enabled India to assert its autonomy in a world dominated by the two superpowers. Some of Nehru's less successful policies may be explained by his excessive idealism and a belief in the power of reason, persuasion and goodwill.

The main parameters of Indian foreign policy were laid down by Nehru, who as the prime minister doubled as India's foreign minister. The element at the core of his policy was based on what subsequently came to be known as the *pancsheel* or the five principles: mutual respect for each other's territorial integrity and sovereignty, mutual nonaggression, mutual noninterference in each other's internal affairs, equality and mutual benefit, and peaceful coexistence. These principles also

▲ **The Taj Mahal, dwarfs Lord Pethick Lawrence and entourage, 1946.**

poverty). Her electoral victory both enabled her to smash the leftwing Naxalbari insurgents and also freed her hands to pursue her conflict with Pakistan to a decisive conclusion. However, much of Indira's political triumph was undone by severe droughts in 1973 and 1974.

The resulting economic crisis, together with charges of corruption and inefficiency, led to popular movements against which Indira resorted to the indiscriminate use of special powers, and she imposed a state of emergency in 1975. Although Indira acted within the limits of the constitution, the parliament was emasculated, political opponents were repressed and demo-cratic institutions were virtually put in abeyance. However, the excessive concentration of power in the hands of the prime minister and the repression that is inevitable under authoritatian rule, gravely discredited both Indira and the Congress without actually resolving any of the

became the basis of the nonaligned movement of which India was one of the architects. Even though India developed close relations with the Soviet Union and signed the Indo–Soviet Treaty of Friendship in 1971, it was careful not to become too dependent upon the Soviet Union and maintained cordial ties with the West.

But nearer home India's foreign policy was less successful. The partition of India, instead of ending the Hindu–Muslim conflict, merely elevated the intercommunity bloodbath into a rivalry between India and Pakistan. The dispute over Kashmir, itself a legacy of the partition, reactivated all the issues and traumas which the partition was intended to still. After three full-scale wars, the problem was still far from resolved.

Nehru was succeeded by the diminutive Lal Bahadur Shastri, a Gandhian disciple. Shastri displayed remarkable leadership in the war with Pakistan in September 1965. However, his sudden death in Tashkent within hours of a peace settlement with Pakistan once again created a leadership vacuum.

Although a number of senior leaders staked their claim, the relatively inexperienced Indira Gandhi, the daughter of Nehru, was installed as prime minister with the support of the Congress kingmaker Kamraj in expectation that Indira would be more amenable to the manipulation of the Congress "syndicate". However the "syndicate" soon discovered that Indira had a mind of her own. After a comparatively poor showing in the 1967 general election, Indira decided both to rid herself of the influence of the "syndicate" and radically transform the image of the Congress. She sacked Morarji Desai, the powerful deputy prime minister, secured the election of V. V. Giri as the president of India and appealed to the masses by the twin populist measures, nationalization of banks and the abolition of the "privy purse" of the princely rulers. Indira's defiance of the "syndicate" split the Congress but she was able to retain her position with the support of the leftwing parties. In early 1971 Indira called a snap election and won a handsome majority on her socialist promise of *gharibi hatao* (removal of

▲ **J. Pandit Nehru and his mentor, M. K. Gandhi, 1946.**

▶ **Ragpickers in Bombay.**

The Kashmir Problem

The princely states of Jammu and Kashmir were under the rule of Maharajah Hari Singh when British paramountcy lapsed in 1947. In theory the princely state was independent but in practice such independence was ruled out by the British. Like other states, Kashmir was expected to accede to either India or Pakistan in that Muslim majority states contiguous to Pakistan would become part of Pakistan and the rest would to go India.

In Kashmir, however, there were some ambiguities. It was adjacent to Pakistan; and while its ruler was Hindu, the population was predominantly Muslim. But Kashmir also had a common frontier with India, which gave the Maharajah some leverage. However, a popular Muslim uprising frightened him into acceding to India. India sent troops to Kashmir; the battle over, Kashmir was divided, two-thirds in Indian and the rest in Pakistani hands.

India insisted that Pakistan withdraw its forces from Kashmir, and referred the dispute to the UN in the hope of censuring Pakistan but was instead reprimanded for not holding a promised plebiscite. In 1954, when Pakistan signed a pact with the United States, India withdrew its offer of plebiscite, arguing that this pact altered the entire context of South Asian security and, more importantly, the accession of Kashmir to India was ratified by Kashmir's Constituent Assembly. Pakistan and India went to war again, in 1965 and 1971, but the Kashmir dispute remained unresolved.

problems for which the emergency had been ostensibly imposed. And when in January 1977 Indira sought a popular mandate by calling a general election, the people registered their disapproval of her authoritarian rule by voting for the Janata Party under Morarji Desai, the former deputy prime minister Indira had sacked. Indira herself lost her seat.

The Janata party was a ramshackle coalition cobbled together with the avowed purpose of ousting the Congress. Once Indira was removed it withered away and in the election that followed Indira was returned to power. The lasting contribution of the Janata party was to dismantle the emergency regime and put India back on a democratic course.

Indira's second stint was marked by a renewed tension between the center and the states. The situation was made worse by a decline of the Congress's power in several states. Indira sought to undermine the authority of the non-Congress governments. State governments were toppled either by weaning away their members in the state assemblies or simply by imposing direct rule from Delhi. The result was violent agitation and political turbulence in Kashmir, Bengal, Assam and the northeast. But by far the most serious challenge came from the Sikhs in Punjab where Congress's attempt to oust the moderate wing of the Akali Dal became entangled with the Sikh secessionist movement. The subsequent military action against the Sikh militants and the storming of the Golden Temple (the holiest shrine of the Sikhs) at Amritsar led to Indira Gandhi's assassination in October 1984 by her personal Sikh security guards.

The Sikh Problem

The Sikhs were the most adversely affected of all Indian ethnic religious groups by the partition of India. The division of Punjab found the community split between India and Pakistan. Since a homeland for the Sikhs was ruled out, Sikh leaders engineered *en masse* migration of the Sikhs in Pakistan to eastern Punjab. In India the Sikhs were rapidly disillusioned: they were refused a compact state with a special status and in the linguistic reorganization of the states in the 1950s and early 1960s their claim was ignored.

In 1966 a Punjabi suba of the Sikhs was created. This only partially met Sikh claims as the party of the Sikhs could only maintain control by allying with non-Sikhs. The Punjabi suba also fell short of the Sikh expectations because Punjab had to share its state capital (Chandigarh) with the neighboring state of Haryana. In the early 1980s there was a rising fear among Sikhs that their internal cohesion was threatened by the growth of the Hindu-Sikh sect, the Nirankaris. This gave birth to a militant movement for the purification of Sikhism. Indira Gandhi saw in the reform movement an opportunity to split the Sikh party and facilitated the rise to power of the movement's leader, Bhindranwale. Bhindranwale fell out with Delhi when it attempted to undermine Punjab autonomy. The militant Sikhs rallied under the banner of Bhindranwale who made the Golden Temple – the holiest shrine of the Sikhs – into an armed fortress and unleashed a campaign of terror against Hindus in Punjab. Indira Gandhi ordered the storming of the Golden Temple and Bhindranwale was killed. This led directly to her assassination in October 1984.

▼ **A militant Sikh assembly.**

The assassination of Indira sparked off sectarian carnage in which thousands of Sikhs were brutally murdered and their properties destroyed. However, Indira's death was followed by orderly and constitutional succession, showing both the maturity of Indian political institutions and also their dependence upon the single Nehru-Gandhi family. The new prime minister was Rajiv Gandhi, Indira's only surviving son, whom she had been grooming for succession.

Rajiv tried to make a break from his mother's policies. Not only were the advisors and bureaucrats who had been close to Indira removed but Rajiv sought to strengthen the political institutions which had been eroded under his mother, and to revamp the Congress party by the introduction of new blood. In a remarkable reversal of his mother's policy, he made it clear that the Congress would not attempt to unseat noncongress government in the states; and although his party lost the elections to the state assemblies in Punjab and Assam, they were seen as a triumph for Rajiv's attempts to restore decency in politics.

Rajiv could not, however, sustain his early successes. The disappointing performance of the Congress in a number of the state assemblies elections and in the *Lok Sabha* by-elections cast doubt about Rajiv's capacity to win elections. This led to a reversal of his earlier policies: many of the unsavory politicians associated with the previous regime were brought back and Rajiv began to show greater tolerance for corruption and narrow party politics. He did little to curb communal violence as he feared any support of the Muslims would lose him Hindu votes which he hoped to mobilize for the Congress. By the time Rajiv went to the polls in November 1989 his image had been badly tarnished; the Congress failed to win a majority, though it remained the single largest party in the *Lok Sabha*.

The government was formed by V. P. Singh's ad-hoc National Front with only 158 seats, propped by the communists and the Hindu fundamentalists Bharatiya Janata party (BJP). Although Singh headed a minority government, his popularity let him pursue his policies quite independently of the parties propping him up. Singh was followed by another minority government headed by Chandra Shekhar and propped up this time by the Congress. But this government lasted less than five months. It was clear that although the period of one-party dominance was over, the political parties in India were not yet ready for coalition government. In May 1991, the world was shocked by the assassination of Rajiv Gandhi, during the general election campaign following the collapse of Chandra Shekhar's government. His assassin, a Sri Lankan Tamil woman, died with him, along with 15 others. Gandhi's widow, Sonia, refused to take his place as Congress party leader, and eventually P. Rao was selected. The dynasty which had been at the sharp end of Indian government since independence was for the first time not represented in its political life. Although Congress failed to win a majority, it was able to form a government headed by Rao.

Economic and Social trends

With a per capita income of approximately US$260 in 1988–89 India remains one of the low-income developing countries. After initial sluggishness the GDP has been growing steadily at around five percent since 1970. However with a population expanding at 2 percent per annum, the GDP per capita in real terms has grown only by 3 percent. But with all its economic achievements some 40 percent of its population still live below the poverty line.

In the first half of the century, under colonial rule, agriculture was the mainstay of Indian economy. But agriculture was stagnant and output of food was outstripped by population growth. Industries were neglected; only cotton textile and jute industries had made any significant mark; iron and steel production was still in its infancy.

The pace of industrialization increased after independence. With a series of Five Year Plans in the 1950s India embarked upon a program of import substitution and diversification of its industries. By the early 1960s it had become largely self-sufficient in basic manufactures. However, lack of competition made Indian industries relatively inefficient and consequently its products failed to compete in the international market. The deregulation and liberalization of the economy was started in the early 1980s and went some way to making industries more competitive. Agriculture remains the most important economic activity, with nearly three-quarters of the work force still employed in it. The principal agricultural products – grown both by subsistence peasants and large-scale farmers – are food grains, including rice, wheat, maize, millet and pulses. The important cash crops are groundnuts, sugar cane, cotton, jute, tea and tobacco. India's exports have not kept pace with imports and consequently there is a persistent negative trade balance. Despite industrial diversification and economic liberalization the main exports are confined to traditional items: tea, fish, iron ore, garments and precious and semiprecious stones. The main imports are petroleum and petroleum products, chemicals, fertilizers and iron and steel.

▲ Women at the sacred river Ganges at Benares, Hindu holy city.

SRI LANKA

DEMOCRATIC SOCIALIST REPUBLIC OF SRI LANKA

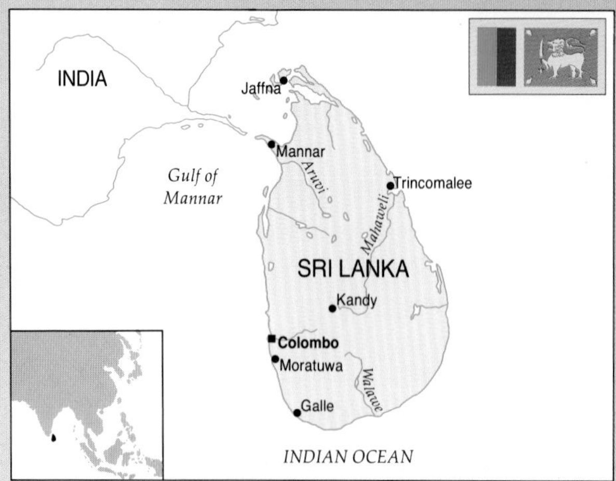

CHRONOLOGY

1948	Proclamation of independence led by Stephen Senanayake, United National party (UNP)
1971	State of emergency declared
1972	Ceylon becomes Republic of Sri Lanka
1977	Government recognizes Tamil as one of official languages
1987	Indian peace-keeping force invited into Sri Lanka
1990	Indian troops withdraw from Sri Lanka

ESSENTIAL STATISTICS

Capital Colombo

Population (1989) 16,855,000

Area 65,610 sq km

Population per sq km (1989) 256.9

GNP per capita (1987) US$400

CONSTITUTIONAL DATA

Constitution Unitary multiparty republic with one legislative house (Parliament)

Date of independence 4 February 1948

Major international organizations UN; Commonwealth of Nations

Monetary unit 1 Sri Lanka rupee (SL Rs) = 100 cents

Official languages Sinhalese; Tamil

Major religion Buddhist

Heads of government since independence (Prime minister) S. Senanayake (1948–52); D. Senanayake (1952–53); Sir J. Kotelawala (1953–56); S. Bandaranaike (1956–59); W. Dahanayake (1959–60) D. Senanayake (Mar–Jul 1960); Mrs S. Bandaranaike (1960–65); D. Senanayake (1965–70); Mrs S. Bandaranaike (1970–77); J. Jayawardene (1977–78); R. Premadasa (1978–88); D.B. Wijeratne (1988–)

The history of Sri Lanka (formerly Ceylon) goes back over 2,500 years. Both the Sinhala and the Tamil inhabitants of Sri Lanka were originally migrants from southern India. The majority of the population are Sinhala, who are also Buddhists; the Tamils are mainly Hindus. Around the 8th century Arab traders – the Moors – settled in the island and brought with them Islam. Christianity followed in the early 17th century when the island came under the domination of Europeans, first the Portuguese and then the Dutch. Toward the end of the 18th century the British established their rule and remained there until the country's independence on 4 February 1948.

Stephen Senanayake, whose United National party (UNP) led Sri Lanka to independence, was conscious of the plurality of the society and sought to reconcile the interests of the Sinhalas and the Tamils. His policy was continued by his son, Dudley Senanayake. But by the mid 1950s Solomon Banaranaike's Sri Lanka Freedom party (SLFP) advocated the "Sinhala only" policy which undermined the ethnic consensus and threatened to relegate the Tamils to an inferior status.

In the late 1960s the political situation became more turbulent with rising unemployment, inflation and scarcity of essential consumer goods. In April 1971 there was a serious insurrection led by Janatha Vimukti Peramuna (JVP), a leftwing group drawn from the militant youth. Although the insurrection was suppressed, it did lead to radical changes, including land reforms and enhanced state control of trade and industries.

The ethnic issue assumed a new dimension after the 1977 parliamentary elections in which the UNP captured 140 seats out of 168, whereas the SLFP lost all but 9 of its 90 seats. More crucially, the Tamil United Liberation Front (TULF) emerged as the official opposition. In August 1977 the government conceded some of the legitimate demands of the Tamils such as the recognition of Tamil as one of the two national languages and the abolition of discrimination against Tamil citizens of Indian origin. But this merely heightened the Sinhala resentment fanned by the SLFP, and in turn provoked the militant Tamils to seek a separate Tamil homeland.

The 50 million Tamils in the Indian state of Tamil Nadu have strong affinity with the Sri Lankan Tamils and offered assistance to fellow Tamils across the Palk Straits. The Tamil insurgency developed into a civil war which threatened to tear the country apart; and the government's attempt to suppress secession provoked widespread allegations of genocide of the Tamils by the Sinhalese forces. Pressures began to grow among the Indian Tamils for India to intervene, and when India's various efforts to mediate failed, Indian prime minister Rajiv Gandhi compelled the Sri Lankan President Junius Jayawardene to sign the Indo-Sri Lanka accord on 27 July 1988 whereby Colombo agreed to the Tamil demand for autonomy within the framework of a united Sri Lanka.

Economic and social trends

Like its South Asian neighbors Sri Lanka is a low-income developing country and its economy displays some features of underdevelopment: low income, limited domestic market, excessive reliance on agriculture and export confined mainly to primary products. But in some respects Sri Lanka is very different from other low-income countries. It has achieved 87 percent adult literacy, life expectancy at birth is nearly 70 years and infant mortality is about 20 per 1,000 live births.

BANGLADESH

PEOPLE'S REPUBLIC OF BANGLADESH

CHRONOLOGY

1905	Region known as Bengal is partitioned
1911	Partition of Bengal is reversed
1947	Pakistan is created
1958	Military coup alienates Bengalis
1971	Bangladesh becomes independent under Sheikh Mujibur Rahman
1975	Military coup leads to killing of Sheikh Mujibur Rahman
1990	Student-led movement overthrows government
1991	Popularly elected government installed

ESSENTIAL STATISTICS

Capital Dhaka

Population (1989) 110,290,000

Area 143,998 sq km

Population per sq km (1989) 765.9

GNP per capita (1987) US$160

CONSTITUTIONAL DATA

Constitution Unitary multiparty republic with one legislative house (National Assembly)

Date of independence 26 March 1971

Major international organizations UN; Commonwealth of Nations; IDB

Monetary unit 1 Bangladesh taka (Tk) = 100 paisa

Official language Bengali

Major religion Muslim

Heads of government since independence (President) Sheikh Mujibur Rahman (1972–75); Khandakar Mushtaque Ahmed (Aug–Nov 1975); Abusadet Mohammed Sayen (1975–77); Maj.-Gen. Z. Rahman (1977–81); Abdus Sattar (1981–82); (Military dictator) Gen. H.M. Ershad (1982–83); (President) Gen. H.M. Ershad (1983–90); Shahabuddin Ahmed (1990–)

Before its war of liberation in 1971, Bangladesh was a part of Pakistan. Bengal was itself partitioned in 1905. The new province of Eastern Bengal, predominantly Muslim, roughly covered the same areas as the present-day Bangladesh. In 1947, when the British abdicated power in India, the Bengali Muslims opted for Pakistan to escape from the exploitation of the landlords who were mainly Hindus. But they had not turned their backs on the composite Bengali culture, which had evolved through centuries of Hindu and Muslim coexistence. Immediately after the creation of Pakistan it became clear that Islam was an inadequate bond to hold together such diverse people. The eastern wing of Pakistan, the home of the Bengalis, was separated from the rest of Pakistan by over 1,500 kilometers. They dressed and ate differently, and had different languages, customs, literature and way of life.

The sense of cultural alienation was heightened by the political and economic deprivation of the Bengalis. The military coup in 1958 meant that the Bengalis, who constituted the majority in Pakistan, were effectively excluded from political participation because of their virtual absence from the armed forces. Moreover, exclusion from political decision-making also increased economic discrimination against Bengalis.

A movement for constitutional autonomy emerged under the leadership of Sheikh Mujibur Rahman. The military dictator, Ayub Khan, was ousted by a spontaneous popular uprising in 1969; and in the election that followed Mujibs Awami League was returned with an absolute majority. But the military junta refused to transfer power to the Bengalis and instead nullified the election results. In the ensuing civil war the Bengali *Mukti Bahini* (freedom fighters), backed by India, defeated the Pakistan army and Bangladesh emerged as an independent state on 16 December 1971.

The postindependence history of Bangladesh is turbulent and tragic. Within three years of independence, a disaffected faction in the army staged a bloody coup, killing Sheikh Mujibur and much of his family. Between August 1975 and December 1990, during which there were numerous coups, counter-coups and assassinations, Bangladesh was ruled by a succession of military rulers. The four "fundamental principles" enshrined in the constitution – socialism, secularism, linguistic nationalism and democracy – were discarded.

A popular movement, led by the students, overthrew the government of General Hossain Mohammed Ershad in December 1990. In the first "free and fair" election in the country since 1975, the results were a surprise. The populist Awami League was trounced by the relative newcomer, Bangladesh Nationalist party. A popularly elected government was installed in March 1991, and in August the constitution was amended to revert to a parliamentary form of government.

Economic trends

Bangladesh is easily one of the poorest countries in the world. It suffers from overpopulation and chronic unemployment. Almost the entire population is malnourished, and especially the women and children. Development is constrained by meager natural resources. Industries contribute a mere 15 percent to the GDP and employ less than eight percent of the work force; the contribution of agriculture to GDP remains about two-fifths of the total. Periodic floods and cyclones add to the existing problems and prevent food self-sufficiency.

BAHRAIN

(State of Bahrain) **Capital** Al Manamah **Population** (1989) 489,000 **Area** 691 sq km **GNP per capita** (1987) US$6,610 **International organizations** UN, LAS, IDB.

Bahrain entered the 20th century with British protectorate status (since 1861), ostensibly to prevent Egyptian, Persian, Russian or German influence. Even before the discovery of oil in 1932, Bahrain was an importance commerce and shipping center in the Persian Gulf. The country is itself only a small oil producer, and its oil reserves are not expected to last beyond 2000; however, the large reserves of natural gas may last longer. In 1968, Britain decided to withdraw all its forces from the Persian Gulf, and Sulman Al Khalifah declared Bahrain independent in 1971, when Sheikh Isa (of the all-powerful Khalifah family) was designated emir of Bahrain. Tensions between Shi'ite and Sunnite Muslims increased at this time, with the Shi'ites pressing for greater participation in government. The system of government evolved was a mixture of traditional Arab autocracy and benign oligarchy, much influenced by the Western system: a constitution promulgated in 1973 created a national assembly but this was dissolved in 1975 after a period of political agitation, and the system reverted to the more traditional Arab *majlis*, whereby petitions were presented to the emir by citizens. Until 1970, Iran occasionally made claims to sovereignty over Bahrain but these were always rejected. Increased pollution in the Gulf, especially that caused by the Gulf War (1991), has almost killed off the fishing industry. In 1989 Bahrain established diplomatic relations with China and in 1990 with the Soviet Union and Czechoslovakia.

BHUTAN

(Kingdom of Bhutan) **Capital** Thimphu **Population** (1989) 1,534,122 **Area** 47,000 sq km **GNP per capita** (1987) US$160 **International organizations** UN.

During the 19th century, Bhutan had been plagued with internal strife as the territorial governors (penlops) and governors of forts (jungpens) fought each other for influence in the country. In 1907 the strongest of the penlops, the penlop of Tongsa, Druk Gyalpo, became the hereditary king of Bhutan. His successors ruled Bhutan from this time. The political, social and economic development of Bhutan occurred under the rule of Druk Gyalpo's grandson Jigme Dorji Wangchuk, who at the same time sought to minimize the discarding of traditional values. His son Jigme Singye Wangchuk continued these enlightened policies after his succession in 1972. In 1974 a plot by Tibetan refugees, who numbered around 7,000, coming to Bhutan after the Chinese invasion of Tibet in 1949, was discovered to assassinate the king. The Tibetans were regarded as a security risk, and moves were made by the Bhutanese National Assembly, and backed by the king, to extradite those Tibetans without Bhutanese citizenship to China. Bhutanese culture and language is nearly identical with that of Tibet, with the exception of south and southwestern Bhutan where there is a Nepali majority who practice Hinduism. Otherwise lamaistic Buddhism is the dominant religion. The king has supreme executive power, though a council of ministers shares administrative responsibility. There are no political parties in Bhutan. In 1983 Bhutan co-founded SARC (the South Asian Regional Cooperation organization).

MALDIVES

(Republic of Maldives) **Capital** Male **Population** (1989) 209,000 **Area** 298 sq km **GNP per capita** (1988) US$410 **International organizations** UN, Commonwealth, IDB, Colombo Plan, SAARC.

A large archipelago of tropical atolls, 1087 islands in all, of which 215 are inhabited, the Maldives were settled before 4500 AD by Buddhists from Ceylon (now Sri Lanka). A large proportion of the population remains Sinhalese, as does the basic language structure. In 1153 Arab settlers brought Islam to the islands, and it is now the State religion, although the Maldives have been distinguished among Muslim nations by the relative freedom of women. The Maldives were taken over as a protectorate in 1887 from the Dutch by the British, who never colonized. In the 1930s the king was deposed and a new king elected; he resigned during World War II and a regency was established until the proclamation of a republic in 1953 under President Amin Dada; but he was deposed and killed the following year and the monarchy restored, with an elected king who ruled until 1978, when a new republic was formed with the former prime minister, Nasir, as president. Britain had acknowledged Maldivian independence in 1965, but retained an airbase there on lease from the government until 1976. In 1978 Nasir resigned and fled to Singapore; he was tried in absentia on charges of misappropriation of funds. Maumoon A. Gayoom took over as president and prime minister. In 1982 the islands joined the Commonwealth; a tourist boom in the 1980s brought about improvements in the infrastructure and a demand for vocational training. Gayoom was reelected in 1983. In 1985 the Maldives was a founder member of SAARC. A coup attempt in 1988 by Sri Lankan mercenaries was quashed with Indian military support; later that year the president was reelected for a third term. The nation's economy is still based on tourism; the literacy rate, at 93 percent (1986) is the highest in Southeast Asia, and the infant mortality rate just below the level considered to indicate endemic malnutrition (1987). Violence is extremely rare and the country is unique in having no prisons; nor are there organized political parties; government takes place through the Majlis (House of Representatives). In 1989 the Maldives was one of several island states concerned with the threat of extinction in the event of a projected rise in ocean levels caused by intensification of the "greenhouse effect", as laid out in a UN report that year.

MONGOLIA

(Mongolian People's Republic) **Capital** Ulan Bator **Population** (1990) 2,094,000 **Area** 1,566,500 sq km **GNP per capita** (1986) US$1820 **International organizations** UN, CMEA.

After Uighur (Turkic) rule (4th–10th century AD), Genghis Khan, a charismatic and brilliant strategist, united the vying nomadic Mongolian peoples and conquered much of Asia, winning the title Ruler of All the Mongols. His son and successor, Ogoden (1229–41), overthrew the Chinese Chin dynasty, and Kublai Khan (1259–94), established a Mongol ("Yuan") dynasty in China for 100 years, and the largest land empire in history. But from the Ming dynasty, China struck back and Mongolian unity was destroyed. The Manchus took first Inner Mongolia and eventually Outer (Northern)

Mongolia; after their downfall in 1912, the Japanese fomented Mongolian nationalism, and Outer Mongolia became independent under the Jabtsandamba Khutagt (the "living Buddha"), with limited support from the politically circumspect Russians, until the Chinese, under Japanese influence, returned in 1917. In 1920, fleeing White Russians invaded, subdued the Chinese and oppressed the Mongolians; they were repelled, and the Chinese expelled, by Mongolians under the popular hero Damdiny Suhbaatar, with Bolshevik help. On the death in 1924 of the Jabtsandamba Khutagt, Outer Mongolia became a People's Republic, under the Revolutionary Party, economically dependent on the Soviet Union. In 1939 a Japanese attack was repulsed by national and Soviet troops. Mongolian demands for independence were finally granted in a Sino–Soviet treaty of 1950, and universally recognized in 1961. Ironically, this coincided with a period of erosion of many national characteristics; the Mongolian alphabet had been lost in the 1940s, and the spread of urbanization was accelerating. The constitution of 1960 lays down government by the Khural, a group of elected deputies, one for every 2,500 people, who elect a Presidium and a Council of Ministers, in a one-party state. In 1986 relations with China improved and treaties were signed in 1986, 1987 and 1988. The late 1980s saw wide public demand for political and economic reforms. In 1989 the Democratic party was founded. The Soviet Union agreed to withdraw its military, and in 1990 the MPRP (Communist) party presidium resigned and gave way to a five-person politburo. A multiparty system was legalized; however, in that summer's elections the new president, Ochirbat, was the former (MPRP) Presidium chair. He promised to introduce a market economy, pursue a nonaligned foreign policy and improve welfare and income levels. Food shortages in 1991 necessitating rationing were attributed to the failure of central planning. Wages and benefits were doubled (so did prices, except on some goods which remained controlled). Despite the economic and cultural changes of the 20th century, traditional customs have remained, like the 2000-year-old Three Games of Men, an annual series of athletic contests, and the New Year national horseriding competitions, dating from the Bronze Age.

NEPAL

(Kingdom of Nepal) **Capital** Kathmandu **Population** (1989) 18,699,884 **Area** 147,181 sq km **GNP per capita** (1988) US$170 **International organizations** UN.

Nepal entered the 20th century under British "guidance" regarding its foreign policy, under the terms of a treaty of 1860 safeguarding the kingdom's independence. In return for this "protection", which was occasionally a slight irritation to both parties, Nepal allowed the recruitment of Nepalis for the British Indian Gurkha units, and was granted virtual domestic autonomy. Nepal maintained good relations with China and Tibet however, so as to balance British ascendancy in South Asia. After British withdrawal from India in 1947 Nepal was suddenly left exposed to political instability, which came first from groups in India hostile to the Rana family, who had effectively ruled Nepal since 1846; these groups formed an alliance with the Nepalese royal family led by King Tribhuvan, and revolted against the Ranas in 1950. This had the result of restoring the monarchy, and the revolutionary forces, under the banner of the Nepali Congress Party, dominated the administration of the country. Democracy was difficult to establish in a country accustomed to autocratic rule, and it was not until 1959 that a constitution was approved. In general elections during that year the Nepali Congress Party won a large majority, but disagreement between the Cabinet and King Mahendra (reigned 1955–72) led the latter in 1960 to dissolve the government and arrest its leaders, abolishing the constitution in 1962 and promulgating a new constitution by which all authority was vested in the monarch. He was succeeded in 1972 by his son Birendra who maintained the nonparty system while trying to liberalize the economy. These efforts were disappointing, and forced Birendra in 1980 to hold a referendum on the issue of returning to a multiparty system. 55 percent were in favor of retaining the nonparty system. As a compromise, Birendra decided to maintain the 1962 constitution but to liberalize the political system by allowing for popular election of the National Assembly, and to allow formerly illegal parties to function with only small constraint. This still failed to satisfy either side. After the success of the prodemocracy movement in the face of violent suppression, political parties were unbanned in 1990, and the king approved a new constitution transferring political power to an elected government. In the 1991 election the Nepali Congress, under G. P. Koirala, came to power.

OMAN

(Sultanate of Oman) **Capital** Muscat **Population** (1989) 1,200,000 [no census ever taken] **Area** 300,000 sq km **GNP per capita** (1988) US$5,070 **International organizations** UN, LAS, IDB, GCC.

Oman's ruling Al Bu Sa'id dynasty was supported by the British throughout the late 19th century against revivals of the Ibadite imamate in the Arabian interior. The British negotiated an agreement between the belligerent imam-ruled tribes and the Omani sultan Taymur ibn Faisal in 1920. In 1951 Oman achieved full independence. The imam-sultan relationship broke down in 1954 on Imam Mohammad al-Khalili's death, when his successor, Ghalib, gained Saudi Arabia's support against the Omani sultan Sa'id ibn Taymur. Saudi Arabia had occupied an area administered jointly by Oman and Abu Dhabi in the hope of striking oil. After arbitration over this issue broke down in 1955 the British moved in and expelled the Saudi occupiers, and a British-led regiment moved into the Omani interior to suppress the imamate's forces, finally succeeding in 1959. In 1968 a rebellion in Dhofar by mountain tribes controlled by the Marxist Popular Front for the Liberation of the Occupied Arab Gulf (later the Popular Front for the Liberation of Oman, the PFLO), with Chinese, then Soviet support, caused Sultan Sa'id to be overthrown in 1970 by his son Qabus ibn Sa'id, who invited British military aid to quell the rebellion, which was finally defeated in 1975. Qabus in 1970 began a program of modernization of Oman, creating a modern governmental structure, and formed close diplomatic ties with other Gulf states – Oman was a founding member of the Gulf Cooperation Council (GCC) in 1981. It maintained neutrality during the Iran-Iraq war owing to its strategic position on the Strait of Hormuz.

QATAR

(State of Qatar) **Capital** Doha **Population** (1989) 427,000 **Area** 11,400 sq km **GNP per capita** (1988) US$11,610 **International organizations** UN, LAS, I-ADB, IDB, OPEC.

In 1916 the powerful Al Thani family signed an agreement with Britain giving the latter full control of Qatar's foreign policy, in return for British protection. Britain played a key role in the country until independence in 1971. After independence, Qatar changed its treaty with Britain to one of friendship, and in 1981 joined five other Arab Gulf states in forming the Gulf Cooperation Council to protect against the threat of the Islamic revolution in Iran and against a spillover of instability from the Iran–Iraq war. Qatar is dominated by the ruling Al Thani, who constitute around forty percent of the population, and the hereditary emir (from this family) appoints a council of ministers to administer the country. Justice is based on Koranic principles, following the lead of Saudi Arabia, adhering to the conservative Hanbali school of jurisprudence. Oil was discovered in Qatar in 1939, and this resource is expected to last only until around 2020 at the rate of production during the mid 1980s.

UNITED ARAB EMIRATES

(United Arab Emirates) **Capital** Abu Dhabi **Population** (1989) 1,827,000 **Area** 77,700 sq km **GNP per capita** (1988) US$15,720 **International organizations** UN, LAS, IDB, OPEC, GCC.

UAE is the desert federation comprising Abu Dhabi, Dubai, Sharjah, Ajman, Umm al Qaiwain, Ras al Khaimah, and Fujairah, and is one of the world's wealthiest states, formerly the Trucial States, so called as a result of three treaties with Britain, the last one signed in 1892, giving Britain exclusive trading rights. Oil was found in Abu Dhabi in 1958, and in Dubai in 1969; both states were transformed and eventually developed welfare states as a result of oil revenues. The oil-based economy centered on Abu Dhabi, which now provides over 60 percent of the national income. In 1971 the British withdrew from the Gulf, and the Trucial States became the UAE (except for Ras al Khaimah, which joined in 1972.) But in 1973 Abu Dhabi moved to pull out of the federation, until several Abu Dhabi ministers were elected to federal government. Dubai retained its own army; the appointment in 1979 of its premier to federal premiership kept it in the federation. All decisions need the approval of at least five member states including Abu Dhabi and Dubai. To add to the potential instability, 80 percent of the population are foreign workers. UAE has tried to reduce its dependence on oil exports, and has almost achieved self-sufficiency in fruit and vegetable production. Abu Dhabi has also begun to use natural gas. The UAE, with five other states, formed the Gulf Cooperation Council (GCC) in 1981 in response to the Iran–Iraq war. The UAE gives aid to other states and has mediated in local conflicts. Iraq's invasion of Kuwait in 1990 caused panic, for Iraqi president Saddam Hussein had accused the UAE, too, of over-production; in the ensuing conflict, Allied forces were stationed there.

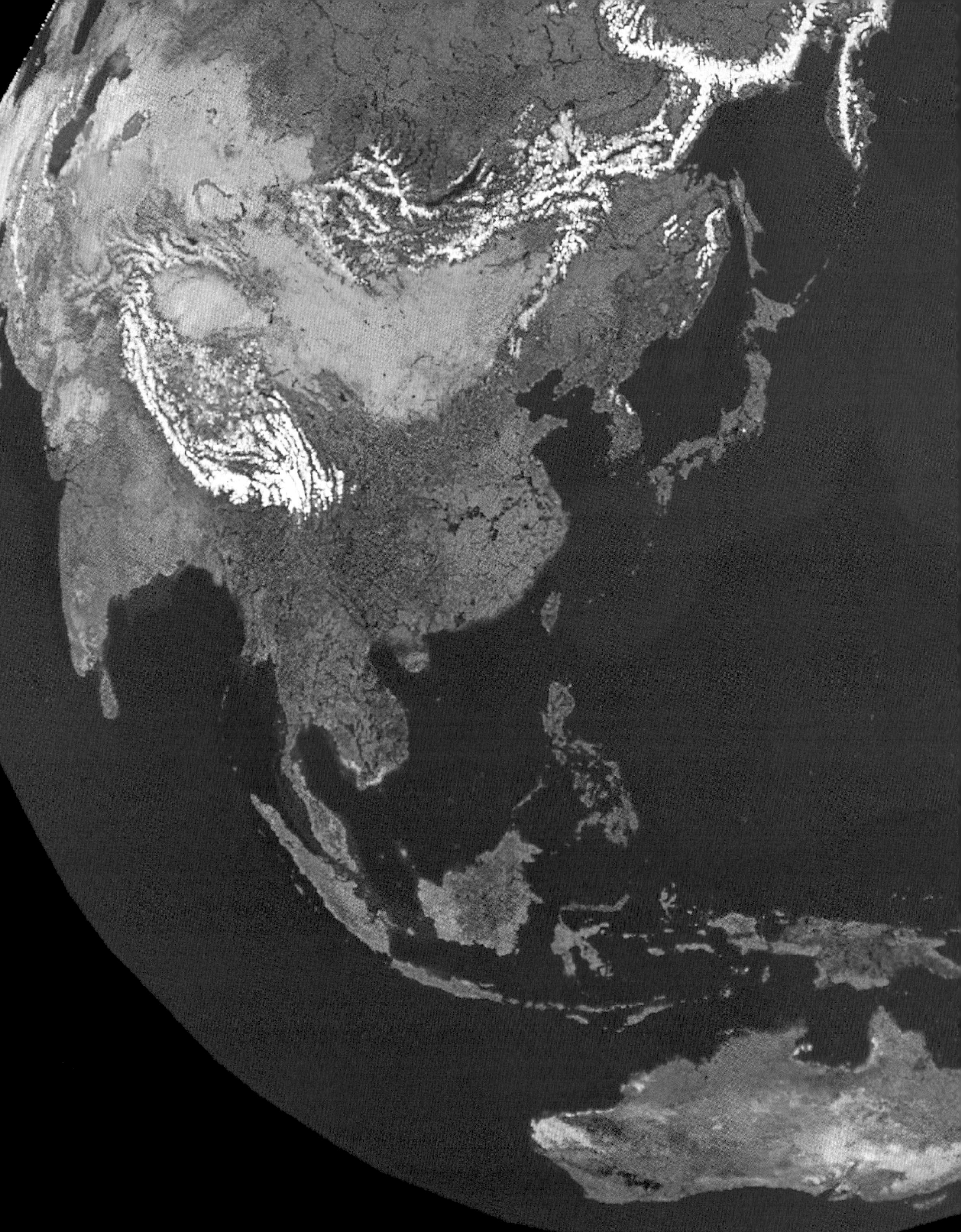

EAST ASIA AND OCEANIA

INDEX OF COUNTRIES

CHINA

PEOPLE'S REPUBLIC OF CHINA

CHRONOLOGY

1900	Government-inspired "Boxer Rising" breaks out
1911	Fall of the Manchu dynasty
1921	Chinese Communist party is founded
1949	Proclamation of the People's Republic of China
1958	The "Great Leap Forward" attempt to achieve economic modernization
1966	The "Great Proletarian Cultural Revolution" increases Mao's popularity
1982	New state constitution is declared
1989	Tiananmen Square massacre of over 1,000 people

ESSENTIAL STATISTICS

Capital Beijing

Population (1989) 1,104,275,000

Area 9,572,900 sq km

Population per sq km (1989) 115.4

GNP per capita (1987) US$300

CONSTITUTIONAL DATA

Constitution Single-party people's republic with one legislative house (National People's Congress)

Major international organizations UN

Monetary unit 1 Renminbi (yuan) (Y) = 10 jiao = 100 fen

Official language Mandarin Chinese

Major religion Chinese folk-religionist

Heads of government (President) Yuan Shikai (1912–16); Sun Yat-sen (1917–25); Jiang Jieshi (1925–49); Mao Zedong (1949–76); Hua Guofeng (1976–78); Deng Xiaoping (1978–90); Li Peng (1990–)

In the course of the 19th century the serene self-confidence of the Chinese imperial government was severely shaken by the aggression of technologically superior Western powers and a series of massive rebellions which rendered large areas of the country beyond the control of central government for years at a time. The two currents came together in 1900 in the so-called "Boxer Rising", an antiforeign, anti-Christian mass movement which provoked armed intervention by the imperialist nations. The Rising confirmed the division of most of the more economically advanced coastal areas of China into "spheres of influence" under the control of the major foreign powers. Confused attempts at domestic reform, including even the abolition of the centuries-old examination system of bureaucratic recruitment, came too late to save the Manchu dynasty from revolutionary overthrow in 1911. In 1912 a republic was proclaimed under Sun Zhong San (also known as Sun Yat-Sen), the foreign-educated leader of the Guomindang (Nationalists).

The new republic, however, commanded even less of the allegiance of the entire country than had the Manchus. By the time of Sun's death it controlled a southern heartland around Guangzhou (Canton), while the north was contested by warlords. Such circumstances favored the emergence of a military leader and Sun's successor was the commander of the Nationalist forces Jiang Jieshi (Chiang Kai-shek). He campaigned vigorously in the north but soon found himself challenged by a new political force – the Communists. The Chinese Communist party (CCP), inspired by the Bolshevik model and strategy, had been established in Beijing in 1921 and, following the Russian example, attempted to foment revolution by capitalizing on the discontent of the urban poor. Communist-led city risings were, however, brutally crushed by the Nationalists and by 1928, the year in which Jiang Jieshi established his capital at Nanjing, the outlook for the CCP seemed bleak indeed. The Communist leader Mao Zedong recognized, however, that effective political action must be based on a realistic understanding of its social and economic context and that the mass backing needed to drive through a revolutionary transformation in China must come from the peasantry who made up the overwhelming majority of the population. Quite simply, he grasped that Russia and China were profoundly different.

Mao's strategy, therefore, concentrated not on attempting to launch a single decisive blow to seize control of a key center of power, but on establishing "liberated areas" which could serve both as secure bases from which to launch further offensive operations and as model communities prefiguring the social vision for which the struggle was being fought.

Beginning from his home province of Hunan in south-central China, Mao moved into the remote fastnesses of Jiangxi where a Chinese Soviet Republic was proclaimed in 1931. Alarmed, Jiang drove the Communists from their base in 1934, precipitating the epic "Long March", during which 100,000 Communists toiled and fought their way 9,000 kilometers (5,600 miles) to safety in the northwestern province of Shaanxi. Only a third survived the ordeal.

Having marginalized the Communists, Jiang turned to face the Japanese, who had seized resource-rich Manchuria in 1931 and in 1937 attempted a fullscale conquest of China, quickly seizing the coastal areas. After accepting Communist cooperation in resisting the invaders, the Nationalists proved unable to reestablish their control, even with United States support. A Soviet invasion of Manchuria in August 1945 boosted the CCP's prospects and in effect marked the opening of a civil war which was not averted even by determined American efforts. The better-armed and US-supported Nationalists were eventually undermined by Mao's guerrilla strategy and by 1948 the Communists were able to face their opponents in set-piece battles. Between November 1948 and January 1949 a series of decisive encounters was fought out in which the Nationalists lost some half a million men, and were forced to flee to the offshore stronghold of Taiwan. On 1 October 1949 Mao proclaimed the establishment of a People's Republic.

Initially reliant on the Soviet Union for both military and economic assistance, China began to question the value of this link as a result of border disputes and concern about Soviet influence in neighboring Vietnam. The deterioration in Sino-Soviet relations was expressed in terms of ideological differences over the nature of Communism and through China's determination to develop an independent nuclear program. It reached its logical conclusion in a rapprochement with the West, which led to China's admission to the United Nations (1971) and official visits to China by the United States president Nixon and Japanese prime minister Tanaka.

Domestically, Chinese affairs under Mao followed a seesaw pattern of consolidation and radicalization. In the immediate aftermath of the civil war the priority was to extend the authority of the central government over the whole country and stabilize a situation of economic chaos. A brief flirtation in 1957 with open debate ("let a hundred flowers bloom...") was followed by a crackdown and the launching of the frenetic and abortive Great Leap Forward (1958–60), intended to achieve economic modernization at a single step. Temporarily restrained by more pragmatic comrades, Mao rebounded with a Cultural Revolution of 1966–68 which took his personality cult to new heights and launched youthful Red Guards into a frenzied assault on the CCP hierarchy, accusing it of corrupting both the ideology and the practice of the revolution. The aftershocks of this immensely damaging upheaval continued after 1976, when Mao died, with the trial in 1980–81 of the Gang of Four, an act of retribution against the ultraleftists on behalf of their victims. In 1981 a party document severely criticized the policies of Mao's declining years and in 1982 new party and national constitutions were promulgated to guard against a further eruption of his volatile political legacy.

▼ Boxers (radical nationalists) fighting westerners in 1900.

The Third Plenum of the CCP's Eleventh Central Committee in 1978 marked a major turning-point as Deng Xiaoping, a survivor from the days of the Long March, assumed effective direction of the leadership and set the country on a more moderate course of cautious modernization at home and conciliation abroad. The United States and China normalized diplomatic relations on 1 January 1979.

Economic liberalization soon began to raise living standards dramatically but was accompanied by a growth in Western cultural influences disturbing to old-guard revolutionaries. Even more alarming, it gave rise to demands for a more open and democratic political system. These pressures culminated in a massive student sit-in at the nation's symbolic heart, Beijing's Tiananmen Square, in the summer of 1989. Pushed to the limit of their political imagination, the aged party leaders ordered that the demonstration should literally be crushed by the tanks and weapons of the "People's Army". Zhao Ziyang, general secretary of the party, was blamed for being too conciliatory and stripped of his power. Ex-security chief Qiao Shi was the coming man. International protests at the brutality of the repression were ignored, prompting fears that, however startling its apparent transformation since the early 1970s, some things in China had changed very little.

Economic and social trends

Western visitors to China at the turn of the century noted the striking contrast between the presence of modern industrial technology and communication systems in the great coastal cities, and the prevalence in rural areas of centuries-old agricultural and craft techniques.

Ironically, World War I, which devastated the European powers which had so long preyed upon China, gave a *de facto*

protection from foreign competition, and this boosted China's emerging manufacturing sector. During the chaotic interwar years industrial development continued in the commercial metropolis of Shanghai and in Japanese-occupied Manchuria, but agriculture languished.

After 1949 the main Communist priorities were land reform, to secure peasant support for the new regime, and control of runaway inflation. By 1953 the party felt able to commit itself to an outline Five Year Plan which gave priority to heavy industry and was to be financed by the export of primary products to offset crucial Soviet machinery imports. Although high rates of investment were achieved, the delayed pay-off from such ventures meant they had little impact on living-standards, which were anyway depressed by the relative

▼ **The advances of the 1950s and 1960s were based on human muscle.**

Tibet

▲ **The 14th Dalai Lama, Tibetan spiritual and state leader in exile.**

Bleak, mountain-ringed Tibet had little difficulty in maintaining its traditional seclusion until the end of the 19th century. Too inhospitable to attract a conqueror for its own sake, it accepted nominal Chinese overlordship and served as a useful buffer between the contending imperialisms of the Manchu Dynasty and British India. Thus secured, the Tibetans could adhere to their own language and traditional culture, based on a distinctive variant of Buddhism which made a monk out of perhaps one in six of the adult male population.

A punitive British expedition in 1904 was followed by British recognition of Chinese suzerainty, which was then enforced by armed Chinese intervention for the first time in a millennium. Then the Manchu Dynasty collapsed and Tibet reasserted its independence (1912), defending it in border clashes until as late as 1931.

Spiritual and nominally political leadership of the nation rested with the Dalai Lama. The death of the 13th occupant of this post in 1933 was followed by a drift back into the Chinese orbit. The ending of China's civil wars was swiftly followed by an invasion of Tibet in October 1950 to reassert Chinese territorial claims. Flouting pledges to respect local autonomy and the traditional authority of the Tibetan lamas, the Chinese curbed monastic powers and initiated land reforms. This provoked a widespread nationalist revolt in March 1959 which was harshly suppressed and led to mass emigrations and the flight of the Dalai Lama to exile in India. Frontier frictions then led to a brief border war with India which decisively confirmed Chinese claims.

Chinese rule in Tibet became increasingly heavy-handed as the public practice of Buddhism was forbidden, agriculture collectivized and an attempt made to remodel society along Chinese lines. Foreign visitors were banned altogether between 1963 and 1971. This harshness gradually gave way to a more relaxed policy, tolerating some religious celebrations, fostering economic development and even encouraging tourism. An outbreak of popular demonstration against Chinese rule in 1987 showed clearly, however, that the Tibetans had by no means accepted the new order. The Chinese response was brutal repression. Although the Chinese had built highways and started air services, Tibet's rich mineral and forest resources still remained largely inaccessible and unexploited.

neglect of agriculture – which in turn jeopardized the ability to pay for imported technology. The country suffered a significant loss of entrepreneurial talent as those who preferred the capitalist way made good their escape.

Opting for a more radical strategy, the party began the collectivization of agriculture and nationalization of industry in 1955–56 and then moved even further from the Soviet model with the Great Leap Forward. It organized the rural population into some 3,000 large, production-oriented, quasi-autonomous "communes" and looked to ideological mobilization rather than conventional economic incentives to provide the momentum for modernization. Longer hours, frenzied effort and inspired ingenuity led to the construction of thousands of "backyard furnaces" and other low-technology, smallscale enterprises, which boosted gross industrial output but, from the point of view of quality or efficiency, represented a grotesque misallocation of resources. The communes, meanwhile, produced a spectacular fall in agricultural output. Two disastrous harvests, compounded by the withdrawal of Soviet technical aid, led to a fundamental revision of development objectives. The previous orders of priority, with agriculture put first and industry directed to serve its needs, were now reversed. Financial incentives and managerial authority were restored. Private plots of land were restored to the peasants and surplus labor transferred from industry to the farms. These tendencies outraged revolutionary purists and were severely attacked during the Cultural Revolution, which canceled bonuses, displaced much managerial talent and disrupted transport, trade and industry.

During the early 1970s development priorities were restated as the Four Modernizations – of agriculture, industry, defense, science and technology – a program originally articulated as far back as 1964. The launching of China's first satellite in 1971 was a dramatic symbol of what could be achieved.

After 1978 Deng Xiaoping's pragmatism promoted economic management through more realistic pricing, wage differentials, wider managerial discretion and a far greater openness to foreign trade and investment. The advent of the new regime was clearly marked by the passage of novel legislation relating to the protection of foreign patents, contracts and the administration of joint ventures. Four freer "special economic zones" were established in the south; in 1984 14 cities were opened to direct involvement in the international economy.

Agriculture was stimulated by a sharp increase in farm prices in 1979 and a progressive dismantling of collective farming in favor of a return to family enterprises. Change in the urban sector, however, proved more problematic, partly owing to its inherent complexity, but more because bureaucratic control had become so deeply entrenched and the disruption of vested interests in the existing order carried the danger of political turmoil, as did the dismantling of subsidies that distorted the pricing system. Light industry and service trades opened up to competition between collectives, cooperatives and private entrepreneurs. But the emergence of a more market-oriented system, while offering the urban consumer unprecedented choice, appeared to have been accompanied by a resurgence of crime and corruption.

The changing pattern of China's foreign trade encapsulates many of the broad trends of its recent economic development. As late as 1950, 70 percent was with the non-Communist world. By 1954 the Communist share was 74 percent. The withering of Soviet influence reduced this to 30 percent by 1965 and by the 1980s it had shrunk to less than ten percent, while more than half was directed to developed capitalist economies such as the United States and Japan. The fact that China's overall balance is comfortably in surplus is another endorsement of the more open policy followed since Mao's death.

The rapid population growth of the early 19th century was slowed by the rebellions, wars and natural disasters of the following hundred years. Improvements in public order, sanitation, medical care and the distribution of food after 1949 brought such a renewed surge in population growth and consequent pressure on food supplies that government-sponsored birth-control programs were initiated as early as 1955. By the 1960s late marriage and the use of contraceptives were being publicly advocated. A decade later a strong element of compulsion was being exerted and by 1979 the one-child family was being exalted less as an ideal than as a required norm.

Immediately after coming to power the Communists launched a major campaign to raise the literacy rate among adults as well as children. The Great Leap Forward pressed for the politicization, and proletarianization, of technicians that they might better become both "red and expert". The Cultural Revolution went much further, disrupting virtually the entire postelementary education system, terrorizing many teachers and exalting ideological purity over academic excellence as the criterion for scholastic advancement. Pragmatic considerations after the late 1970s reinstated elite secondary schools and competitive examinations. Even more significant perhaps was the development of overseas study programs for postgraduates and the importation of foreign academics.

Cultural trends

In terms of population China is over ninety percent Han Chinese, but there are also over fifty identifiable minority groups spread over some two-thirds of the total area of the country, notably in the frontier areas of the northeast, northwest, west and southwest. Major minority groups include the Ching-chia, Tai, Tibetans, Miao-Yao, Mongolians and various Turkic-speaking Muslim peoples such as the Uighurs, Kazakhs and Kirgiz. More than half of these groups had no written form of their language before the accession to power of the Communists, who protected their cultural identities as well as promoting their material well-being. The ethnically homogeneous Han majority shares a common culture, traditions and written language, but is divided by the spoken forms through which that language is expressed. Even the official tongue – known in the West as Mandarin, in China as *p'u-t'ung hua* (common speech) – has three main variant modes, relating to different geographical regions.

The Cultural Revolution damaged and destroyed cultural treasures of inestimable value and attempted to outlaw the practice of many ancient arts and crafts. After Mao's death cultural liberalization fostered a diversification of expression which embraced both a greater openness to Western influences and a reassertion of traditional forms. And the care and pride with which recent remarkable archeological finds have been displayed suggests both a renewal of cultural self-confidence and a shrewd appreciation of the commercial potential inherent in a heritage so vast, varied and distinctive.

Physical artifacts can be vandalized or conserved. Artistic expression can be patronized or persecuted. Social customs are, however, less open to ready manipulation. If female foot-binding was gladly abandoned as a barbaric vestige of the past other aspects of tradition have proved more tenacious as the reiteration of official disapproval implicitly confirms. By 1990 exposure of female infanticide or bride-purchase in some remote rural area might now be occasion for public scandal but reverence for ancestors and a marked preference for male offspring was still widespread.

Hong Kong

A convention signed by the British and Chinese governments in 1898 assigned to the former a 99-year lease on the "New Territories", a 1,000-square-kilometer (400-square-mile) hinterland for Hong Kong island and the Kowloon peninsula, which Britain already held. Britain gave the colony constitutional government though not democracy, and this, plus the economic opportunities offered by a free port, proved enough to attract successive waves of migrants from mainland China throughout the century.

The Japanese occupation of December 1941 to August 1945 reduced the colony's population from 1,600,000 to 650,000 but numbers recovered with an influx of refugees from the Chinese civil war. Industrial discontent briefly turned into political violence in 1967 under the influence of the Cultural Revolution but was relieved by social programs.

On 19 December 1984 the British and Chinese governments came to an agreement on the future of Hong Kong. From 1 July 1997 the colony would revert to Chinese sovereignty but its existing social and economic system would be guaranteed for 50 years by a Hong Kong Special Administrative Region directly under central government control. The 1989 massacre of students and protestors in Beijing's Tiananmen Square made this seem an uncertain prospect for a borrowed place living on borrowed time. Between 1984 and 1997 the British undertook to introduce a program of democratization in Hong Kong, to discourage widespread emigration through fear of the Chinese political system. In the 1990s the situation was complicated by the pressure in Hong Kong of thousands of "boat people" who had escaped from Communist Vietnam. They were generally denied refugee status and returned to Vietnam.

Hong Kong is smaller than many cities and its only resources are its superb harbor and its people. Together they made it the world's biggest container port and 13th largest trader. Industry remained underdeveloped until the 1951 UN embargo on China and North Korea greatly damaged entrepôt trade. Cheap labor and refugee mainland entrepreneurs encouraged light manufactures, especially textiles. By the 1970s rising skill standards facilitated the development of electronics but by the 1980s industry was overshadowed by the colony's new status as the world's fourth largest financial center.

Hong Kong's split personality is reflected in the dual status of English and Chinese as official languages, and the observance of traditional Chinese festivals alongside Western public holidays. Cultural life developed vigorously with the establishment of two universities (1911 and 1963) and two polytechnics (1972 and 1984) and the inauguration in 1977 of an international Film Festival which celebrates the enormous productivity of local studios.

▲ **Hong Kong Stock Exchange, a growing financial center.**

▼ **"Boat people", refugees from Vietnam to Hong Kong.**

NORTH KOREA

DEMOCRATIC PEOPLE'S REPUBLIC OF KOREA

CHRONOLOGY

1910	Annexed by Japan
1948	Democratic People's Republic of Korea established
1950	War breaks out with South Korea – ending in 1953
1958	Collectivization of agriculture
1972	New constitution inaugurates executive presidency
1980	Kim Jong-il designated as official successor to Kim Il-Sung
1983	Assassination of South Korean Cabinet delegation in Rangoon (Burma)

ESSENTIAL STATISTICS

Capital Pyongyang

Population (1989) 22,418,000

Area 122,400 sq km

Population per sq km (1989) 183.2

GNP per capita (1986) US$910

CONSTITUTIONAL DATA

Constitution Unitary single-party republic with one legislative house (Supreme People's Assembly)

Date of independence 9 September 1948

Major international organizations UN affiliations

Monetary unit 1 won = 100 chon

Official language Korean

Major religion Traditional beliefs

Heads of government since independence (Party leader) Marshal Kim Il Sung (1948–72); (President) Marshal Kim Il Sung (1972–)

Korea, renowned as "the hermit kingdom" in its isolation under the long-lasting Yi dynasty (1392–1910), sought to realize the ideal of Confucian social order: cultured scholar-gentlemen (the *yangban* class) benevolently administered a diligent peasantry, while keeping the necessary evil of commerce in a subordinate and marginal position. Japan took the country over following the Russo-Japanese war of 1904–05; between 1905 and 1945 the brutal Japanese rule, and the Japanese policy of systematically destroying Korean culture, followed by the immense physical destruction, partition and the internationally-sponsored civil war (1950–53) destroyed this vision for ever. Yet, in a curious way, the self-sealed Communist regime which eventually replaced it, represents a restatement of the traditional ideal in a modern form. The Democratic People's Republic of Korea (otherwise known as North Korea) was established, with Soviet backing, on 9 September 1948. It was dominated from its foundation by ex-Soviet army major Kim Il-Sung, first as premier and, after the inauguration of the 1972 constitution, as president. As head of the central people's Committee he controlled the supreme decision-making body and became the focus of a personality cult. In 1980 his son, Kim Jong-Il, already a holder of key party posts, was designated as his official successor.

Economic and social trends

Japanese colonial rule forced economic modernization, including the establishment of railways, the exploitation of coal and other mineral resources, the construction of hydroelectric plants along the Yalu river and the building of fertilizer factories (which could later be converted to explosives production). The emphasis on the development of heavy industry continued after the Communist takeover. In the late 1980s manufacturing employed a quarter of the labor force but a scarcity of consumer goods persisted. Minerals, especially coal, earned much-needed foreign exchange. Although two-fifths of the labor force was still employed in agriculture (after 1958 in cooperatives modeled on Chinese communes), and individual households were allowed small plots of land, foodstuffs were still subject to rationing and occasional shortages.

Despite the potential of its mineral wealth and disciplined workforce, North Korea remained economically retarded owing to huge defense expenditures, a mismanaged external debt and a lack of advanced technology.

The population of North Korea is 99.8 percent Korean. The only significant minority consists of some 50,000 Chinese. The war years saw not only the expulsion of foreign influences (such as missionaries) and the general disruption of civil society, but also the deliberate destruction of centuries-old lineage records to undermine the established kinship system and create discontinuity with the past. The country's youthful age-structure means that by 1990 three-quarters of the population had never known anything but an all-embracing Communist regime; nevertheless, about one person in three still adhered to some form of religious observance.

The welfare achievements of the regime were impressive. The literacy rate was over ninety percent and life expectancy had been raised to about seventy years. Medical care, like education, was free, but trained personnel scarce. Despite vigorous efforts parasitic diseases persisted in rural areas, reflecting the lack of basic services in villages.

SOUTH KOREA

REPUBLIC OF KOREA

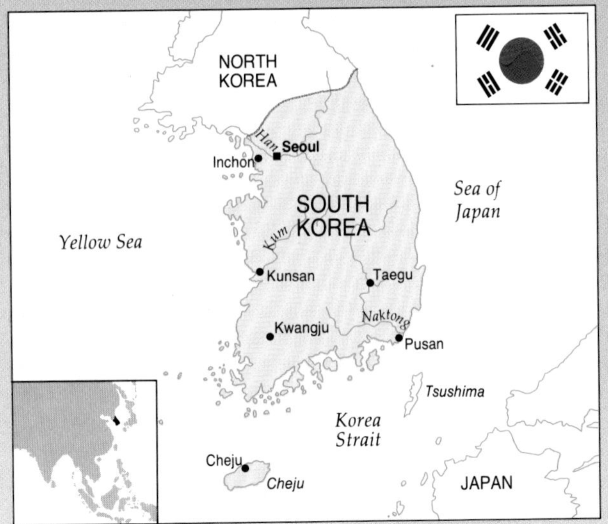

CHRONOLOGY

1910	Annexed by Japan
1919	Demonstrations for independence
1948	Republic of Korea proclaimed independent
1950	War between North and South Korea until 1953
1962	Beginning of export-led growth
1987	Direct election of President Roh Tae Woo
1988	Olympic Games held in Seoul

ESSENTIAL STATISTICS

Capital Seoul

Population (1989) 42,380,000

Area 99,237 sq km

Population per sq km (1989) 427.1

GNP per capita (1976) US$2,690

CONSTITUTIONAL DATA

Constitution Unitary multiparty republic with a National Assembly

Date of independence 15 August 1948

Major international organizations UN affiliations

Monetary unit 1 won (W) = 100 chon

Official language Korean

Major religion Buddhist

Heads of government since independence (President) Synghman Rhee (1948–60); Chang Myon (1960–61); Park Chung Hee (1961–79); Choi Kyu-hah (1979–80); Chun Doo Hwan (1980–88); Roh Tae Woo (1988–)

During the 19th century the peace of Korea was troubled by widespread social discontent which provided a pretext for intervention by Korea's more powerful neighbors. Outright annexation of Korea by Japan came in 1910. Koreans declared their independence on 1 March 1919 but, despite widespread popular support, the effort was easily beaten down by the Japanese occupation forces. A quarter of a century of oppression followed, until the Japanese were disarmed in 1945 by Soviet forces moving in from the north and Americans from the south. The peninsula, for the first time in a millennium, was divided. The establishment of a Republic of Korea (also known as South Korea) was declared on 15 August 1948.

On 25 June 1950 Communist forces from the north invaded to attempt reunification by force. Three years and four million casualties later, despite intervention from China and the United Nations spearheaded by United States forces, the division of the peninsula was reconfirmed roughly where it had been before, along the 39th Parallel.

Politically speaking, postwar Korean history was a story of strong-arm rule. Veteran nationalist Synghman Rhee held the reins until overthrown by student riots in 1960. Park Chung Hee seized power in a military coup in 1961, initiated a successful program of rapid economic growth but fell victim to rivalries within the government elite and was assassinated in 1979. Chun Doo Hwan, another ex-general, continued the policy of economics before politics until, in 1986–87, popular discontent forced a process of constitutional revision. The keenly-fought presidential election of December 1987 gave victory to ex-general Roh Tae Woo, thanks to the division of his opponents. A successful, if uneven, transition to a more open political system had at last been set in train.

Economic and social trends

Korea's mineral resources are concentrated in the north half of the peninsula, and the development efforts of the Japanese colonial administration in the south focused on increasing the output of the main staple, rice. Significant increases were achieved but for export to Japan, while local consumption actually declined. The negative effects of systematic exploitation were then compounded by the devastation which accompanied partition and war, during which time some 40 percent of existing industrial facilities and a third of the housing stock were destroyed. Reconstruction was hampered by Korean unwillingness to cooperate with fast-recovering Japan. A program of export-led growth was initiated in 1962; this raised annual average per capita income from US$100 to over US$5000 in the course of a quarter of a century. In the 1980s the country's international financial status was transformed as it shifted from being the world's fourth largest debtor to becoming a net creditor, with heavy industry, textiles and electronics sectors. National planners targeted 1992 as the year for joining the OECD and becoming one of the world's ten largest trading nations. As the new decade opened these goals did not seem unattainable.

The meticulously organized opening ceremony of the 1988 Olympic Games brilliantly symbolized Korea's postwar achievement. Literate, prosperous, healthy and hardworking, the Koreans entered the last decade of the 20th century to face a novel challenge – enjoying the fruits of unsought trauma and unstinted effort.

JAPAN

JAPAN

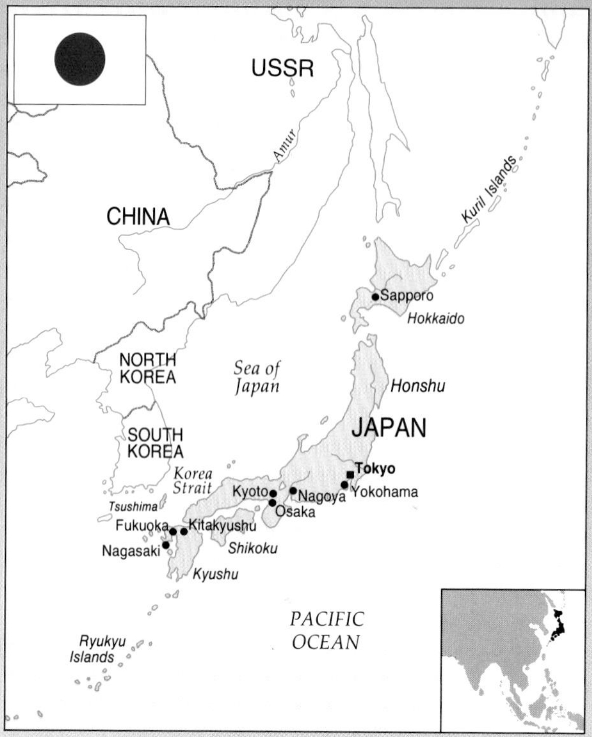

CHRONOLOGY

1904	Russo-Japanese war
1933	Japan withdraws from League of Nations
1941	Attack on US naval base at Pearl Harbor
1945	Dropping of atomic bombs on Hiroshima and Nagasaki
1952	End of Allied Occupation
1955	Formation of Liberal Democratic Party
1989	Japan becomes world's largest foreign aid donor and capital supplier

ESSENTIAL STATISTICS

Capital Tokyo

Population (1989) 123,120,000

Area 377,835 sq km

Population per sq km (1989) 325.8

GNP per capita (1989) US$15,030

CONSTITUTIONAL DATA

Constitution Constitutional monarchy with a national diet consisting of two houses (House of Councillors; House of Representatives)

Major international organizations UN; I-ADB

Monetary unit 1 yen (y) = 100 sen

Official language Japanese

Major religion Shinto and Buddhism

Heads of government since 1900 (Emperor) Mitsuhito (1868–1912); Yoshito (1912–26); Hirohito (1926–89); Akihito (1989–)

Japan entered the 20th century uncertain of its place in the international order and determined to improve on it. It had accomplished the modernization of its armed forces, the reconstruction of its national institutions and the creation of an advanced industrial sector in less than half a century. Imperial expansion was the next logical step on the path to undisputed "great power" status. In 1894–95 Japanese forces had thrashed those of decaying Qing China and by 1898 Japan had acquired the underdeveloped island of Taiwan as the spoil of war. The signing of an alliance with Britain in 1902 meant that France would not intervene in any war undertaken by Japan. This enabled Japan to attack Russia in 1904–05, destroying the Russian fleet at Port Arthur.

The outbreak of World War I permitted Japan to seize German possessions in East Asia and the Pacific and to retain most of them as part of the postwar settlement. But Japanese attempts to force strife-ridden China into accepting "Twenty One Demands" in 1915 were thwarted by the intervention of other powers. The attempt to insert a racial equality clause into the charter of the new League of Nations was likewise frustrated. American limitations on "oriental" immigration in 1924 compounded the insult.

Japan in the 1920s had cities that were vigorously modern, and a countryside still visibly traditional. However, strains lay beneath a surface impression of bustle contrasted with tranquility. Violent "rice riots" had swept the nation in 1918. And, if the extension of the franchise to all adult males in 1925 seemed to presage the establishment of a more democratically based parliamentary system, it was accompanied by an enlargement of police powers of surveillance over political activity. When Hirohito assumed the throne in 1926 he chose as the name for his reign "Showa" – "enlightened peace". It proved an inappropriate choice. Radical nationalists were soon calling for a "Showa Restoration", a deliberate echo of the dramatic 1868 coup known to history as the Meiji Restoration, which had initiated Japan's great program of change. What they now looked for was the purging of liberalism, the curbing of capitalism and the resumption of expansionism. Their motive force came from the distress of the peasantry, who had more than paid for Japan's advances in every sense but scarcely benefited from them.

The collapse of world trade after 1929 made distress acute and provided the context for a brutal initiative by the Kwandong army, stationed on Chinese territory to safeguard Japanese railway and mineral concessions. Seizing the vast and resource-rich province of Manchuria, the overseas military left the domestic politicians to face the diplomatic consequences of their actions. Japan rejected the condemnation of the League of Nations and withdrew from membership in 1933. In 1934 a puppet state of Manzhouguo was established, headed by Henry Pu Yi, the last of the Chinese Manchu dynasty.

An attempt by junior officers to assassinate leading cabinet ministers and take power in the Emperor's name in February 1936 was aborted by the uncharacteristic intervention of the Emperor himself, a scrupulous constitutionalist. In 1937, however, the military began a fullscale offensive in China, and as a consequence the militarization of Japan proceeded apace.

The outbreak of war in Europe made the French and Dutch colonies in Southeast Asia more vulnerable, tempting the Japanese to advance one step further. Spurning American protests, Japan chose a gambler's strategy, staking all on a preemptive strike on the US naval base at Pearl Harbor, in Hawaii. The operation of 7 December 1941 achieved total surprise and inflicted spectacular damage but failed to destroy either vital facilities or the crucial aircraft carriers, which were absent from their moorings. The year 1942 saw the high tide of Japanese expansionism, as the last great infantry army in history swept through mainland Southeast Asia, captured the key British base at Singapore and threatened the borders of India itself. But the battles of the Coral Sea and of Midway Island with the US Navy so crippled Japanese naval power that Allied forces could adopt an "island-hopping" strategy until they were within bombing range of the Japanese home islands. A three-day fire-bomb raid on Tokyo in March 1945, which inflicted 140,000 deaths, and the loss of another 100,000 lives in the capture of the island of Okinawa, provided a yardstick by which the price of a full-scale invasion might be measured. The dropping of atomic bombs on Hiroshima and Nagasaki on 6 and 9 August 1945, coupled with Russia's declaration of war on Japan and the fear of territorial partition along German lines, enabled the Emperor to break a political deadlock and call upon his people to surrender.

At the order of General Douglas MacArthur, Supreme Commander of the Allied Power (SCAP) occupation forces, Japan acquired a new, democratic constitution which assigned to the Emperor, who no longer claimed divinity, a purely symbolic role, guaranteed civil liberties and, in its unique Article IX, renounced the right of belligerency. Drafted in English, and effective from 3 May, 1947, this alien document provided the framework for postwar political life without amendment. Occupation preferences for decentralization in such matters as education and police were, however, largely undone by the "reverse course" of the 1950s. Politics, business, the military and education were purged of their prewar leaders, but much continuity was provided throughout by the highly competent bureaucracy which was scarcely purged at all, since its cooperation was essential for the successful implementation of the occupation reforms. Apart from a brief period under socialist-led coalition in 1948, postwar Japan has been ruled by conservative parties. Capitalist and pro-American, funded and led by big business and supported electorally by the countryside and smaller cities, the Liberals and Democrats merged in

▼ **Hiroshima, devastated by the atom bomb, August 1945.**

1955 to meet the challenge of a merger of socialist parties. The Liberal Democratic party (LDP) has won every general election ever since, though scandals, mostly financial, have often tarnished its image. The LDP's sustained electoral success, however, is a tribute to more than its skill in distributing largesse. The opposition parties have been unable to present the electorate with a coherent coalition alternative. And the LDP has delivered what most people wanted most, a consistently rising standard of living, if not an unambiguously improved quality of life.

During the Occupation Japan's foreign relations were controlled by the United States and on the same day that Japan regained its sovereignty it signed a security treaty with the United States which has remained the cornerstone of its diplomacy ever since. Japan was admitted to the UN in 1956. Revision of the US Security Treaty in 1960 provoked violent extraparliamentary protests, cutting short the career of premier Kishi; but its renewal on more equal terms implied an upgrading of Japan's status in the alliance to that of partner.

Japanese confidence in its "special relationship" with the United States was severely jolted in the early 1970s by the "Nixon shokku", arising from the unilateral imposition of tariffs on Japanese goods and the abrupt recognition of "Red China" after 20 years' refusal to do so. In neither matter had Japan been consulted. The "oil shokku" of 1973 prompted reappraisal on overreliance of a single external partner.

By the 1980s Japan's global economic presence and financial strength led its government to pledge itself to a policy of "*kokusaika*" – internationalization. This vague phrase embraced such varied measures as enhanced aid to developing countries, redressing structural imbalances in trade, promoting cultural exchange and taking initiatives to resolve regional conflicts in areas such as Southeast Asia and the Middle East.

Economic trends

Before 1853, Japan's economy had been virtually cut off from international trade, relying on traditional craft technologies and overwhelmingly dependent on agriculture, forestry and fishing. By 1900 it had become innovative, dynamic and exposed to international pressures and opportunities. Yet, despite the eager absorption of Western industrial technology, agriculture was still clearly the basis of the state, although agriculture itself was changing as farmers were urged to adopt chemical fertilizers and improved crop strains developed through systematic scientific research. Labor was plentiful, however, and little attention was given as yet to the introduction of machinery. Improvements in agricultural productivity thus enabled Japan to keep pace with the demands of a population which was not only growing rapidly but increasingly urbanized.

Manufacturing at this time was dominated by two sectors – textiles and an emerging complex of shipbuilding and associated engineering industries. Within the textile sector cotton manufacture, based on imported Western technology, was expanding rapidly to meet domestic and regional export demand; but silk, still primarily rural and smallscale in its organization, was the main source of foreign exchange. Shipbuilding and engineering had been developed primarily for strategic motives and were stimulated by the wars of 1894–95 and 1904–05. Significant developments of the first decade of the present century included the opening of Japan's first major steelworks, the growing exploitation of hydroelectric power and the nationalization of railroads to establish a more integrated transport infrastructure. Overseas, the acquisition of Taiwan (1898) and Korea (1910) opened up new markets and sources of raw materials.

World War I enabled Japan to move into Asian markets neglected by the combatant Western powers. The interruption of supplies from Germany also helped to stimulate the domestic chemical industry. But these gains were accompanied by rapid inflation, which provoked disruptive labor disputes. The collapse of the war boom in 1920 brought bankruptcies and lay-offs but these were soon overshadowed in 1923 by an earthquake which destroyed or severely damaged three million homes in the Tokyo area.

The worldwide economic crisis of the 1930s brought disaster to the Japanese countryside as the collapse of the silk trade deprived farmers of a crucial source of earnings. A measure of popular misery can be seen in the fact that two-fifths of all deaths were clearly linked with inadequate diet and poor living conditions. Overseas expansion provided an obvious avenue of relief but the "Greater East Asia Co-Prosperity Sphere", allegedly intended to exploit the complementarity of industrialized Japan and its neighbors, simply became an instrument for exploitation. In the later stages of World War II Japanese living standards plunged to subsistence level as the overburdened economy strained to match the performance of an enemy with a productive potential more than ten times as

◀ **Students and police stage a battle in the 1970s.**

▼ **Traditional arts such as calligraphy retain their popularity.**

Japan's "Invisible Race"

The renowned homogeneity of Japanese society obscures the survival of an outcast social category whose very existence has periodically been disavowed. Once known as *eta*, a highly derogatory term denoting filth and pollution, they have more normally been referred to in the present century by the more neutral term *tokushu burakumin* ("special hamlet people"), the hamlets in question being, in effect, ghetto communities. The antecedents of this group are uncertain and variously related to prisoners of war disgraced by capture, physical mutants rejected by their families or simply grossly impoverished loners. Whatever their origins the *burakumin* became associated with occupations regarded by the prevailing religions of Buddhism and Shinto as degrading and polluting because of their association with death, such as butchery, tanning and the manufacture of leather goods.

During the Tokugawa era (1603–1867) the *burakumin* were segregated from mainstream society by law and obliged to signal their separation by means of distinctive appearance and the observance of curfews and rituals of deference and avoidance. Legal discrimination was abolished by an emancipation decree in 1871; however, this did nothing to tackle social prejudices.

Self-help began in 1922 with the establishment of Suiheisha, a national organization which adopted militant tactics such as school boycotts and the nonpayment of taxes. In 1946 it was succeeded by the All-Japan Committee for Burakau Liberation whose radicalism provoked the establishment of a breakaway moderate rival, Dowakai (Society for Integration) in 1960.

In the late 1980s *burakumin*, indistinguishable from Japanese in appearance, still suffered from widespread discrimination in matters of employment and were generally obliged to marry within their own community.

great. Allied bombing destroyed not only most industrial facilities but also much of the infrastructure and every major city except historic Kyoto.

By 1945 Japan was literally on the edge of starvation. The population of the capital had halved to three millions as its inhabitants fled to the countryside to barter possessions for food. From that level Japan, "the Asian Phoenix", arose to become the world's leading producer in turn of steel, ships, vehicles and electrical goods. The immediate postwar years were a period of continuing hardship as rampant inflation, black marketeering and the repatriation of six million Japanese from the lost overseas empire threw the economy into confusion. Initial occupation policy was hostile to industrial reconstruction. This attitude was, however, reversed as the Communist takeover of mainland China prompted the United States to reassess Japan as an outpost of democracy, to be nurtured and stabilized. Agrarian reform, which transferred the land to the peasants who worked it, was seen as fundamental to this process, providing both economic incentives for increased output and a political stake in the new social order.

In 1960 prime minister Ikeda announced to a skeptical world that Japan would double average real incomes within ten years. In fact the target was achieved in seven. By 1964 Japan was sufficiently advanced to join the OECD, the "rich nations' club"; by 1968 it was the free world's second largest economy. Breakneck growth was dramatically checked in 1973 when the OPEC oil crisis quadrupled the price of Japan's biggest single import item. Renewed growth was achieved at the cost of massive reductions in energy-intensive industries such as steel, chemicals and aluminum-smelting. Restructuring industry toward more complex products, such as optical and precision instruments and office equipment, had the dual advantage of requiring less imported energy and raw materials and utilizing the highly educated labor force. Japan became an acknowledged world leader in robotics, biotechnology and computers.

The second oil price rise, in 1979, was accommodated with relative ease and by the early 1980s Japan's trade balance had moved into such massive surplus that frictions with the United States and the European Community were to dominate their relations for the rest of the decade. By the end of the 1980s Japan had become both the world's largest creditor and, in absolute dollar terms, its most generous donor of foreign aid.

Social and cultural trends

Japan's constant effort to maintain and develop its distinctive identity, while incorporating foreign influences, is well illustrated by the early history of its cinema. The Japanese first saw moving films in 1897. Three years later they had manufactured their first film camera, opened a processing laboratory and despatched a crew to film the Boxer Rising in China. Over the next 20 years they struggled to master the new medium – combining (all male) actors from the traditional *kabuki* theater, with plots drawn from translations of Western novels and the intrusive "assistance" of *benshi*, the dramatic narrators whose commentaries had long accompanied stylized stage-plays. *Souls on the Road* (1921) marked a breakthrough by introducing a more naturalistic style of acting, spurning the studio for natural settings and focusing on the real problems of real people. But it was not until 1951, when Kurosawa's medieval enigma *Rashomon* was awarded the Grand Prix at the Venice Biennale, that Japanese cinema, by then half a century old, was finally "discovered" by Western intellectuals.

The 1920s saw the heyday of the *mobo* (modern boy) and *moga* (modern girl) as Japan's urban youth flirted with Western frivolousness in music and fashion. The militaristic and economically depressed 1920s saw a reversion to a strident cultural chauvinism. The 1940s saw another complete cultural somersault as the military occupation was accompanied by an enthusiastic interest in things American. A new legal code enshrined novel notions of individual liberty and equality, ending the head of household's traditional authority and eroding the sense of mutual responsibility between members of extended families. Many hailed these changes as the final death of feudalism; others saw in them the seeds of a new selfishness and irresponsibility. Women acquired the vote but almost half a century later Japan appeared to Western visitors as a male-oriented society in which women occupied few commanding positions in business or public affairs.

Within a generation the Japanese had become the most highly educated of all the major industrial nations, with over ninety percent completing 12 years of schooling and almost forty percent going on to higher education. Paradoxically the national obsession with education both increased familiarity with Western culture and, coupled with rising affluence, enabled the mass of the population to participate on an unprecedented scale in the elegant pastimes of the traditional elite, such as tea-ceremony, calligraphy or classical music and drama. Education also created a voracious reading public which treated men of letters as national celebrities but also devoured *manga* (comic books) by the hundreds of millions each year. The brilliant staging of the 1964 Tokyo Olympics and the inauguration for the event of the superfast *shinkansen* ("bullet-train") service caught the imagination of the world.

▶ 1980s Tokyo, one of the world's most intensively urban environments.

住友銀行
ショウルーム

TAIWAN

REPUBLIC OF CHINA

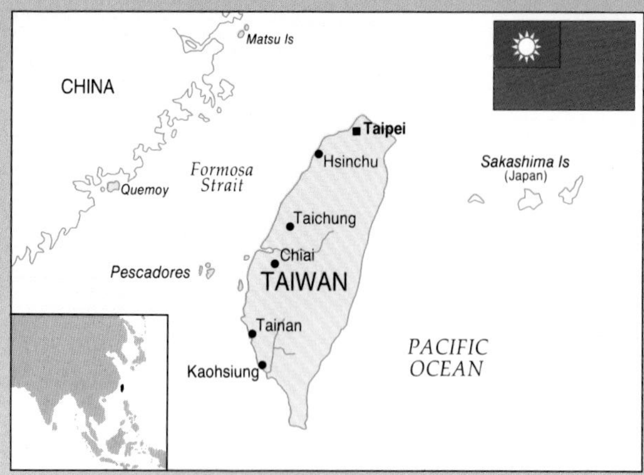

CHRONOLOGY

1895	Island ceded to Japan by China
1945	Taiwan returned to China
1949	Chinese state in Taiwan is established enforcing martial law
1971	Taiwan expelled from United Nations
1979	US end diplomatic relations
1987	Martial law is lifted
1988	Public demonstrations are legalized
1989	Ban on opposition parties is dropped

ESSENTIAL STATISTICS

Capital Taipei

Population (1989) 20,024,000

Area 36,000 sq km

Population per sq km (1988) 556.2

GNP per capita (1988) US$6,020

CONSTITUTIONAL DATA

Constitution Unitary republic with a National Assembly

Date of independence 25 October 1945 (from Japan)

Major international organizations none

Monetary unit 1 New Taiwan dollar (NT$) = 100 cents

Official language Mandarin Chinese

Major religion Chinese folk-religions

Heads of government since independence (Presidents) Jiang Jieshi (1945–49); General Li Zongren (1949–50); Jiang Jieshi (1950–75); Yen Chia-kan (1975–78); Jiang Ching-kuo (1978–88); Lee Teng-hui (1988–)

Taiwan, a neglected possession of the dying Chinese Qing empire, was ceded to Japan after it defeated China in 1894–95. Local resistance was soon crushed and from 1898 onward a program of colonial modernization was imposed, bringing substantial improvements in local health, curbing consumption of opium and establishing the infrastructure necessary for further development. Japanese rule, if authoritarian, did succeed eventually in gaining the active collaboration of the ruled as well as doubling the population in half a century and extending the area under cultivation to its optimum extent. In September 1945, following Japan's defeat, the island was occupied by the Chinese Nationalist (Guomindang) troops of Jiang Jieshi. The harsh rule of his representative, Chen Yi, provoked a revolt which claimed 20,000 lives and left a legacy of lasting bitterness between Taiwan and mainland China. As China's civil war turned against Jiang's forces, he transferred his seat of government, and by 1950, two million of his followers to Taiwan which henceforth became "the Republic of China". Until his death in 1975 Jiang maintained that his was the sole legitimate government of all China and that the reconquest of the mainland would inevitably occur in the aftermath of a spontaneous uprising against Communist rule. This fantasy was gradually abandoned by his son and successor as party leader, then president, Jiang Ching-kuo. In 1987 martial law (in force since 1949) was finally lifted, as was the ban on travel to mainland China. Jiang Ching-kuo died in January 1988, the month in which public demonstrations were at last legalized. He was succeeded by Vice-President Lee Teng-hui who was a native Taiwanese. Having no power-base in the army or the security services, Lee needed to develop genuine popular support and a more collective style of leadership. In January 1989 the formal ban on opposition parties was finally dropped as the country continued to edge towards a more open political system. Elections for a new National Assembly were announced for December 1991.

Economic and social trends

In the early years of the century, Taiwan's role in Japan's colonial empire was to export rice and tropical crops (sugar, pineapples and bananas) to feed the homeland, and to import manufactures to support the homeland's industries. This required the construction of roads and harbors, and the development of human skills. An agricultural research station was established at Taipei as early as 1898, and in the 1930s Kao-hsung was developed as a modern port, complete with its own shipyards and refineries. The authorities' concern to spread Japanese culture required an expansion of educational provision and by 1943 over 90 percent of children were receiving elementary education. Knowledge of the Japanese language was to be a useful commercial asset a generation later.

Even so, Taiwan in 1950 was an unlikely setting for an economic miracle. Three-quarters of the island is uncultivable mountain. The resource base is poor and domestic sources of energy meager. Typhoons batter the coastline and earthquakes shake the interior. Postwar Taiwan also suffered the burden of an overblown military establishment, the loss of 400,000 repatriated Japanese technicians and managers and the hostility of the indigenous population towards a government of strangers. But the American perception of Taiwan as an "unsinkable aircraft carrier off the China coast" led to the disbursement of

$US1465 million in foreign aid between 1951 and 1965 and, if Taiwan was poor in resources, it was rich in educable labor. While the island's economic transformation can be attributed in part to low wages, indifference to environmental damage and an eager appetite for pirating foreign designs and technologies, credit should also be given for the positive achievements of the government – a series of Four-Year Economic Development plans, beginning in 1952; a land reform which put the farms in the hands of the peasants and gave fair compensation to the landlords; and the systematic extension of power, transport and irrigation systems. By 1967 the total overseas trade of Taiwan amounted to almost half that of mainland China, which had a population 50 times greater. And over the next 20 years per capita income was to rise 20 times. The basic engine of growth was export-led manufacturing, a sector dominated by small, labor-intensive firms, 85 percent of which employed fewer than fifty people. Heavy industry was limited to shipbuilding and steel, chemicals and building materials for domestic use. Light industry, headed by textiles and electrical appliances, produced a wide range of consumer goods, making Taiwan the world's largest exporter of footwear, bicycles, sports rackets, fans and umbrellas.

By 1987 Taiwan had built up massive foreign exchange reserves of $US75 billion, a hoard second only to Japan's in size. Faced with threats of protectionism from the US, its largest single market, and the constant competitiveness of its rivals, Hong Kong and South Korea, Taiwan set itself to increase investment in research and development, to encourage automation and the application of information technology and to develop high-technology, value-added goods such as precision tools, computer peripherals and medical equipment.

Taiwan now has its own foreign aid program, which supports the relocation of labor-intensive industries like textiles, on the mainland of Southeast Asia, where labor costs and raw materials are significantly cheaper. Further indicators of maturity are the state's commitment to curbing pollution and introducing nationwide medical insurance and other labor benefits and its application to join the General Agreement on Tariffs and Trade (GATT).

Cultural trends

True to its Confucian heritage, Taiwan honors learning and the arts, allocating one-seventh of its national budget for their promotion. In 1945 the island had but one university and three private colleges, with a total enrollment of just over 2000 students. By 1988 it had over a hundred institutions of higher education with more than 460,000 students. Approved fields of endeavor include not only such traditional activities as painting, poetry, calligraphy, Beijing opera and seal-carving but also modern pastimes such as Western classical music.

▼ **Taiwanese industry relies on "nimble fingers".**

BURMA

UNION OF MYANMAR

CHRONOLOGY

1885 Annexed by the British

1937 New colonial constitution established

1948 Independence declared

1962 Ne Win coup inaugurates military rule

1988 Armed forces seize power

ESSENTIAL STATISTICS

Capital Rangoon

Population (1989) 40,810,000

Area 676,577 sq km

Population per sq km (1988) 60.3

GNP per capita (1986) US$ 200

CONSTITUTIONAL DATA

Constitution Military regime – pending elections

Date of independence 4 January 1948

Major international organizations UN

Monetary unit 1 Myanmar kyat (K) = 100 pyas

Official language Burmese

Major religion Buddhist

Heads of government since independence (Prime minister) U Nu (1948–58); Gen. Ne Win (1958–60); U Nu (1960–62); (Military ruler) Gen. Ne Win (1962–88) (Prime minister) General Saw Maung (1988–)

In 1885 Britain annexed the deeply Buddhist country of Burma and integrated it with India. This was much resented in Burma, and the young Men's Buddhist Association established by London-educated lawyers in 1906 set up schools to promote Burmese culture. In 1920 national consciousness was further strengthened by protests against British domination of Rangoon's new university and against proposals to exclude Burma from constitutional reforms for India. In 1937 Burma was separated from India and granted a new constitution.

During World War II, ex-student activist Aung San led "30 Comrades" in establishing a Burmese army with Japanese support. After the Japanese invasion of early 1942, he switched sides to help the Allies evict the Japanese, a task that was completed in 1945. Aung San then founded a broad "Anti-Fascist People's Freedom League" (AFPFL) which negotiated a peaceful handover of power from the British. This resulted in a deep division within the AFPFL, and Aung San was assassinated. On 4 January 1948 the Union of Burma became a sovereign independent republic outside the Commonwealth; the veteran nationalist U Nu was its first premier. Opting for neutrality in the Cold War, Burma pursued a minimalist foreign policy and preoccupied itself with counterinsurgency operations against Communists and ethnic minorities seeking autonomy. In 1958 U Nu, nettled by AFPFL splits, passed the premiership to ex-"30 Comrades" General Ne Win who staged the February 1960 election. This gave U Nu a firm majority, but in March 1962 Ne Win led a coup pledged to building a socialist state under the Burma Socialist Program party (BSPP).

During the next 25 years, Burma was ruled by the military or as a one-party state: it was preoccupied by unending anti-guerrilla campaigns which absorbed 40 percent of the budget while the economy slid into chaos. Mass protests in 1988 were repressed bloodily; Ne Win stepped down but continuing confusion led to an army-backed "State Law and Order Restoration Council" (SLORC) under General Saw Maung. Elections of 1990 showed victory for the opposition, although the SLORC continued to rule, pending "fresh elections".

Economic integration with British India introduced Burma's peasant farmers, accustomed to barter, to a money economy in a brutal fashion. The Irrawaddy delta was rapidly cleared of mangroves to become a rice-growing area for export. The establishment of modern roads and railroads accelerated the extraction of timber and mineral resources such as oil and rubies but did little to develop indigenous enterprise. To the Burmese, it appeared that the sole beneficiaries were the British ruling elite and their accomplices, the Chinese merchants and Indian moneylenders and laborers. After independence Burma made slow but steady progress in repairing the ravages of war. Then 25 years of socialism transformed the resource-rich country from a moderately prosperous exporter of primary products to one of the poorest states in Asia. It had to negotiate "least developed country" status with the UN to qualify for increased aid and relied on a massive black market to supply its daily necessities. After an economic crisis in 1987–88, the government vigorously promoted cooperation with foreign governments and companies to develop trade, offshore fishing, onshore oil exploration and tourism.

Some 85 percent of the population still follow Theravada Buddhism and the Burmese language retains its dominance in a country of a hundred different tongues.

LAOS

LAO PEOPLE'S DEMOCRATIC REPUBLIC

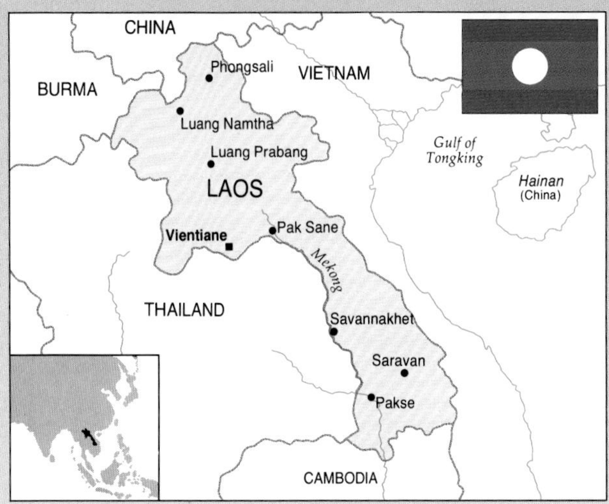

CHRONOLOGY

1893	French protectorate established
1945	Proclamation of independence
1949	Self-government within French Union granted
1954	Full independence recognized by Geneva Conference
1958	Elections lead to hostilities between Pathet Lao movement and rightwing government
1973	Ceasefire agreed
1975	Pathet Lao abolishes monarchy and establishes Communist regime
1977	Twenty-five-year friendship treaty with Vietnam
1986	Laos moves towards more open economic system

ESSENTIAL STATISTICS

Capital Vientiane

Population (1989) 3,936,000

Area 236,800 sq km

Population per sq km (1988) 16.6

GNP per capita (1987) US$ 140

CONSTITUTIONAL DATA

Constitution Unitary single-party People's Republic with one legislative house (Supreme People's Assembly)

Date of independence 23 October 1953

Major international organizations UN

Monetary unit 1 new kip (KN) = 100 at

Official language Lao

Major religion Buddhist

Overseas territories none

Heads of government (King) Sisavang Vong (1904–59); Savang Vatthana (1959–75); (Prime minister) Kaysone Phomvihane (1975–)

Laos, divided into three rival kingdoms in the 18th century, fell under Siamese domination, which was replaced by a French protectorate between 1893 and 1907. In March 1945 colonial rule was terminated by the Japanese, who proclaimed an independent Laotian state. When the French returned they recognized Laotian autonomy in 1946 and granted self-government within the French Union in 1949. This arrangement was challenged by the Pathet Lao (Lao Country) movement under Prince Souphanouvong, who established a firm grip on the northeast of the country.

The 1954 Geneva Conference recognized the complete independence of Laos and from 1955 to 1958 a broad front government, including Pathet Lao elements, held precarious sway. The 1958 elections, however, exposed deep political divisions which led to open hostilities between the Pathet Lao and a rightwing, United States-backed government based in Vientiane. A reconvened Geneva Conference attempted to reconcile the antagonists in 1961, and in 1962 the moderate Prince Souvanna Phouma (half-brother to Souphanouvong) formed another broad coalition. The search for reconciliation was, however, bedeviled by the escalation of the conflict in neighboring Vietnam. A ceasefire agreement in February 1973 was followed by the inauguration of a "Provincial Government of National Unity" in April 1974. The better organized Pathet Lao, encouraged by Communist takeovers in South Vietnam and Cambodia, finally established control over the whole country as their royalist opponents fled to exile in Thailand.

A "Lao People's Democratic Republic" was established in December 1975 under Prince Souphanouvong, with the Pathet Lao finally revealed as a cover for the 20-year-old Laotian Communist party. Following the negotiation of military and economic agreements in 1977 the new regime became in effect a client-state of Vietnam. In 1986 it moved towards a more open economic system linked with dynamic Thailand, a prospect clouded by conflicts arising out of a border dispute. A draft constitution was published in 1990.

Economic and social trends

Landlocked but unprotected by natural frontiers, underpopulated and ravaged by the rivalries of more powerful neighbors, Laos in 1990 was one of the world's least developed countries. The French introduced coffee-growing and laid out a basic network of roads, but the country still has not a single railroad. Three-quarters of the labor force was engaged in farming, though natural calamities threatened even the basic staple, rice. Collectivization was however accompanied by a doubling of the area under cultivation and a more than doubling of rice output between 1976 and 1984. Illegally traded opium remained an important secondary crop. Of the country's substantial mineral resources only tin was exploited commercially. The country's major hydroelectric plant at Nam Ngum produced far more power than the consumer goods industries could use and 90 percent was exported to Thailand, making electricity the largest single source of foreign exchange.

The Laotian peoples can be divided into four basic groups, distinguished from one another by language and lifestyle, but most following Theravada Buddhism as their religion, often supplemented by spirit and ancestor worship. Music, dance, literature, sculpture and other crafts still bear the strong imprint of Buddhist and Hindu tradition.

VIETNAM

SOCIALIST REPUBLIC OF VIETNAM

CHRONOLOGY

1945	Proclamation of independence from France
1946	French renew armed conflict
1954	Geneva Conference provides ceasefire – Vietnam divided
1963	Coup in Saigon leads to military regime
1975	North Vietnamese troops capture Saigon
1976	Formal reunification of north and south
1978	Vietnam joins COMECON
1980	Refugee exodus reaches three quarters of a million
1986	Retirement of "old guard" revolutionary leadership

ESSENTIAL STATISTICS

Capital Hanoi

Population (1989) 64,747,000

Area 331,688 sq km

Population per sq km (1989) 195.2

GNP per capita (1987) US$198

CONSTITUTIONAL DATA

Constitution Socialist Republic with one legislative house (National Assembly)

Date of independence 2 September 1945

Major international organizations UN; COMECON

Monetary unit 1 dong (D) = 10 hao = 100 xu

Official language Vietnamese

Major religion Buddhist

Heads of government (President of Democratic Republic) Ho Chi Minh (1945–69); (Prime minister) Le Duan (1969–76); (Party leader of Socialist Republic) Le Duan (1976–86); Nguyen Van Linh (1986–91); Do Muoi (1991–)

The French conquest of Vietnam began in 1858, but the colonial state was created by Paul Doumer, governor-general from 1897. He initiated a wide-ranging program of public works to aid the rapid exploitation of local resources for France. The peasants were denied civil liberties, oppressed by a collaborationist elite of local landowners, and had virtually no access to schooling or medical care. Revolts in 1916, 1917 and 1930 were all ruthlessly suppressed. In 1925 the Revolutionary League of the Youth of Vietnam, the nucleus of the Vietnamese Communist party, was founded by Nguyen Ai Quoc (known from 1943 as Ho Chi Minh). The Communists became the best organized and disciplined of all the groups opposed to French rule. During World War II Ho Chi Minh's "Viet Minh" guerrillas were armed by the United States to assist their harassment of the Japanese occupation army. In the confusion that accompanied the ending of hostilities, Ho Chi Minh proclaimed the independence of Vietnam, on 2 September 1945. The French soon reconquered the southern half of the country and then attempted to negotiate a power-sharing agreement with the Viet Minh. Armed conflict was renewed in November 1946 with a French bombardment of the port of Haiphong which caused 6,000 civilian casualties. Conflict ended in May 1954 when Viet Minh forces under General Vo Nguyen Giap overwhelmed the French at Dien Bien Phu. A Geneva conference provided for a ceasefire and the country was divided at the 17th Parallel of latitude, pending national elections (which were never held).

For a decade the war-ravaged North dedicated its energies to reconstruction – and supporting Communist-led insurgency in the South. In Saigon an anti-Communist regime under Ngo Dinh Diem struggled, with lavish American support, to stabilize a chaotic situation by strong-arm methods. Diem was killed in a 1963 coup which, two years and nine governments later, led to a military regime under pistol-packing Air Vice Marshal Nguyen Cao Ky. Fearing the complete overthrow of the shaky Saigon regime, the United States increased its involvement from 17,000 military advisors in 1963 to 510,000 combat troops by 1968. Despite massive bombing of the North, the United States failed to break the Communists' will and by 1973, following negotiations in Paris, withdrew its ground forces. After further fierce fighting the Communists took Saigon on 30 April 1975, subsequently renaming it Ho Chi Minh City. Formal reunification of the country was marked by the proclamation of the Socialist Republic of Vietnam on 2 July 1976. Socialist "reforms" then produced an exodus of some three-quarters of a million refugees by 1980. Meanwhile Vietnam sent 200,000 troops into Cambodia to oust the Khmer Rouge regime in 1979 and fought a border war with the Chinese. These military adventures hampered reconstruction and perpetuated the country's diplomatic isolation. The death of veteran party boss Le Duan in 1986 was, however, accompanied by a significant weakening of the old guard and a willingness among reformers to build a more cooperative relationship with the non-Communist world in the interests of economic recovery.

Economic and social trends

The French colonial regime was organized to denude Vietnam of its rice, coal, rare minerals and rubber. Industry was restricted to the production of simple consumer goods and building materials for local use. French manufactures

dominated the import trade in the interwar years but the French colonial elite and its favored landowning class of 6,500 families could not constitute a viable major market. Irrigation projects quadrupled the rice-growing area between 1880 and 1930 but rice consumption among the peasant cultivators actually fell, as did standards of literacy. After World War II, despite 40 years of warfare, there were significant periods and pockets of retrenchment. Vietnam has the potential for prosperity. Its population is largely literate. There are vast areas of cultivable land even after the devastation of prolonged chemical warfare waged especially by US forces between 1968 and 1973, rich forests, varied mineral deposits and a long coastline with excellent harbors.

After the 1976 reunification, however, recovery was hampered by the inherent problems of integrating a command and a free enterprise economy, by damaged infrastructure and shortages of key equipment and by the removal of skilled and educated non-Communists from positions of authority. Vietnam faced the loss of Chinese entrepreneurial talent (and the accompanying 5000 fishing boats) and the continuing burden of high military expenditure.

In the late 1980s the primary sector still employed some three-quarters of the population and had scarcely been touched by modern methods. Citrus, seafood and timber were potential export earners. State control of economic life was pervasive, with rationing the normal method of distribution and free provision of education, health care and transport to and from work. Vietnam joined Comecon in 1978 and later qualified for assistance from the World Bank and International Monetary Fund (IMF). In 1988 Vietnam took a more generous approach to foreign investment. Vietnam's proximity to the fast-growing Association of Southeast Asian Nations (ASEAN) countries could bode well for the 1990s but greater economic integration with their success depended on the adoption of less threatening military postures.

A thousand years of Chinese rule has left an indelible mark on the culture of the ethnic Vietnamese, from literature to architecture. The Cham and Khmer, descendants of mighty empires, exhibit an Indian heritage. The many tribal peoples of the highlands were sufficiently awakened by French missionary contacts during the colonial period to have become fiercely attached to their traditional cultures and suspicious of lowland infiltration and control. Religious observance in Vietnam ranges from animism through Taoism to Mahayana Buddhism and Christianity (purged of foreign priests) plus pockets of Islam and residual Hinduism. Poetry remains the most widespread and prestigious cultural form but there are also lively followings for traditional "Chinese opera" and a local variant, *cai luong*, a sort of satirical musical comedy whose performers are carefully controlled employees of the state. Among crafts high-quality lacquerware is preeminent, while carving and weaving still flourish among the mountain peoples.

▲ The Vietnamese suffered greatly in the wars of the 1960s and 1970s.

CAMBODIA

STATE OF KAMPUCHEA

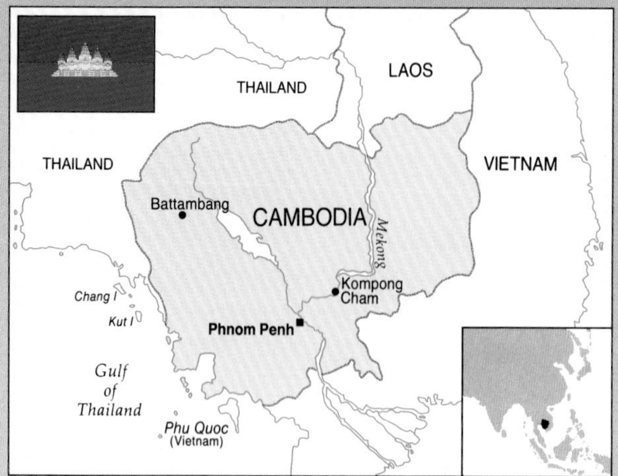

CHRONOLOGY

1897	French takeover of monarchy
1953	Full independence achieved
1975	Lon Nol overthrown by Khmer Rouge
1978	Vietnamese install Heng Samrin government
1982	Formation of tripartite resistance coalition
1990	Withdrawal of Vietnamese troops

ESSENTIAL STATISTICS

Capital Phnom Penh

Population (1989) 8,055,000

Area 181,035 sq km

Population per sq km (1989) 44.5

GNP per capita (1981) US$90

CONSTITUTIONAL DATA

Constitution Single-party people's republic with one legislative house (National Assembly)

Date of independence 9 November 1953

Major international organizations UN

Monetary unit 1 riel = 100 sen

Official language Khmer

Major religion Buddhist

Heads of government since independence (King) Norodom Sihanouk (1941–54); Norodom Suramarit (1954–60); (Head of State) N. Sihanouk (1960–70); Lt.-Gen. Lon Nol (1970–72); (President) Lon Nol (1972–75); (Head of State) N. Sihanouk (1975–76); (President) Khieu Samphan (1976–77); (Party leader) Pol Pot (1977–79); (President) Heng Samrin (1979–)

In the late 19th century France established a protectorate over Cambodia, as a byproduct of its efforts to conquer Vietnam. In 1897 the French representative in Phnom Penh, the Cambodian capital, became chairman of the Cambodian Council of Ministers, reducing the king to a splendid cipher. Under Sisowath (1904–27) and Monivong (1927–41) the monarchy remained a potential but inert focus for nationalist sentiment. In 1941 the Japanese invaded, but left French administration in theoretical control. The French placed the teenage Prince Norodom Sihanouk on the throne, assuming that he would prove as easily influenced as his predecessors. When the French administration collapsed throughout Indochina in March 1945, Sihanouk declared independence, then accepted the reassertion of French control of what became an "autonomous state within the French Union". French failure to regain power over Vietnam, coupled with radical internal Cambodian opposition to the policies of the throne, prompted Sihanouk to move decisively in January 1953, dissolving parliament, declaring martial law and asserting his country's independence. France accepted the *fait accompli* of this "Royal Crusade" in November 1953.

In May 1954 an international conference at Geneva recognized the Sihanouk administration as Cambodia's sole legitimate government. Determined to outflank the radicals, Sihanouk abdicated in favor of his father in 1954 and established a new mass movement – Sangkum Reastr Niyum (People's Socialist Community) – which won every seat in the elections that September. For the next 15 years Sihanouk ran Cambodia, assuming the position of chief of state, rather than monarch, after his father's death in 1960. Suspicion of US hostility led Sihanouk to abandon his policy of neutrality after 1965, in favor of an alignment with the Communists. The consequent use of Cambodian territory by Vietnamese Communists alienated the conservative urban elite and officer corps. Though still a semidivine figure to the peasant masses, Sihanouk now had enemies on both left and right.

In March 1970, while Sihanouk was out of the country, rightwingers led by General Lon Nol deposed him and declared a "Khmer Republic". Setting up a government in exile in Beijing, Sihanouk proclaimed himself head of a National United Front of Cambodia. In May 1970 US and Vietnamese forces entered Cambodia to support Lon Nol, who faced armed opposition from both the Vietnamese and indigenous leftwing "Khmer Rouge" revolutionaries, nominally loyal to the exiled Sihanouk but in fact controlled by his past opponents, such Khieu Samphan. Despite continuing American aid the Lon Nol regime eventually succumbed to the revolutionaries' strategy of attrition as they took control first of the countryside, then of provincial capitals, and finally of Phnom Penh itself on 17 April 1975. Sihanouk returned as titular chief in September and a new constitution for "Democratic Kampuchea" was established in January 1976. Sihanouk gave way to Khieu Samphan in April, with Pol Pot being named as premier. In 1977 all governmental power was vested in the Communist party.

The Pol Pot regime's attempts to eradicate prerevolutionary society cost at least a million lives, perhaps twice as many – which would be a third of the entire population. Border clashes with Vietnam led to a Vietnamese-backed guerrilla invasion and the installation of a government headed by Heng Samrin in 1979. Little stability ensued as the new regime fought the

Khmer Rouge, the *Armée Nationale Sihanoukist*, which was likewise led from exile, and the anti-Communist Khmer People's National Liberation Front, led by ex-premier Son Sann. In 1982 these three formed a common front, which retained international recognition as the legitimate government of Cambodia. Meanwhile hundreds of thousands of refugees fled to Thailand. In 1988 Vietnam announced its intention to withdraw its troops by December 1990, leaving its protégé government to stand alone. As the deadline approached the prospects for a lasting peace appeared remote, with the Khmer Rouge still in existence as a major power though unacceptable to many Cambodians and to much international opinion.

Economic and social trends

Cambodia is a typical monsoon rice-growing country, and self-sufficiency in this staple crop was the rule until recent tragic times. The French introduced rubber cultivation for export, but otherwise the economy remained one of the least developed in the entire region. The lack of any significant mineral or energy resources limited other possible incentives for change. The civil war of 1970–75 severely damaged the country's infrastructure and reduced foreign trade to a trickle. The Pol Pot regime then attempted to chart a new course of total self-sufficiency by the brutal but simple expedient of emptying the capital to swell the ranks of forced laborers building industrial facilities and constructing massive irrigation systems to facilitate multiple cropping. Widespread conflict from 1978 onward negated the possible benefit of these efforts and led to general famine in 1979–80. By then only wartorn Afghanistan had higher levels of child mortality. The economy of this former rice-exporting country had been reduced to subsistence level. In 1987, following similar changes in Vietnam, the government made tentative moves toward the introduction of a more liberal system of economic management, and Soviet aid made possible the restoration of international telephone and telex links for the first time since 1975. Tourists began to visit the impressive ruins of the Buddhist temple complex of Angkor Wat. But the work of reconstruction had scarcely begun.

In a region noted for its diversity Cambodia exhibits a relatively high degree of homogeneity, over 90 percent of the population being ethnic Khmers and just under 90 percent Theravadan Buddhists. Expulsions of Vietnamese, Cham-Malay and European minorities have reinforced this homogeneity. Savage and systematic attacks were made on organized Buddhism in the period 1975–78. Modern Cambodian culture developed in the shadow of the greatness of the medieval Khmer empire which at its height (between the 11th and 13th centuries AD) sprawled over the Indo-Chinese peninsula and left an enduring impact on the whole region out of all proportion to Cambodia's present size and standing. Pride in this heritage led the government, after gaining independence from France in 1953, to foster a revival of national arts by establishing schools of ballet, theater and music and a university of fine arts. The devastation of the country's education system by two decades of war and persecution make the prospects for a renewal of this renaissance dim indeed.

▼ **A museum records the horrors of the Pol Pot regime.**

THAILAND

KINGDOM OF THAILAND

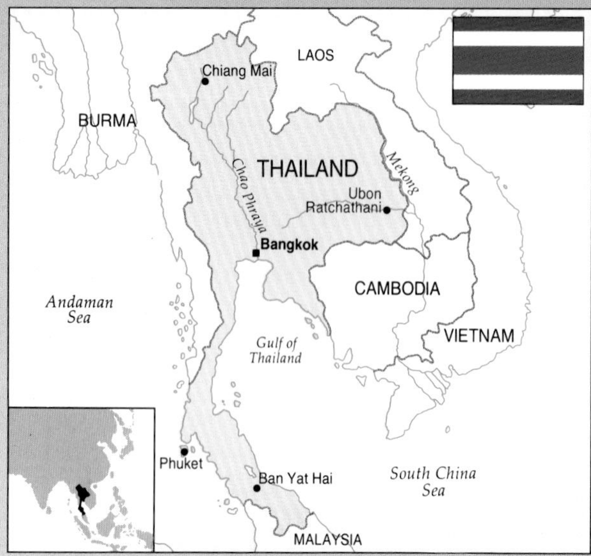

CHRONOLOGY

1917	First western-style university opened
1932	Bloodless coup ends absolute monarchy
1939	Official name changed from Siam to Thailand
1941	Japanese occupation begins, ending in 1944
1973	Riots followed by interlude of civilian rule
1980	General Prem Tinsulandona inaugurates era of civilian rule

ESSENTIAL STATISTICS

Capital Bangkok

Population (1989) 55,258,000

Area 513,115 sq km

Population per sq km (1989) 107.7

GNP per capita (1987) US$840

CONSTITUTIONAL DATA

Constitution Constitutional monarchy with a multiparty National Assembly (Senate; House of Representatives)

Major international organizations UN; ASEAN

Monetary unit 1 Thai Baht (B) = 100 stans

Official language Thai

Major religion Buddhist

Heads of government (Kings) Chulalongkorn (1868–1910); Vajiravudh (1910–25); Prajadhipok (1925–35); Ananda Mahidol (1935–46); Bhumipol Adulyadej (1946–); (Prime minister) Marshal Luang Pibul Songgram (1938–44); Kuang Aphaiwongse (1944–45); Nai Thawi Bunyakat (1945–46); Kuang Aphaiwongse (Jan–Mar 1946); Nai Pridi Phanomyang (Mar–Aug 1946); Luang Dhamrong Nawassat (1946–47); Marshal Luang Pibul Songgram (Nov 1947); Kuang Aphaiwongse (1947–48); Marshal Luang Pibul Songgram (1948–57); Marshal Sarit Thanarat (1957–63); Thanom Kittikachorn (1963–73); Sanya Dharmasakti (1974–75); Seni Pramoj (Feb–Mar 1975); Kukrit Pramoj (1975–76); Seni Pramoj (Apr–Oct 1976); Thanin Kraivichien (1976–77); Gen. Kriangsak Chomanan (1977–80); Gen. Prem Tinsulanonda (1981–88); Gen. Chatichai Choonhavan (1988–91); Anand Panyarachun (1991–)

The Kingdom of Siam entered the 20th century under the progressive rule of Chulalongkorn (Rama V; 1868–1910), who oversaw the reform of the legal code and the establishment of Western-style centralized bureaucracy with ministries distinct from the royal household. In foreign relations he set the pattern for his successors, fending off colonial powers by pragmatically ceding marginal territories to both the French (1907) and the British (1909). King Vajiravudh (Rama VI; 1910–25) continued the policy of Westernization by making education compulsory, introducing the Western calendar and joining the war on the Allied side in 1917.

In 1932, a bloodless coup forced King Prajadhipok (Rama VII; 1925–35) to accept constitutional limitations on what had hitherto been an absolute monarchy. The two youthful Western-educated coup leaders, Pibul Songgram and Pridi Phanomyang, dominated politics for the next quarter of a century. The mood of the 1930s was chauvinist, aggressively rejecting the position of non-Thais in the country's economic and political affairs. Hoped-for schemes of social reform failed to materialize and severe restrictions were imposed on the commercially active Chinese minority. The influence of the army in political life increased markedly and in 1938 the militarist Pibul became premier, espousing a Pan-Thai expansionist policy. He changed the country's official name in 1939 from Siam to Prathet Thai (Country of the Free), a proud celebration of the fact that it was one of the very few countries in the region that had avoided colonial rule. The new name was also an unintentionally ironic comment on his own regime and the country's immediate future. Taking advantage of the defeat ofFrance in 1940, Pibul revived old claims to parts of the French colonies in Laos and Cambodia, and received them as a result of Japanese "mediation". In December 1941 Thailand, after token resistance, accepted Japanese occupation and in 1942 Pibul declared war on the United States and Britain, receiving territories in Burma and Malaya as a reward in 1943. Pridi, however, formed an anti-Japanese resistance with support from the Allies.

Pibul stepped down in 1944, making way eventually for Pridi; a period of political confusion and economic dislocation ensued during which inflation ran riot, the territories acquired during the war were returned and King Ananda Mahidol (Rama VIII; 1935–46) was found shot dead in mysterious circumstances. Pridi restored the name Siam to signify rejection of Pibul's policies. Meanwhile Bhumipol Adulyadej assumed the throne as Rama IX (1946–). Pibul soon ousted Pridi and restored the name Thailand in 1949. In 1951 he was obliged to concede much power to General Sarit Thanarat, though he remained premier until exiled in 1957, when Sarit took full control of the government after a military coup. Suspending the constitution and dissolving parliament and all political parties, he ruled by decree until his own death in 1963. Army rule, of varying degrees of directness, continued under Thanom, justified in part by fear of Communist insurgency. The United States was allowed to develop local bomber bases to support its war-effort in Vietnam in return for military and financial aid. In 1973 a bloody suppression of student demonstrations brought royal protests and a split in the army, ushering in an interlude of civilian rule.

Renewed student disorders in 1976 provided the pretext for a new hardline government under Thanin Kraivichien, who was in turn deposed by a coup in 1977. In 1980 a coup displaced Kriangsak Chamanand in favor of Prem Tinsulanonda who, suppressing attempted coups in 1981 and 1985, stabilized the nation both politically and economically as the region came to terms with the legacy of Vietnam and the emerging significance of cooperation through the Association of Southeast Asian Nations (ASEAN). With royal support Prem not only wooed Communist dissidents out of the forests but promoted a phased introduction of democracy. This culminated in the 1988 election of his successor, Chatichai Choonhavan, whose cabinet, which consisted mainly of ex-businessmen rather than ex-soldiers, symbolized the emergence of a new political style and agenda. Chatichai worked to improve relationships with the other countries of the region, replacing years of hostility with the policy of "turning Indochina's battlefields into a marketplace". After 26 coups, attempted coups and counter-coups and 13 constitutions it began to look as though the country had at last found a durable political formula.

Economic and social trends

Throughout the century Thailand's economy has been based on the exploitation of its agricultural and mineral wealth; since 1984 manufacturing has contributed more to GDP than agriculture, and by 1990 tourism had developed to the point at which it made the largest single contribution to the balance of payments, surpassing textiles in 1987. Thailand's most important mineral is tin, of which it is the world's third largest producer. Rice, which as an export boosted postwar recovery, remains the staple crop but agriculture diversified to include fine breeds of cattle and pigs introduced from the West. Rubber, which had been introduced in the 19th century, is still a major export but the logging of teak was banned in 1989 to prevent further deforestation. Industry developed largely out of processing minerals, foodstuffs and forest products, but also includes oil-refining and gem-cutting; electrical goods and vehicle assembly were introduced by Japan, the country's major trading partner. Integrated circuits now account for more than five percent of all exports. With the lowest foreign debt in southeast Asia, a large pool of cheap and youthful labor and abundant reserves of natural resources, Thailand was well-placed in 1990 to sustain its own recent rapid growth and even to act as the pacemaker of development for its neighbors.

While Thai monarchs have favored modernization, they have also epitomized the continuity and conservation of national identity. King Vajiravudh opened the country's first Western-style university in 1917 – and named it after his father. He promoted an immense literary outpouring which stressed the need for Thai unity, pride and loyalty to king and country. The nation remains 95 percent Buddhist and in 1990 there were still 32,000 monasteries and some 200,000 monks. Symbolic, perhaps, of the cultural significance of the monarchy has been the growth of the royal capital Bangkok, to the point where its population of eight million is now 40 times larger than any other Thai city. Population growth slowed during the 1970s and 1980s but the dramatic impact of nationwide family-planning campaigns was offset by the influx of refugees from neighboring Laos and Cambodia. Conservation of forests is now matched by a planned policy of restoring key archeological sites as "Historical Parks", partly as a matter of national pride and partly as an investment in tourist appeal.

Thai unity has been strengthened in the 20th century by the nationwide spread of the speech of the central plains through mass education, and by a reversal of the 1930s policy toward integrating the Chinese through intermarriage and the ready granting of citizenship. Marked regional disparities of wealth and lifestyle remain, with average annual income in the capital eight times higher than that in the impoverished northeast. The cosmopolitan, not to say glamorous, erotic, nature of Bangkok was emphasized by its growing popularity as a tourist center in the 1980s. Breakneck economic development runs the risk of aggravating such differences rather than relieving them but the establishment of the country's first social security system in 1989 was an encouraging sign for the future.

▼ A Hindu temple in Bangkok.

MALAYSIA

MALAYSIA

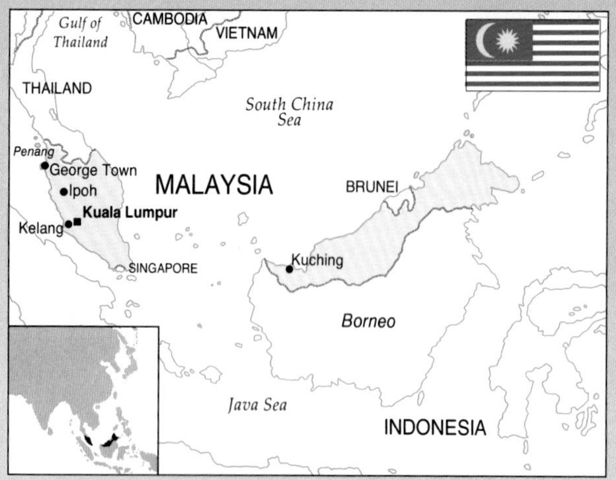

CHRONOLOGY

1941	Malayan Communist Party (MCP) is formed
1946	United Malayan National Organization (UMNO) is formed
1948	MCP armed uprising
1957	Independence of Malayan Federation
1963	Establishment of Federation of Malaysia
1966	Ending of "confrontation" with Indonesia
1969	Race riots lead to suspension of parliamentary rule
1987	Autocratic government leads to arrest and detention of opposition leader

ESSENTIAL STATISTICS

Capital Kuala Lumpur

Population (1989) 17,421,000

Area 330,442 sq km

Population per sq km (1989) 52.8

GNP per capita (1987) US$1,800

CONSTITUTIONAL DATA

Constitution Federal Constitutional Monarchy with two legislative houses (Senate; House of Representatives)

Date of independence 31 August 1957

Major international organizations UN; Commonwealth of Nations; ASEAN; IDB

Monetary unit 1 ringgit or Malaysian dollar (M$) = 100 cents

Official language Malay

Major religion Muslim

Heads of government since independence (Prime minister) Tunku Abdul Rahman (1957–59); Abdul Razak (Apr–Aug 1959); Tunku Abdul Rahman (1959–70); Abdul Razak (1970–76); Datuk Hussein bin Onn (1976–84); Dr Mahathir Mohammad (1984–)

The history of resource-rich Malaysia is largely the product of a Malay-Chinese rivalry within a framework established by British colonial rule. The opening of the Suez Canal in 1869 greatly increased European commerce in the region and heightened interest in Malaya's vast, recently-discovered tin deposits. This quickened immigration from southern China to work in the tin mines and encouraged the establishment of coffee and rubber plantations – which in turn led to Indian immigration. Britain secured its interests by the introduction of "residents" to supervise the administration of the petty states of the Malay peninsula, four of which were federated in 1896, with Kuala Lumpur as their capital, under the supervision of Sir Frank Swettenham. Other Malay states gradually moved out of the Siamese sphere of influence and came within the orbit of the British, who favored the continuance of traditional institutional forms while at the same time promoting economic development which necessarily challenged their relevance.

Japan's 70-day campaign in 1941–42 to conquer the peninsula and Singapore, bastion of British power in the East, led to a harsh occupation and the formation of a Malayan Communist party (MCP) among the Chinese community to fight a guerrilla resistance. After the war was over, Britain tried to rationalize previous treaty arrangements to make a single colonial state, the Malayan Union; however this provoked a massive nationalist reaction, expressed in the establishment in 1946 of the United Malaya National Organization (UMNO). In 1948 the ten states of the Malay peninsula and the island state of Penang were federated. In the same year the MCP began an armed insurgency leading to a period of "emergency" in which the British colonial authorities fought the uprising with a combination of military counter-measures and political concessions. High Commissioner General Templer fostered cooperation between UMNO and the Malayan Chinese Association (MCA) with a view to *merdeka* (independence).

The landslide victory of the anti-Communist, anticolonial alliance in the 1955 national election led to the attainment of *merdeka* on 31 August 1957, though the emergency did not end until 1960. When the Chinese-dominated island state of Singapore, self-governing since 1959, sought union with the peninsula, Malaya offset the further growth of Chinese influence by suggesting that the British colonies in North Borneo, Sabah and Sarawak join them in a new federation, Malaysia. This was established on 16 September 1963. The Sultanate of Brunei declined admission. Although Singapore detached itself in 1965, the new federation, with British military support, proved sturdy enough to resist armed harassment from Indonesia in 1963–66, which attempted unsuccessfully to destabilize "Eastern Malaysia".

Thereafter threats to stability were as much internal as external. Malay-Chinese ethnic clashes in May 1969 led to a suspension of parliamentary rule until February 1971. The suppression of Communist forces was a preoccupation of the 1970s, to be followed by a resurgence of Islamic fundamentalism among students under the impact of the missionary "Dakwah" movement and tensions created by the arrival (from 1979 onwards) of "Vietnamese boat people" refugees, who were mostly ethnic Chinese. After 1981 politics was dominated by Dr. Mahathir Mohammad, the first commoner to become prime minister. His "Look East" policy, based on admiration of Japan's economic modernization, aimed to foster economic diversification

through industrialization. Faltering economic performance weakened his hold on the UMNO-led National Front coalition government and led to a major challenge to his leadership in 1987, which he survived only at the cost of dividing his party.

Economic and social trends

Britain's political unification of the Malay peninsula was accompanied by economic integration through the construction of railroads, first inland from the coast to the tin fields, then down the peninsula to link with the great entrepôt port of Singapore. Tin and rubber remained the mainstays of the economy until independence and Malaysia is still the largest producer of these commodities. Key resources that were subsequently of major importance include crude petroleum, palm oil and timber which by the 1980s, with tin and rubber, accounted for a sixth of all exports. Industrialization, beginning with import substitution, developed to a level of international competitiveness which meant that electronic components accounted for a sixth of all exports. Nevertheless over half the labor force remained in the primary sector and, although industry was the fastest-growing sector of the economy, many manufactures were still imported, especially from Japan. Confidence in the value of producing steel, cement and a "national car" faltered in the late 1980s. Postindependence economic policy was interventionist, using such instruments as tariffs, tax-holidays and government-sponsored research institutes and industrial estates, as well as promoting infrastructural development. Measures to increase economic participation and control by ethnic Malays (*bumiputras*) were a particular concern. These were down-graded by the New Development Policy of June 1991, which favored privatization and foreign investment to achieve an ambitious eight-fold increase in national income by 2020.

Malaysian society is notable for its diversity. Muslim Malays constitute about half the population and the Chinese a third, while peoples from the Indian subcontinent account for another ten percent. At least another 25 ethnic groups are discernible, as well as Europeans, Arabs and Thais. Music and dance, literature and the decorative arts have resisted Westernization. Painting and architecture, essentially alien cultural forms, attracted younger artists eager to experiment and produce a distinctively Malaysian style.

▲ **Rubber growing and processing is a major industry.**

INDONESIA

REPUBLIC OF INDONESIA

CHRONOLOGY

1926	Communist-inspired revolt is crushed
1927	Indonesian Nationalist Party (PNI) established under Ahmed Sukarno
1933	Abortive naval mutiny leads to increasingly authoritarian regime
1942	Japan invades Indonesia
1945	Japanese surrender leads to PNI declaration of independence
1946	British-sponsored truce signed
1949	Formal transfer of Dutch sovereignty
1959	Sukarno imposes dictatorship, under the name "Guided Democracy"
1963	Accession of western New Guinea (Irian Jaya) and "confrontation" with Malaysia
1965	"Anti-Communist" massacres carried out by armed forces
1967	General Suharto takes over as President
1976	Annexation of east Timor

ESSENTIAL STATISTICS

Capital Djakarta

Population (1989) 177,046,000

Area 1,919,443 sq km

Population per sq km (1989) 92.2

GNP per capita (1987) US$450

CONSTITUTIONAL DATA

Constitution Unitary multiparty republic with two legislative houses (House of Representatives; People's Consultative Assembly)

Date of independence 17 August 1945

Major international organizations OPEC; IDB; ASEAN; UN

Monetary unit 1 Indonesian rupiah (Rp) = 100 sen

Official language Bahasa Indonesia

Major religion Muslim

Heads of government since independence (President) Ahmed Sukarno (1945–67); General Suharto (1967–)

In 1900 the 65 million people of the Indonesian archipelago represented the core of the Dutch colonial empire, then third largest in the world. Embryonic nationalist sentiment began to coalesce around Budi Utomo (High Endeavor), an elitist cultural society which attracted Indonesian student support. Other early nationalist groups included the "Sarekat Islam" (Islamic Association) which achieved a mass membership, the Eurasian "Indische Partij" and the Indies Social Democratic assocation, the forerunner of the Indonesian Communist party (PKI). A Communist-inspired revolt in 1926 was easily crushed, removing the Communists as a major contender for power for 20 years. In 1927 the Indonesian Nationalist party (PNI) was formed; it grew rapidly under the energetic leadership of the multilingual Ahmed Sukarno. The Dutch response was imprisonment and exile for PNI leaders and, after an abortive naval mutiny in 1933, an increasingly authoritarian regime.

A Japanese invasion in 1942 soon swept aside Dutch resistance and led to active collaboration between the PNI and the occupiers. Following the Japanese surrender, the PNI proclaimed Indonesia's independence in Djakarta on 17 August 1945. The Dutch returned in force and fighting ensued until a British-sponsored truce was signed on 15 November 1946. The attempt to develop power-sharing within a United States of Indonesia soon broke down, and in July 1947 the Dutch began a "police action" so brutal that international pressure forced a ceasefire that December. A Communist rising against the PNI in September 1948 gave the Dutch the opportunity for a decisive blow and on 19 December they bombed Jogjakarta, the PNI headquarters. This resumption of "police action" brought renewed international pressure and a conference at The Hague, which led to a transference of power over all former Indonesian territories (except western New Guinea) on 27 November 1949, though vestigial links with the Dutch Crown were retained until 1956.

Plans for a federation were soon shelved in favor of a unitary state which would inevitably be dominated by densely populated Java. In due course the ambitious Sukarno, figurehead president since independence, seized direct power in a pseudo-constitutional coup in 1959. Disputes with the Dutch continued until they handed over western New Guinea (Irian Jaya) in 1963, at which point Sukarno launched a *confrontasi* (confrontation) with the newly established Malaysian federation with the aim of annexing the former British colonies of Sabah and Sarawak. The only significant result of this campaign was to weaken the Indonesian economy.

The year 1965 proved a major turning-point as the increasingly powerful army gave up on Sukarno's theatrical and chaotic administration and launched into a violent suppression of Communists, real or imagined, which led to the massacre of some half million or more people and the destruction of the PKI. Army minister General Suharto finally ousted Sukarno as president on 12 March 1967 and, abandoning his predecessor's grandiose visions of regional hegemony, proclaimed a "New Order" embracing the depoliticization of domestic affairs, currency stabilization and the promotion of rapid economic development. He did, however, annex Portuguese Timor in 1976 and defied UN condemnation of this act. By creating Golkar, in theory a council to represent the main groupings of Indonesian society but in practice a government party, Suharto was able to mobilize popular support at the polls with increasing success. Nevertheless he continued to suppress opposition. By 1990 the greatest external achievement of his regime was the creation and development of the Association of Southeast Asian Nations (ASEAN). The possibility of Suharto's retirement after 1993 sparked off political speculation.

Economic and social trends

In 1901 the so-called "Ethical Policy" was introduced, by which Dutch government funds were to be used to promote health, education and agricultural improvement among the Indonesian population. The aims were noble but the achievement meager. By the 1930s the literacy rate was still only six percent and agricultural productivity had failed to keep pace with an inexorable increase in population. Under Suharto a stable framework for economic management was established, even if corruption remained an endemic problem. Literacy, life expectancy and personal incomes all rose and there were major successes in promoting family planning and achieving national self-sufficiency in rice. The exploitation of oil, gas and tin for export was the main engine of growth but in the late 1980s agriculture still accounted for a quarter of the national product, employed more than half the population and provided major exports in rubber, spices, oil palm and other plantation products, a clear legacy of colonial days. Manufactured goods were still mostly imported, though the nation's rich and varied mineral deposits provided the basis for an expansion of metal-processing industries. The textile and garment industries were of growing importance but relied on imported raw materials. An overblown bureaucracy had held back progress in the past but weakening oil prices in the late 1980s obliged the government to look more favorably on foreign cooperation, whether through aid or private investment. There were ambitious plans to develop aircraft and electronics industries though, given the country's massive foreign debt and shortage of technical skills, tourism and handicrafts seemed more certain ways of attracting foreign currency.

Binneka Tunggal Ika – Unity in Diversity – is an apt national slogan for a country with 250 distinct languages and 300 identifiable ethnic groups scattered over an archipelago 5,000 kilometers (3,000 miles) long. The population base is fundamentally Malay with centuries of Arab, Indian, Chinese and European admixture. Sukarno's most lasting positive achievement, and no mean one, was the promotion of *Bahasa Indonesia*, rationalized Malay, as a unifying national speech, which by 1990 was used throughout the education system. Most people professed Islam; either traditional or with fundamentalist strands. The government consistently sought to head off potential religious conflicts, notably by promoting another legacy of the Sukarno era, the syncretic ideology known as *Pancasila* – five principles: belief in one god; humanitarianism; nationalism; democracy; social justice. Indonesian diversity is well reflected in its ancient culture, which is rich in heritage and myriad in forms, from the immense Buddhist monument at Borobudur or the Hindu temple complex at Prambanan, to the delicacy of *wayang* (shadow puppet) plays, or the brilliance of Javanese batik-dyed textiles, and the many and varied styles of carving and dance.

▶ **Ahmed Sukarno proclaims Indonesia's independence.**

SINGAPORE

REPUBLIC OF SINGAPORE

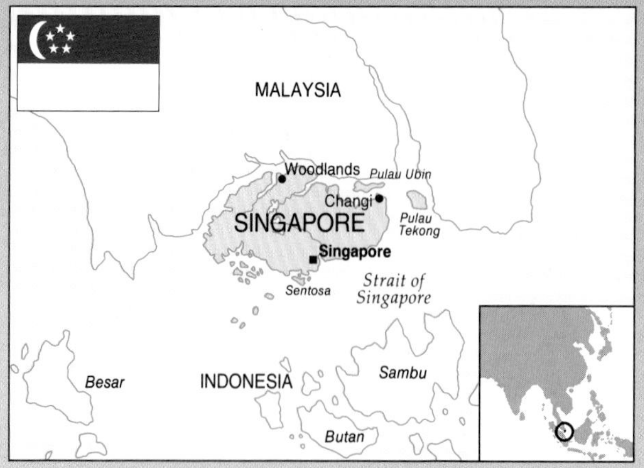

CHRONOLOGY

1867	British Crown colony established
1942	Occupied by Japanese troops
1959	Internal self-government is achieved
1963	Singapore becomes member of Federation of Malaysia, leaving in 1965
1971	Closure of British naval base

ESSENTIAL STATISTICS

Capital Singapore

Population (1989) 2,644,000

Area 622 sq km

Population per sq km (1989) 4,299

GNP per capita (1987) US$7,940

CONSTITUTIONAL DATA

Constitution Unitary multiparty republic with one legislative house (Parliament)

Date of independence 9 August 1965

Major international organizations UN; Commonwealth of Nations; ASEAN

Monetary unit 1 Singapore dollar (S$) = 100 cents

Official languages English; Malay; Mandarin Chinese; Tamil

Major religion Buddhist

Heads of government since independence (Prime minister) Lee Kuan Yew (1959–90); Goh Chok Tong (1990–)

A British Crown Colony from 1867, Singapore benefited commercially from the increase in Far Eastern trade after the opening of the Suez Canal in 1869 and the advent of steamshipping. The ending of the Anglo-Japanese alliance in 1921 led to the establishment of a major naval base there, as the chief bastion of British power in the region. It fell to the Japanese in a week in February 1942. The island remained in Japanese hands until September 1945.

After 1946 the British ensured that Singapore, as a predominantly Chinese community, pursued a path of constitutional development separate from neighboring Malaya. By 1955 responsibility for government, except for defense and foreign affairs, was in the hands of elected ministers and a legislative assembly with an elected majority. By 1959 complete internal self-government had been achieved under the leadership of Lee Kuan Yew and his People's Action Party (PAP).

In October 1971, 152 years of British military presence came to an end and Singapore transferred its security arrangements to a five-power defense arrangement which brought in Australia, New Zealand and Malaysia.

Economic and social trends

Virtually devoid of natural resources, too low-lying for hydroelectric power and dependent on its neighbor Malaya for water as well as food, Singapore built its wealth on its position and its people. By 1924 railways connected it with the Malay peninsula, enabling it to develop as an outport and processing center for primary products. The British naval base helped to develop a ship-repairing industry. Even in disadvantaged agriculture the tiny, swampy island achieved its successes – self-sufficiency in pig, poultry and egg production, and the development of high-value products such as orchids and rare aquarium fish. Manufacturing expanded especially rapidly, moving from food-processing and oil-refining to electronics and computer peripherals. Financial services, commodity exchanges and their associated communications facilities also became of major significance as Singapore looked to the uncertainties of 1997, when Hong Kong would be returned to Chinese rule.

In 1823 Singapore had a population of 10,000; by 1931 it had grown to 560,000 and by 1985 to 2,556,000. Of these three-quarters were Chinese, one-seventh Malay and about one in twelve "Indian". None of these groups was linguistically homogeneous. As a result, postindependence Singapore gave official status to no less than four languages – English, Malay, Tamil and Mandarin Chinese. English, the language of education of more than half the island's children, was the main medium for government and commerce. A third of all children were taught in Mandarin, China's national language, which transcends dialects. Learning a secondary language in school was compulsory. In a continent renowned for soaring population growth, sprawling slums and chaotic traffic, Singapore became a byword for its cleanliness and orderliness: its strict litter laws were legendary. A vigorous family-planning program was so successful that the economy suffered from labor shortages, as the government had set its face against immigration as a solution. Massive public construction programs did away with the old housing problem, while high-quality medical services raised health standards to those of a developed country.

PAPUA NEW GUINEA

PAPUA NEW GUINEA

CHRONOLOGY

1884	Southeastern quadrant (Papua) declared a British protectorate
1906	Papua becomes Australian territory
1942	Armed conflict begins as Japanese troops invade
1949	Union of northern and southern territories under Australian administration
1972	Beginning of copper exports
1975	Full independence achieved

ESSENTIAL STATISTICS

Capital Port Moresby

Population (1988) 3,592,900

Area 462,840 sq km

Population per sq km (1989) 7.8

GNP per capita (1987) US$730

CONSTITUTIONAL DATA

Constitution Constitutional monarchy with one legislative house (National Parliament)

Date of independence 16 September 1975

Major international organizations UN; SPC; ACP; Commonwealth of Nations

Monetary unit 1 Papua New Guinea kina (K)=100 toea

Official language English

Major religion Protestant

Heads of government since independence (Prime minister) Michael Somare (1972–80); Sir Julius Chan (1980–82); Michael Somare (1982–85); Paias Wingti (1985–88); Rabbie Namaliu (1988–)

In 1900 Papua New Guinea was claimed – controlled is too strong a word – by three imperial powers. The northeastern quadrant was in German possession and the western half held by the Dutch. The southeastern quadrant had been claimed by Queensland in 1883, made a British protectorate in 1884 and became Australian territory in 1906. War in 1914 led to Australian takeover of the German portion, which remained under military administration until the League of Nations mandated it to Australia as a trust territory after the war. During World War II both Papua proper and the trust territory were subject to Japanese occupation. In 1949 the two territories were combined under a restored Australian administration, which acknowledged indigenous participation and devolved the practical work of government to local districts. Papua New Guinea became selfgoverning on 1 December 1973 and a fully independent member of the Commonwealth on 15 September 1975. The transition to independence, however, brought an immediate threat of fragmentation as the offshore, copper-rich island of Bougainville asserted its own independence. A year was required to quell the secessionist attempt. In the first decade of independence Papua New Guinea's government changed hands no less than four times. The dominant leaders have been Michael Somare of the Pangu Pati and Paias Wingti of the People's Democratic Movement. In 1985 Somare was ousted by Wingti at the head of a five-party coalition. The 1987 elections confirmed Wingti in power with a slender majority. Seeking to break out of the confines of past relationships Wingti pledged a strengthening of ties with the United States, China, Japan and other major states.

New Guinea is monsoonal, humid and swampy, or mountainous and thickly forested. The island is rich in natural resources which have yet to be definitively explored, let alone exploited. Most of the major agricultural products are consumed locally but there are exports of copra, coffee, tea, cocoa, rubber, prawns, tuna and marine shells, chiefly to Japan and Australia. Copper exporting began in 1972 and by 1990 accounted for a third of all exports. Natural gas in commercially usable quantities has been prospected and there is also the possibility of oil as well as hydroelectric potential. Significant deposits of gold and silver also exist and pearl culture has been encouraged. Industry is diversifying from brewing, tobacco-processing, boat-building and the production of concrete and paints for local needs, to include more sophisticated ventures such as the manufacture of industrial gases and the assembly of electrical appliances. Shortages of educated labor and difficulties of communication remain major obstacles to further development.

The achievement of national unity will be, to say the least, a challenge for a country in which 700 languages are spoken and communications are so poor that it takes three weeks to complete the voting in a general election. Pidgin English evolved from the Melanesian speech of the coasts and islands to serve as a common tongue. Despite the nominal prevalence of Christianity, belief in spirits and magic is still general and boys are invariably initiated into men's secret cults. Education is free but not compulsory and is under-resourced. The survival of malaria and malnutrition show the need for further extension of basic health care. Unruly gangs of unemployed youths have created a major problem of violent crime as bonds of tribal authority loosen.

PHILIPPINES

REPUBLIC OF THE PHILIPPINES

CHRONOLOGY

1898	First declaration of independence
1935	Establishment of "Commonwealth"
1946	Full independence achieved
1949	Rebellion by Communist-led Hukbalahap group
1972	Martial law is declared
1983	Assassination of opposition leader Benigno Aquino
1986	People overthrow Marcos following dubious election

ESSENTIAL STATISTICS

Capital Manila

Population (1989) 59,906,000

Area 300,000 sq km

Population per sq km (1989) 199.7

GNP per capita (1987) US$590

CONSTITUTIONAL DATA

Constitution Unitary republic with two legislative houses (Senate; House of Representatives)

Date of independence 4 July 1946

Major international organizations UN; ASEAN

Monetary unit 1 Philippine peso (P) = 100 centavos

Official languages Pilipino; English

Major religion Roman Catholic

Heads of government since independence (President) Manuel Roxas y Acuna (1946–48); Elpidio Quirino (1948–53); Ramon Magsaysay (1953–57); Carlos P. Garcia (1957–61); Diosdado Mascaoagak (1961–65); Ferdinand Marcos (1965–86); Corazón Aquino (1986–)

In 1896 the Spanish authorities in their colony of the Philippines executed the reformist intellectural José Rizal and provoked a rising by the Kapitunan, a nationalist movement. A year of fighting led to a truce, Spanish pledges of reform and the exile of rebel leader Emilio Aguinaldo. The outbreak of war between Spain and the United States in 1898 enabled Aguinaldo to return home with American help. He liberated an area south of Manila and declared the independence of the Philippines on 12 June 1898. A congress at Malolos drafted a republican constitution and in January 1899 a government was established, with Aguinaldo as president. Local American sympathy was not, however, in line with US national policy, which ignored Filipino independence. Two years fighting followed until Aguinaldo was captured and persuaded to urge acceptance of American rule. The Filipinization of government under United States tutelage did, however, proceed quickly, supported by a massive expansion of educational opportunity.

The 1916 Jones Act established a 24-member senate, almost wholly elected, on a franchise of all literate adult males. The Tydings–McDuffie Act (1934) set a 10-year target for full independence and a "commonwealth" was inaugurated in 1935 for the intervening period, during which the United States would continue to control defense (which accounted for no less than a quarter of the national budget) and foreign policy, which was overshadowed by fears of Japanese expansionism. The perfection of democracy was shaken by bouts of violent agrarian unrest and then cut short by the Japanese conquest of December 1941–April 1942. While an "Executive Commission" of the landed elite cooperated with the occupiers, guerrilla resistance proliferated; many groups were pro-American but the Hukbalahap was Communist-led. United States General Douglas MacArthur liberated the country in 1944 and promoted excollaborator Roxas to become first president of the fully independent Republic of the Philippines in July 1946. The fraudulent election of his successor, Quirino, led to the Hukbalahap rebellion (1949–53) which was suppressed with US aid. Reform attempts by the charismatic Magsaysay (1953–57) were frustrated by the obstruction of entrenched interests.

Ferdinand Marcos, president from 1965, wrestled with an economy sliding out of control, won reelection in 1969, then attempted to revise the constitution to perpetuate his power. When this provoked protest he declared martial law (1972). His carrot-and-stick policies included land reform and a war on crime, accompanied by the arrest of political opponents and the prohibition of strikes. Despite increased efforts Communist insurgency continued to spread. Tarnished by allegations of financial corruption, Marcos was accused of political corruption too after the long-postponed interim National Assembly elections of 1978 failed to reward opposition leader Benigno Aquino with representation that corresponded in degree to the support he had won during the election campaign. With Aquino in exile, Marcos won a new six-year term in 1981. Aquino returned to the country in 1983 and was immediately assassinated; this provided a focus for popular discontent which rattled Marcos into calling an early presidential election in February 1986. When Marcos was declared the winner after dubious electoral proceedings the indignation of "people power" overthrew him a month later, to install Cory Aquino, widow of the murdered leader and candidate of the united opposition, as the new president.

Marcos died in exile while the new government attempted vainly to recover a proportion of the money alleged to have been removed from the country by him and his flamboyant wife, Imelda – the "Iron Butterfly" – during their two decades of power. "Cory" survived five coup attempts in her first 18 months in office and inaugurated a new constitution overwhelmingly endorsed by popular referendum (1987). Meanwhile she battled with a perilous legacy of economic mismanagement. However, the Communist guerrillas of the "New People's Army" remained in control of approximately one-fifth of the nation's 42,000 villages and Muslim separatists in the southern islands of the archipelago continued to defy government efforts at conciliation.

Economic and social trends

The constitution of the Malolos republic (1899) decreed that church-held lands would be nationalized, but ever since the desire for land reform has been continually frustrated by the power of the landed oligarchy which came to dominate Filipino politics. Large landowners promoted plantation sugar production to meet export demand, while the percentage of farmers under share tenancy agreements doubled between 1900 and 1935. Imports of tariff-free American manufactures meanwhile deterred the growth of local industry. Burdensome defense spending, the harsh effects of Japanese occupation and a destructive liberation campaign further retarded development. The grievances of tenant farmers fueled rural insurgencies while widespread corruption, smuggling and tax evasion hampered sporadic governmental drives for responsible economic management. Confusion reached a critical level in the last years of the Marcos regime. By 1985 GDP per head was no higher than its 1975 level. Agriculture remained the mainstay of the economy with rice the staple crop (self-sufficiency was achieved in 1968), and coconut products, sugar, timber and hemp provided major exports. The industrial sector concentrated on the exploitation of the country's rich mineral resources and the production of light manufactures such as garments, plastics and electrical goods. With government encouragement the basically free-enterprise economy benefited in the late 1980s from a diversification of its international trading partners away from a stifling dependence on US markets. But a foreign debt of US$26 billion from the Marcos years remained a major burden, while land reform prospects depended on the need both to satisfy domestic grievances and not to alarm potential foreign investors. Another political problem with profound economic implications was the future of the long-contested US military bases. The termination of the American presence when the current agreement ended in 1991 would gratify nationalist feelings but would also deprive the country of a significant, if resented, source of hard currency.

Cultural trends

The shaping forces of the Filipino cultural inheritance have been described as "300 years in the convent and 50 years in Hollywood". However, this does less than justice to the strength of indigenous Asian traditions, which flourish in such fields as myths and legends and the music and dance and wood-carvings which often embody them. The Filipino peoples are a mixture of Malay, Chinese, Spanish and American descent. The aboriginal inhabitants of the islands, pygmy Negritos, now constitute only a tiny fraction of the population. Almost 70 different languages are spoken, divisible into eight major groups. The most important are Pilipino, the national language, based on Tagalog, spoken around Manila, and English, the other official language which remains dominant in public life, business and higher education. The replacement of Spanish by English as the leading European language is a testament to the effectiveness of the educational crusade undertaken by the American colonial authorities. Under the Spanish fewer than 20 percent of those who actually went to school achieved a competent degree of literacy. The United States drafted in hundreds of American teachers and then trained Filipino replacements so quickly that by 1927 they accounted for 99 percent of the 26,000-strong teaching force. School attendance quintupled in a generation and by 1939 more than a quarter of the entire population could speak English, a larger percentage than any of the local tongues. This not only strengthened national unity and gave broader access to Western technology and culture but also opened up new career opportunities, making it easier to staff the goverment services by Filipinos. The Philippines remains the only mainly Christian country in the region and the Catholic Church commands the loyalties of the bulk of the population.

▲ **Antigovernment guerrillas in the late 1980s.**

AUSTRALIA

COMMONWEALTH OF AUSTRALIA

CHRONOLOGY

1945	Social security system introduced
1951	Australia signs ANZUS treaty
1960	Social security benefits extended to Aborigines
1967	Aborigines given vote
1988	Bicentenary of first settlement celebrated

ESSENTIAL STATISTICS

Capital Canberra

Population (1989) 16,804,000

Area 7,682,300 sq km

Population per sq km (1989) 2.2

GNP per capita (1987) US$10,900

CONSTITUTIONAL DATA

Constitution Federal parliamentary state with two legislative houses (Senate; House of Representatives)

Date of independence 1 January 1901

Major international organizations UN; Commonwealth of Nations; SPC

Monetary unit 1 Australian dollar ($A) = 100 cents

Official language English

Major religion Christian

Overseas territories Cocos (Keeling) Islands; Christmas Island; Norfolk Island

Heads of government (Prime minister) E. Barton (1901–03); A. Deakin (1903–04); J. C. Watson (1904); G. H. Reid (1904–05); A. Deakin (1905–08); A. Fisher (1908–09); A. Deakin (1909–10); A. Fisher (1910–13); J. Cook (1913–14); A. Fisher (1914–15); W. Hughes (1915–23); S. Bruce (1923–29); J. Scullin (1929–31); J. Lyons (1931–39); R. Menzies (1939–41); A. Fadden (Aug 1941); J. Curtin (1941–45); F. Forde (Jul 1945); J. Chifley (1945–49); R. Menzies (1949–66); H. Holt (1966–67); J. McEwen (1967–68); J. Gorton (1968–71); W. McMahon (1971–72); E. Whitlam (1972–75); J. Fraser (1975–83); R. Hawke (1983–)

Australia announced its nationhood and constitutional democracy on the first day of the 20th century. Political hegemony was proclaimed in terms of a federation of states that took in the former colonies of New South Wales, Victoria, Queensland, South Australia, Tasmania and Western Australia. The prime minister, Edmund Barton, hailed federalism as the system by which Australians would assume full political responsibility for the continent.

Australia's European influence exists alongside its geopolitical location in the Asia-Pacific region. To Australians, Europe represents cultural roots but implies domination; Asia is economically Australia's natural region, but politically it is perceived as a threat. Usually the tensions of nationalism have drawn Australia toward Europe and, since World War II, North America. Australia's foremost ally is the United States through the ANZUS (Australia, New Zealand and United States) treaty but it is also part of SEATO (South East Asia Treaty Organization) and has contributed to the Colombo Plan, for Asian/Pacific economic development.

Political practice in Australia is based on the Westminster model with a prime minister and cabinet, but the structure of parliament took lessons from many sources. The House of Representatives is the legislature while the Senate was designed to protect the rights of the states. Central government was consolidated with the passing of time and especially during the two world wars and the 1930s' depression. In practice the states have become the second tier of government. At the third level, local government is very weak.

The modern two-party system in politics dates from 1910 when the conservative groupings in parliament joined forces in opposition to Labor. Thereafter, Australian politics have been characterized by populist Labor and Conservative parties. In 1911 electoral enrollment was made compulsory and voting became compulsory in 1924. A system of preferential voting was also introduced whereby voters are required to list their preference for all candidates on the ballot paper. The Australian Labor Party (ALP) is the oldest (continuous) political party in federal politics though conservatism has had the larger share of electoral results. The minority Country party, which was formed in 1920, has traditionally maintained a conservative coalition with the major non-Labor parties, whose names have been variously Nationalist, United Australia party, and, since 1944, Liberal. The Country party, renamed the National party in the 1970s, has not really developed as a third force in Australian politics but represents what was the traditional demarcation between the freetraders and protectionists.

Unlike the House of Representatives, the Senate is elected on a proportional voting system which has tended to encourage the smaller parties and independents and which has meant that the major parties have rarely achieved a majority. The formation of the liberal-left Australian Democrats in 1977 signaled the emergence of a new political force in the Senate where it has held the balance of power from time to time. From the early 1980s the Green party and Nuclear Disarmament party contested vacancies.

Economic trends

Australia is a primary-producing nation whose main exports have been wool, wheat and minerals, while manufacturing has tended toward the local market. The wool industry is a major source of export revenue and with large-scale improvements in refrigeration and shipping the sheep-meat trade has grown apace. A live sheep trade began in the 1920s with shipments to Singapore and in the late 1960s this was expanded to include the lucrative Middle East market. The cattle industry also profited, especially with exports to the United States, and Australia accounts for approximately 25 percent of the world's export beef trade.

With federation and the removal of trade barriers between the states, Australia's internal commerce flourished and, after the 1920s, manufacturing grew behind tariff walls erected as protection against the international trade. At the turn of the century Britain was Australia's largest trading partner and up to this time all trade agreements relating to Australia had been negotiated in London. With the passing of the century, continental Europe, North America but especially Asia became more important.

Australia is a mineral-rich country with major deposits of lead, copper, iron, nickel, zinc, uranium, gold, silver, diamonds and zircon. It also has major reserves of oil and gas. The most important and sustained growth in mineral exploitation in the 20th century was underway by the mid 1950s and continued through to the 1970s during which period the stock exchanges reached ever new high points. Mineral exploration continues to be one of the most important facets of Australia's economy, notably in Western Australia, Queensland and the Northern Territory.

Successive Australian governments have seen their major role as fostering the development of industries and technology in a range of ways, from financing the construction of roads, railroads, and infrastructure to providing shipping, air transport, insurance services, postal services, electricity and telecommunications. Government maintained its interventionist policy despite free market political rhetoric in the late 20th century and even entered into direct partnership with companies. It offered tax incentives for mining and exploration, subsidized the cost of fertilizers, formed boards to regulate the prices of most primary goods, erected tariff walls to protect manufacturing, and provided substantial material and financial inducements for overseas companies.

Social and cultural trends

Australia's greatest enterprise this century has been war. Contingents were sent to New Zealand, China, the Sudan and South Africa in the 19th century but in the 20th century Australian Armed Forces have fought in Europe, Africa, Asia and the Middle East. The nation's official national day is Australia Day which celebrates the arrival of Europeans in Australia in 1788; however, the unofficial national day is Anzac Day which commemorates the Australian and New Zealand Army Corps' participation in World War I when more than 200,000 Australian casualties were suffered.

▼**Bondi Beach, near Sydney, symbol of the outdoor Australia.**

By 1901 Australian social policy headed the world; the welfare state was a fact of life, but it was achieved at a cost. The state was financed by massive overseas loans and social and cultural institutions were maintained by strict immigration regulations – the White Australia Policy or "Australia's Monroe Doctrine" as it was sometimes called. Women had the vote and other civil rights but were unequal in citizenship; Aborigines, Asians and Africans had no rights and were not citizens.

In the mid 1970s the White Australia Policy was officially relinquished. Australia now takes immigrants from all over the world without reference to race or ethnicity and the monoculturalism of the first 50 years has given way to multiculturalism. Along with Canada, Australia is now one of the most ethnically diverse nations in the world – in the late 1970s governments introduced multicultural and multilingual radio and television – and multiculturalism has become official government policy since the 1980s.

In the area of health and welfare Australia has pioneered many reforms from pension services for the elderly, the invalid, widows and soldiers through to maternity benefits and child endowments. In 1907 the basic wage became fundamental to industrial policy by guaranteeing a minimum male (family) wage, with no reference to employers' abilities to pay. Women did not share in the family rate and earned between a half and two-thirds the male rate. The basic wage was the basis of all wage arbitration until officially scrapped at the federal level in 1967 around which time all forms of discrimination at the workplace were outlawed.

In the mid 19th century a dual system of state and denominational schooling was introduced and universal compulsory primary education had been achieved by the turn of the century. Later reforms included distance education, including traveling schools and Schools of the Air, which pioneered correspondence lessons on the radio.

Literacy was virtually universal by 1901. Not surprisingly, then, literary culture has been perceived as the standard bearer of Australian culture. More recently art, film, television and music have made important contributions. In the 1890s writers such as Henry Lawson, Australia's favorite literary son, Joseph Furphy and "Banjo" Paterson turned their backs on middle-class European literature and created in Australia a distinctive type of national-proletarian writing. The work of these writers, who congregated around the Sydney weekly, the *Bulletin*, continue to have an influence.

In the 1930s writing moved away from short fiction and verse toward the novel as the dominant literary form but the egalitarianism remained. The two decades to the 1950s also encouraged women's fiction which emerged in the forefront of radical literary politics in the 1970s and 1980s.

▲ **Sydney's bridge and opera house.**

Australian Aborigines

In 1788 the original inhabitants of Australia made up its whole population. In the late 20th century they represent approximately one percent. Recent population studies suggest their number in 1788 could have been three million. Today there are approximately 160,000 Aborigines.

In 1901 the Australian constitution stated "in reckoning the numbers of people... the Aborigines shall not be counted". The constitution also maintained that the Australian government could pass legislation on any race or ethnic group except for the Aborigines for whom responsibility rested with the state governments. In effect these two clauses meant that Aborigines had no citizenship rights; a factor that was only corrected in 1967 when over 90 percent of the electorate voted in a national referendum to confer citizenship rights to all Aborigines.

Until quite recently, 20th-century methods of dealing with Aborigines were a simple continuation of 19th-century practices. Detribalization was a prime motivation on the part of the Europeans; children were taken away from their parents to be raised on missions and half-castes were forbidden to live on reserves. In 1937 state and federal officials met and a policy of assimilation was adopted by which part-Aborigines were to be assimilated into the white community; selected children were to be educated in white schools and the remainder were to be left on the reserve. The current belief was that the race was dying out.

During World War II, Aboriginal stockmen pressured station-owners to pay wages instead of rations. The industrial campaign succeeded in part and in 1946 northwestern people walked out and went on strike, refusing to work on the nearby stations. Their boycott has lasted over 40 years. Since the 1960s when assimilation was abandoned, Aboriginal Australians have become increasingly vocal in their claims for legal, political, social, cultural and land rights. Yet they remain the most underprivileged section of Australian society, in many instances living in third-world conditions where disease and dietary disorders are endemic. Aborigines make up a distorted percentage of the prison population and in the late 1980s an investigation was launched because of the large numbers of deaths in custody. In the late 20th century Aborigines are reclaiming their cultural and political rights and heritage and are pressuring for changes at all levels of society. In 1988 prime minister Bob Hawke announced that a "treaty" or "compact" would be drawn up to take full consideration of aboriginality, but already Aborigines have made claims for political autonomy including recognition of their separate cultures and nationalities.

▼ **Aborigines protest for their rights, 1988.**

NEW ZEALAND

NEW ZEALAND

CHRONOLOGY

1907	New Zealand achieves dominion status
1919	Women are eligible for parliamentary seats
1930	Unemployment relief introduced
1932	Pay cuts lead to rioting
1947	New Zealand becomes independent
1951	State of emergency declared following dock strike
1988	Cyclone *Bola* causes widespread devastation

ESSENTIAL STATISTICS

Capital Wellington

Population (1989) 3,371,000

Area 267,844 sq km

Population per sq km (1989) 12.6

GNP per capita (1987) US$8,230

CONSTITUTIONAL DATA

Constitution Constitutional monarchy with one legislative house (House of Representatives)

Date of independence 25 November 1947

Major international organizations UN; Commonwealth of Nations; SPC

Monetary unit 1 New Zealand dollar ($NZ) = 100 cents

Official languages Maori; English

Major religion Anglican

Overseas territories Cook Islands; Niue; Tokelau; Ross Dependency

Heads of government since independence (Prime minister) P. Fraser (1940–49); Sir S. Holland (1949–57); K. Holyoake (Sep–Dec 1957); Sir W. Nash (1957–60); K. Holyoake (1960–72); Sir J. Marshall (Feb–Dec 1972); N. Kirk (1972–74); H. Watt (Aug–Sep 1974); W. Rowling (1974–75); R. Muldoon (1975–84); D. Lange (1984–89); G. Palmer (1989–90); M. Moore (1990); J. Bolger (1990–)

New Zealand is a constitutional democracy with a long-standing two-party parliamentary system, based on the Westminster model. It has no single constitutional document, but, rather, a series of adopted English laws and practices. The British monarch is the ceremonial head of government, represented locally by the governor; but real political power resides in the prime minister and cabinet who are drawn from the majority party in the House of Representatives. Parliamentary power is centralized in the national government with the lower house, the House of Representatives, the law-making house. The upper house, or Legislative Council, originally comprised appointed life members, but this was reformed in 1950 after which time elected members sat in the chamber. The House of Representatives is the most important house where governments are made and unmade. Elections are held every three years and, while enrollment is compulsory, voting is optional. Citizenship remains something of a contested arena, despite acts in 1865, 1948 and 1977, but the indigenous Maoris may vote within the parliamentary framework or in one of the four Maori electorates, which have become increasingly powerful.

Since the 1890s the main parliamentary divide has been between left-liberal parties on the one side and conservative groupings on the other. The National and Labor parties have inherited the schism and dominated politics in the second half of the century. The National party was formed after Labor won its first majority in 1935. It drew together a coalition of the United (formerly Liberal) and Reform parties. The Reform party was constituted in 1905 to take account of changed social and political factors brought about by the rise of a mercantile urban middle class who now shared economic hegemony with the farmers. In the 1890s New Zealand moved away from a predominantly rural society to become increasingly urbanized, though the farming lobby remained very powerful. The bourgeoisie did not wish to overthrow conservative politics but did want a share in its spoils. However, it was the Liberal party which dominated politics in the 20 years to 1911. New Zealand Liberals charted a number of social reforms, earning New Zealand the reputation of being a "classic" example of "state socialism". New Zealand continued as a strong and stable democracy and in the 1980s, in a marked maneuver away from early dependence on Britain and, after 1945, the United States, the Labor government of David Lange began to pursue an independent foreign policy, focusing on its own area and into the Pacific where it enjoyed excellent relations with many of the island states. The peace and antinuclear movements are very strong in New Zealand, though the country also has a reputation as a fighting nation in European and Pacific wars. At home racial violence, once endemic, has continued though on a lesser scale.

Economic and social trends

New Zealand is a primary producing nation with chief exports in the pastoral and agricultural industries. With a modicum of manufacturing it relies heavily on the sales of a range of primary goods overseas. Early in the 19th century, New Zealand traded primarily with Australia but this was surpassed by Britain with the expansion of land tenure, improved agricultural and pastoral techniques and transport, including the important refrigerated ships by the 1890s. At the turn of the century Britain was taking over two-thirds of New Zealand exports, a figure which grew to almost ninety percent by the mid-1940s. Since World War II the UK trade has diminished dramatically and by the late 20th century, Australia, Japan and the United States became more important. New Zealand's heavy dependence on the sale of primary goods means that it was susceptible during periods of economic downturn. However, a free-trade agreement with Australia brought it into the orbit of a larger economy. More recently New Zealand entrepreneurs have used Australia as a springboard into the arena of high finance and international trade.

Dual systems of secular and religious education grew up in New Zealand in the 19th century but the Education Act of 1877 moved toward a compulsory state-based schooling system. Secondary and university education expanded in the 1940s and 1960s and free compulsory education from the age of 6 to 15 was introduced in 1964. From the 1970s successive governments have also taken to subsidizing private schools. In science and technology, New Zealand is probably best known for the work of Ernest Rutherford in the field of nuclear physics. Research is supported within the universities but is also given extra government incentive through the Department of Scientific and Industrial Research (DSIR), established in 1926. Concerned largely with agricultural sciences, the DSIR has also pioneered many advances geared toward high technology, secondary industries, geology, physics and chemistry.

Society and culture in New Zealand are dominated by the twin issues of proprietorship over land and Maori–Pakeha (European) relations. Both issues pivot on the Treaty of Waitangi. This treaty, signed in 1840, recognized the sovereignty of British rule, after a negotiated series of conditions relating to Maori rights had been settled. The understanding of the Maoris was that their rights and cultures had been guaranteed under this arrangement and that concessions would be made relating to the land. However, there was some confusion occasioned by the drawing up of two documents, both signed by the Maoris and the Pakeha. The English version differed from the Maori translation which gave the clear impression of conceding more than the original. There is also evidence to show that the original was tampered with, to the detriment of the Maori claims, after the signing. Moreover, Waitangi was not incorporated into statutory law and, while it became a powerful symbol of both Maori and Pakeha nationalism in the 20th century, it lacked legal status. In 1990 a commission of inquiry was established to incorporate Waitangi in New Zealand law, but its implementation was delayed.

Maori culture is intrinsic to New Zealand culture; Maori languages and legends are taught in schools and Maori arts, literatures and oral cultures are officially and popularly sanctioned. Pakeha-New Zealand culture still carries many of the emblems of colonialism. Until 1948 all New Zealanders were British subjects and the last formal associations with British law were broken in 1977. While other kinds of cultural imperialism were also present (notably Australian and American), Pakeha writers such as Katherine Mansfield, Frank Sargeson and Janet Frame, actors such as Sam Neill, rock bands such as Split Enz and Crowded House developed distinctive New Zealand voices which have been heard on a world stage. New Zealand is a highly urbanized consumer society which has given rise to many leisure industries. Its mountain areas and hot springs have made it an impressive tourist destination for Australians, Asians and Americans in particular. In the world of sport the All-Black Rugby Union side is legendary and New Zealand has produced its fair share of athletes, especially long-distance runners such as John Walker.

▼ **The Maori wardance of the New Zealand rugby team.**

GAZETTEER

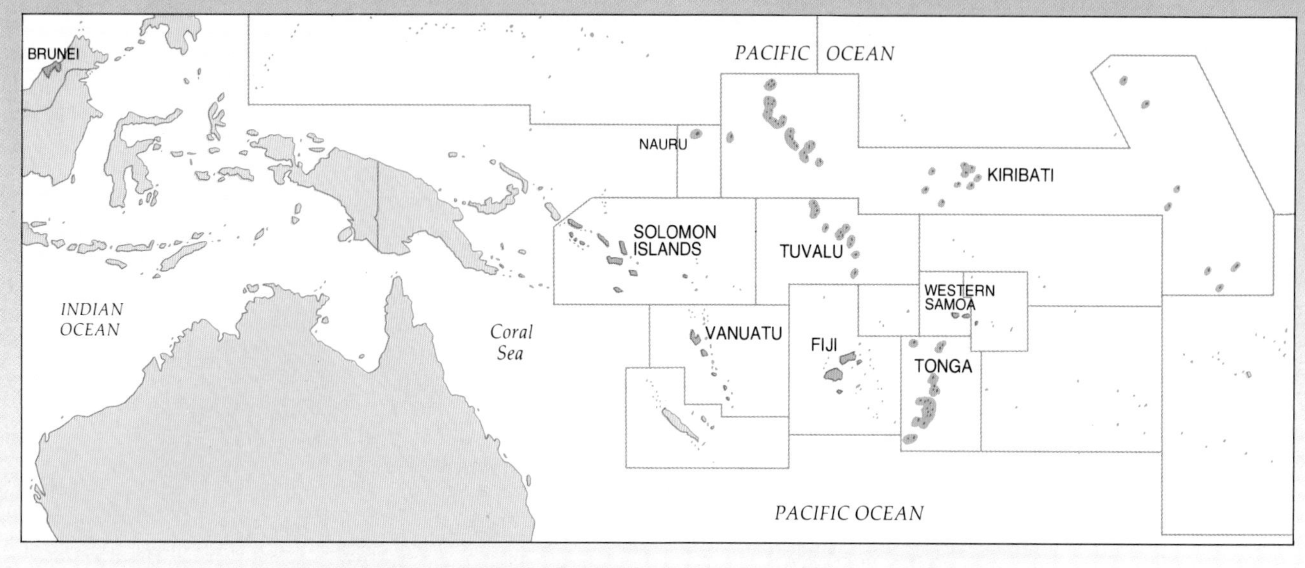

BRUNEI

(Sultanate of Brunei) **Capital** Bandar Seri Begawan **Population** (1989) 249,400 **Area** 5,765 sq km **GNP per capita** (1986) US$15,400 **International organizations** UN, Commonwealth, ASEAN, IDB.

Brunei became a British protectorate in 1888, and administrative power was handed over in 1906 to a British resident, who advised the hereditary Sultan and was responsible to the governor of Sarawak (also a British protectorate) after the Japanese occupation of 1941–45, but only until 1959, when the two administrations were separated and a British high commissioner was appointed for Brunei. In the same year a constitution was promulgated, and in 1962 a Legislative Council with limited authority was set up. A revolt caused the declaration of a state of emergency, and parts of the constitution were suspended; this situation still obtained in 1991. The revolt was eventually put down by British troops at the request of the Sultan. Defense and foreign affairs were still handled by Britain. In 1967 Sultan Sir Omar Ali Saifuddin abdicated in favor of his son Crown Prince Bolkiah, and thereafter Brunei remained stable; its economy was stable owing to its being one of the world's wealthiest (on a per capita basis) oil-producing countries – the Sultan of Brunei is reputedly the richest man in the world, though the wealth of Brunei is not shared equally. It remained a protectorate until 1971, and when in 1976 Britain wished to withdraw its troops the Sultan protested, being suspicious of the territorial ambitions of Malaysia and Indonesia. Both the latter countries insisted that they had no designs on Brunei, and so independence was achieved in 1984.

FIJI

(Republic of Fiji) **Capital** Suva **Population** (1989) 734,000 **Area** 18,274 sq km **GNP per capita** (1988) US$1,540 **International organizations** UN, SPC, ACP, Colombo Plan.

European commercial interest in Fiji developed in the early 19th century with the discovery of sandalwood there. In the 1860s cotton plantations were established, and ensuing financial disputes undermined an already unstable community. In 1874 negotiations between European nations led to Fiji's becoming a British crown colony. The first governor, Sir Arthur Gordon, governed through the traditional political structure, and initiated economic policies designed to protect the Fijians from speculators. Indentured Indian laborers were introduced at this time also, and encouraged to settle in Fiji. This system lasted until 1920, when Indians began to express grievances brought about by the circumscription of their political and economic rights, and racial tension between Fijians and Indians increased. In 1943, during the Allied occupation of Fiji, it was discovered that Indians outnumbered Fijians in the islands, and this heightened the tension further. Developments toward independence began in the 1960s, with a restructuring of the Fijian political structure, including an extension of the franchise to all adults, including Fijians, where it had previously included only Europeans and some Indians. Independence was finally achieved in 1970, and until 1987, Fiji was governed by the Alliance Party on a platform of multiracialism. Only briefly, in 1977, were the Fijians swayed toward nationalism. In 1987, the Fijian-led Labour Party entered into a coalition with the Indian-led National Federation Party, thus creating an Indian majority in government, which was greeted with dismay by Fijians. It was only a few weeks before the new leaders were deposed in a coup by Lieut. Col. Sitiveni Rabuka, who demanded Fijian domination of the country. The governor-general negotiated a compromise with the new military leaders to keep civilian rule, but this only caused Rabuka to stage a second coup toward the end of 1987, and shortly afterward that year declare Fiji a republic. He appointed a new civilian government with Ratu Sir Kamisese Mara as prime minister. Rabuka himself became minister of home affairs in 1988, and in that year the cabinet approved a new system reserving the majority of parliamentary seats and also all key positions for ethnic Fijians, although no date was set for elections. In 1990 Mara forced Rabuka to leave the cabinet, achieving civilian rule, and President Ganilau drew up a new constitution; it was unpopular because, in an attempt to redress the balance in Fijian society, it discriminated against the Indian population in favor of the ethnic Melanesians.

KIRIBATI

(Republic of Kiribati) **Capital** Bairiki **Population** (1989) 69,600. **Area** 849 sq km **GNP per capita** (1988) US$650 **International organizations** Commonwealth, SPC, ACP.

Kiribati is a republic of coral islands in the central Pacific Ocean, consisting of 33 islands, 20 of which are inhabited, with the population mainly concentrated in the 16 Gilbert Islands, which became a British protectorate in 1892. In 1900 the island of Banaba was annexed after the discovery of rich phosphate deposits. Until these were exhausted in 1979, Kiribati's economy depended heavily on the export of phosphate. Coconut palms dominate the landscape of the islands, coconuts being of primary dietary importance. Kiribati society remains conservative and resistant to change: ties to family and traditional land remain strong, conspicuous displays of individual achievement or wealth are strongly discouraged, and traditional pursuits such as building and sailing canoes, dancing and music are widespread. Kiribati depends a great deal on foreign aid, and food accounts for one third of all its imports. The Gilbert Islands and Banaba were linked with the Ellice Islands from 1916, and the colony was subsequently extended to include some other islands also. Administration was through island governments. Resettlement schemes were initiated in response to drought and perceived overpopulation in the 1930s and later. During World War II the Gilberts were occupied by the Japanese and later liberated by American Marines. Only after the war was the colony developed with the introduction of aid-funded programs. An elected House of Representatives was established in 1967. Racial tensions led to the division of the colony in 1975–76, and independence was achieved in 1979 and the House of Representatives renamed Maneaba ni Maungataba. From then onward a high priority was given to economic development, especially the exploitation of marine resources. The UN report of 1989 on the aggravation of the "greenhouse effect" owing to the damage to the Earth's atmosphere from industrial and automotive pollution listed Kiribati as one of the countries in danger of submersion as a result of the projected rise in sea levels.

NAURU

(Nauru) **Capital** (unofficial) Yaren **Population** (1989) 9,000 **Area** 21 sq km **GNP per capita** (1985) US$ 9,091 **International organizations** Commonwealth, SPC

A small Pacific island with a mainly Polynesian populace, today only partly habitable or cultivable and therefore overcrowded, because of the phosphate mining industry on which its economy is based, Nauru was first visited by Europeans in 1798. In the late 19th century internecine strife laid the island open for annexation in 1888 by Germany, as part of the Marshall Islands. In 1899 phosphate was discovered, and mined from 1906 by a British company, who brought in Chinese laborers, and with them dysentery, infantile paralysis, and TB. After World War I Nauru came under British government until 1947, when the UN drew up a trusteeship agreement between Britain, Australia and New Zealand. In 1968 it became the first Micronesian sovereign country under President Hammer de Roburt, with one of the world's highest per capita incomes owing to its phosphate resources. De Roburt was reelected in 1971 and 1973, but the people grew tired of his self-regarding rule, and in 1976 elected Bernard Dowiyogo in his place, reelecting him in 1977. But de Roburt's allies were obstructive and in 1978 Dowiyogo resigned, was reelected, and resigned again on an economic issue. De Roburt was reelected five times in the next five years. After being reelected again in 1987 he began to pursue the matter of seeking compensation from the British Phosphate Commission for Nauru's land areas debilitated by colonial mining; and he set up a commission to examine the issue. The commission proposed that the estimated costs (US$216 million) be borne jointly by the three former trust governments. They were not forthcoming, and Nauru went to the International Court of Justice. Australia agreed to accede to the ICJ's ruling. In 1989 de Robert was dismissed by a no-confidence vote and Kenos Aroi was elected, but illness prevented his standing in December's general elections, when Dowiyogo was returned. Also in 1989 was the issue of the UN report on the greenhouse effect, naming Nauru as one of the islands under threat of submersion as sea-levels rise.

SOLOMON ISLANDS

(Solomon Islands) **Capital** Honiara **Population** (1989) 308,000 **Area** 28,370 sq km **GNP per capita** (1988) US$430 **International organizations** UN, Commonwealth, ACP, SPC.

The Solomons are a South Pacific island chain with over 60 languages and dialects spoken, a subsistence-based agricultural economy, with excessive fish and timber harvesting depleting stocks, and a bauxite mining industry, named by Spanish explorers on the fallacious assumption that they were the source of King Solomon's gold. In 1886 they were divided between the British and the Germans; but by 1900 Germany had ceded most of its share to Britain, which had annexed other islands in the group. In 1942 the Japanese invaded, to be expelled the following year by the USA. During the 1960s the people were given increasing electoral power, and in 1974 the third constitution since 1960 established a single legislative assembly under a governor. In 1978 the Solomons became an independent Commonwealth state, governed by the PPP (Progressive Party), with the PAP (Alliance Party) in opposition, and a native governor general. The first prime minister, Sir Peter Kenilorea, was reelected in 1980, but replaced in 1981 after a no-confidence vote by PAP leader Solomon Mamaloni, who began to devolve power to provincial administrations. In 1984 Kenilorea was re-elected and reversed this process. Two further no-confidence motions failed but Kenilorea resigned in 1986 after allegations that he had secretly taken French aid to repair cyclone damage in his home village. In 1989 Mamaloni was returned to power and initiated progress toward a federal republican system, to be in place by 1992. He suspended aid programs, asking for trade instead. He also advocated cooperation with neighbors Vanuatu and Papua New Guinea, as a possible Melanesian federation; but in 1990 conflict arose with Papua New Guinea, which the Solomons government alleged was interfering with traditional seaways. Opposition leader Andrew Nori criticized Mamaloni's authoritarian style of leadership. In late 1990 Mamaloni resigned as party leader but remained head of government, bringing in four opposition members.

TONGA

(Tonga) **Capital** Nuku'alofa **Population** (1988) 95,300 **Area** 780 sq km **GNP per capita** (1986) US$740 **International organizations** Commonwealth, SPC, ACP.

In 1875 Tonga became an independent, unified constitutional monarchy with a modern administration. George II (ruled 1893–1918), requested British protectorate status in 1900 to protect Tonga from German interests, and from then on all Tonga's foreign affairs were conducted via the British consul, who had extensive powers of veto over Tonga's finances and foreign policy. He was succeeded by Queen Salote Tupou III, who ruled until 1965, when Taufa'ahau Tupou IV became king. Tonga regained full independence in 1970. Tongan culture still flourishes in spite of Western influences: traditional dancing, carving and other handcrafts (often to meet the needs of the tourist trade), as well as the elaborate ritual of drinking kava, a pepper plant-based narcotic.

TUVALU

(Tuvalu) **Capital** Fongafale (Funafuti island) **Population** (1989) 8,900 **Area** 23.96 sq km **GNP per capita** (1985) US$420 **International organizations** Commonwealth (special member), ACP, SPC

The Ellice Islands became a British protectorate in 1892, and formed part of the Gilbert and Ellice Islands colony in 1916. Racial tensions mounted in the 1960s, and the Ellice Islands seceded in 1976 as a separate colony, achieving independence in the Commonwealth – with no political parties – as Tuvalu in 1978 under prime minister Tomasi Puapua, who was reelected in 1985. Tuvalu signed a friendship treaty with the USA in 1978, but relations with France suffered because of persistent French nuclear testing on Mururoa Atoll. In 1989 a new prime minister, Bikenibeu Paeniu, was elected, and a number of new government members appointed. Paeniu promised to support his country's national identity and self-determination, and to reduce dependency on foreign aid. 1989's UN "greenhouse effect" report listed Tuvalu among the island groups at risk of submergence if predicted sea-level rises took place.

VANUATU

(Republic of Vanuatu) **Capital** Port Vila **Population** (1988) 154,000 **Area** 12,190 sq km **GNP per capita** (1988) US$820 **International organizations** UN, Commonwealth, SPC, ACP.

Vanuatu was first mapped as the New Hebrides in 1774 by British Captain James Cook, but it was not until the 1840s that missionaries and sandalwood traders settled in some of the islands. At this time the Europeans had little impact on the indigenous population, but when during the 1860s thousands of ni-Vanuatu people returned from plantation work in Fiji, New Caledonia and Australia, they were able to compete successfully with the European traders and planters in the islands; and so in 1887 the French and British governments set up a Joint Naval Commission to protect the interests of the planters and missionaries, replaced in 1906 by an unwieldy condominium which had little impact on the ni-Vanuatu. New Hebrides was a major Allied base in the Pacific during World War II, and contact with spendthrift black American troops inspired the development of an anti-European political movement. Land ownership became the chief bone of contention postwar – over a third of the New Hebrides being owned by foreigners. In 1979 elections were held and a constitution drawn up to prepare for independence as the Republic of Vanuatu in 1980, during which year an unsuccessful attempt was made by Jimmy Stevens, leader of the Na-Griamel Party, to establish the separate independence of Espiritu Santo island. 1989 saw a rapprochement with France.

WESTERN SAMOA

(Western Samoa) **Capital** Apia **Population** (1989) 164,000 **Area** 2,831 sq km **GNP per capita** (1988) US$580 **International organizations** UN, Commonwealth, SPC, ACP.

In 1873 US special agent Col. A.B. Steinberger, made peace among vying Samoan chiefs, and also drafted a constitution and became the virtual dictator of the islands until he was arrested and deported by the British in 1876. Germany gained Western Samoa as a protectorate from 1899. It was occupied by New Zealand during World War I, and afterward granted to New Zealand as a mandate by the League of Nations. Resistance by the Mau of Pule, a movement formed under German administration, continued, and after a shootout in 1929, the New Zealand government pursued a conciliatory policy as of 1936. The US military used Western Samoa as a base during World War II, thus beginning the development of the islands, which after the war became a UN trust territory administered by New Zealand. Moves toward independence were made: a draft constitution in 1961 was ratified by a plebiscite. Western Samoa became independent in 1962, and a Treaty of Friendship was signed with New Zealand, authorizing that country to handle Western Samoa's foreign affairs on request. Only the *Matai* (male clan leaders) and Europeans were allowed to vote, until universal suffrage was accorded after a referendum in 1990, and a new legislature elected (but only *Matai* could stand for election), under the ruling HRPP (Human Rights Protection party); President Tofilau Eti Alesana then appointed the first woman to a cabinet post.

INTERNATIONAL ORGANIZATIONS

ADB – African Development Bank
Established 1963, commenced operating 1966. Membership: 50 African member states, 25 non-African. Provides loans, advisory services and technical assistance to public and private sectors of African countries. Institution organized to maintain a specifically African character.

ADB – Asian Development Bank
Commenced operations 1966. Membership: 47 countries. Undertakes loan provision and technical assistance with development projects in the region, also assists member states to coordinate economic policies.

Andean Group (Acuerdo de Cartagena, also known as the Andean Pact)
Created 1969. Membership: Bolivia, Colombia, Ecuador, Peru, Venezuela. A cooperative development group set up to accelerate economic and social development through regional integration.

ANZUS Security Treaty
Signed September 1951, ratified April 1952. Membership: Australia, New Zealand, USA (US and New Zealand suspended security relations following a dispute). Established to coordinate members' efforts for collective defense and the preservation of peace and security in the Pacific area. Has no permanent staff or secretariat.

ASEAN – Association of Southeast Asian Nations
Established 1967. Membership: Brunei, Indonesia, Malaysia, Philippines, Singapore, Thailand. Established to accelerate economic growth, social progress and cultural development of the region through joint measures aimed at establishing a peaceful and prosperous community of southeast Asian nations

BIS – Bank for International Settlements
Established 1930 as part of the Hague Agreements. Membership: 24 European countries and 5 others. The central banks' bank, an international financial organization set up to promote the cooperation of the central banks.

CACM – Central American Common Market
Established 1960, ratified September 1963. Membership: Costa Rica, Guatemala, El Salvador, Honduras, Nicaragua. Established to facilitate regional economic integration through liberalization of trade.

CARICOM – Caribbean Community and Common Market
Established 1973. Membership: Antigua and Barbuda, Bahamas, Barbados, Belize, Dominica, Grenada, Guyana, Jamaica, Montserrat, St. Kitts-Nevis, Saint Lucia, Saint Vincent and the Grenadines, Trinidad and Tobago. Established to encourage unity in the region through economic cooperation and coordination of foreign policy and social and cultural development.

CEAO – Communauté Economique de l'Afrique de l'Ouest (West African Economic Community)
Established 1974, replacing the West African Customs Union. Membership: Benin, Burkina Faso, Ivory Coast, Mali, Mauritania, Niger, Senegal. Established to facilitate trade and regional and economic cooperation and integration through community development.

The Colombo Plan for Co-operative, Economic and Social Development in Asia and the Pacific
Established 1950. Membership: 26 countries. Established to accelerate economic and social progress in the region through cooperative effort and development assistance. However planning and finance are organized bilaterally on a government-to-government basis rather through any central funding or organization.

The Commonwealth
Evolved in the 1840's with the introduction of self-government in Canada. The modern Commonwealth began in 1947 with the entry of India and Pakistan. In 1950, India's assumption of republic status while continuing in full membership of the Commonwealth established a new basis for the association whereby allegiance to a common monarch ceased to be a condition of membership. All members of the Commonwealth recognize Queen Elizabeth II as the symbolic head of the Commonwealth but the position is not vested in the British Crown. Membership: 50 countries in 1991 (including Nauru and Tuvalu who participate in functional activities but are not represented at meetings of the Heads of Government). The Commonwealth is a voluntary association without a constitution which aims to establish cooperation, consultation and mutual assistance in a range of diverse areas from international inequality of wealth and opportunity to nuclear disarmament.

COMECON or CMEA – Council for Mutual Economic Assistance
Founded 1949; ended 1991. Membership on dissolution: Bulgaria, Cuba, Czechslovakia, Hungary, Mongolia, Poland, Romania, USSR, Vietnam. Founded to assist socialist economic integration and to achieve rapid economic and technical progress for member states.

Council for Arab Economic Unity
Established 1964. Membership: 12 Arab countries. Supervises the Arab Common Market and facilitates economic development through cooperative effort and enterprise.

The Council of Europe
Founded 1949. Membership: 24 European states. Concludes conventions and agreements which attempt to harmonize national laws, put citizens in member countries on an equal footing and pool resources and facilities. The agreements are effected by Ministers of Foreign Affairs in member countries. Human rights has been an area of major concern to the council.

EC – European Community
Founded 1967. Membership: Belgium, Denmark, France, Germany, Greece, Ireland, Italy, Luxembourg, The Netherlands, Portugal, Spain, UK. Following the reunification of Germany in October, 1990, the former German Democratic Republic immediately became part of the EEC. The umbrella organization, with its own commission, to administer the EEC, Euratom and the ECSC; formed by merging their three commissions.

ECOWAS – Economic Community of West African States
Established 1975. Membership: 16 West African states. Established to promote trade, cooperation and self-reliance in West Africa in economic, social and cultural areas.

ECSC – European Coal and Steel Community
Founded 1952, to integrate member states' coal and steel industries, and based on a proposal in 1950 by French foreign minister Robert Schuman. It removed trade barriers, established common regulatory rules, and fixed prices; infringements were punished with fines. Membership as EC.

EEC – The European Ecomomic Community
Founded 1957, effective from 1 January 1958. Membership as EC: Established to promote the harmonious development of economic activities combined with raised living standards and political and economic stability. The free movement of persons, goods and capital and closer relationships between the member states have become key motivating ideals.

EFTA – European Free Trade Association
Established 1960. Membership: Austria, Finland, Iceland, Norway, Sweden, Switzerland. Aims to bring about free trade in industrial goods and an expansion of trade in agricultural goods between member countries.

EURATOM – European Atomic Energy Community
Effective from 1 January 1958. Membership: as EEC. Founded to facilitate the peaceful development of nuclear energy.

European Space Agency
Established 1975. Membership: Austria, Belgium, Denmark, France, Federal Republic of Germany, Republic of Ireland, Italy, the Netherlands, Norway, Spain, Sweden, Switzerland, UK. Aims to promote the advancement of space research and technology and the implementation of long-term European space policy and coordinate national space programs.

The Franc Zone
Membership: 15 member countries in French West and Equatorial Africa. The zone includes all 6 countries whose currencies are linked with the French franc at a fixed rate of exchange and who agree to hold reserves mainly in French francs.

GCC – Gulf Cooperation Council
Established 1981. Membership: Bahrain, Kuwait, Oman, Qatar, Saudi Arabia, United Arab Emirates. Aims to coordinate and integrate in economic, social and cultural areas.

IBEC – International Bank for Economic Cooperation
Established 1963; ended 1991. Membership: Bulgaria, Cuba, Czechoslovakia, Hungary, Mongolia, Poland, Romania, USSR, Vietnam. Until October 1990, the German Democratic Republic was a member. Acted as the central bank of COMECON and assisted in economic cooperation and development of member countries.

IDB – Inter-American Development Bank
Established 1959. Membership: 44 members, countries from outside the region included from 1976. Main activity is provision of loans as a means of promoting regional economic and social development.

IDB – Islamic Development Bank
Commenced operations 1975. Founded to aid the economic and social development of Muslim countries through the granting of interest-free loans, in accordance with Islamic law, for infrastructural projects.

International Red Cross and Red Crescent Movement
International Committee of the Red Cross (ICRC) formed 1863 to deal with conflict situations. The League of the Red Cross and Red Crescent (the Muslim equivalent) Societies (LRCS) formed 1919 for peacetime activities. ICRC is exclusively composed of Swiss nationals. A private and independent association, it gives legal protection and material assistance to military and civilian victims of war. LRCS has a membership of approximately 250 million people in 148 countries organized into federated societies, and operates to bring relief to victims of natural disasters, refugees and other displaced people.

LAIA – Latin American Integration Association
Established 1980, ratified 1982. Membership: Argentina, Bolivia, Brazil, Chile, Colombia, Ecuador, Mexico, Paraguay, Peru, Uruguay, Venezuela. Aims for regional integration through policies of economic preference and agricultural, scientific and technical cooperation.

LAS – League of Arab States (or Arab League)
Founded 1945. Membership: 21 Arab states. A voluntary association of sovereign Arab states which aims to coordinate economic, social and cultural activities for the common good of all Arab countries.

NATO – North Atlantic Treaty Organization
Founded 1949. Membership: Belgium, Canada, Denmark, France, Germany, Greece, Iceland, Italy, Luxembourg, Netherlands, Norway, Portugal, Spain, Turkey, UK, USA. A collective defense organization based on the premise that an attack against one country constitutes an attack against all. NATO aims to safeguard peace through political solidarity and the maintenance of defense forces at the lowest level needed to deter all forms of aggression.

Nordic Council
Founded 1952. Membership: Denmark (with Faeroe Islands and Greenland), Finland (with Aland Islands), Iceland, Norway, Sweden. Established to facilitate communications and decision-making by national member parliaments over a wide range of political, economic and cultural matters.

OAPEC – Organization of Arab Petroleum Exporting Countries
Established 1968. Membership: Algeria, Bahrain, Egypt, Iraq, Kuwait, Libya, Qatar, Saudi Arabia, Syria, United Arab Emirates. Established to safeguard the interests of members, to co-ordinate the economic activities of the petroleum industry and to ensure the flow of oil to consumer markets. Also attempts to link petroleum research in Arab states. Member states produced 22.4 per cent of total world petroleum in 1988.

OAS – Organization of American States
Founded 1948. Membership: 33 member countries plus 26 permanent observers. Established to promote economic and social development and to foster peace, security and mutual understanding and cooperation amongst the nations of the Western hemisphere.

OAU – Organization of African Unity
Founded 1963. Membership: 50 African states. Established to promote unity amongst African states, raise living standards, defend sovereignty and territorial integrity, to promote political, economic, cultural, scientific and defense policies and to eliminate colonialism in Africa.

OECD – Organization for Economic Co-operation and Development
Founded 1961. Membership: 24 countries. Provides a forum within which representatives of the governments of industrialized democracies discuss and attempt to coordinate economic and social policies.

OPEC – Organization of the Petroleum Exporting Countries
Established 1960. Membership: Algeria, Ecuador, Gabon, Indonesia, Iran, Iraq, Kuwait, Libya, Nigeria, Qatar, Saudi Arabia, United Arab Emirates, Venezuela. Established to link countries whose main source of export earnings is petroleum, OPEC aims to unify and coordinate petroleum policies and to safeguard members' interests. Members accounted for 33.8 per cent of world petroleum production in 1988.

SADCC – Southern African Development Co-ordination Conference
First conference July 1979. Membership: Angola, Botswana, Lesotho, Malawi, Mozambique, Namibia, Swaziland, Tanzania, Zambia, Zimbabwe. Established to harmonize development and reduce the economic dependence of the region on South Africa.

SPC – South Pacific Commission
Established 1947, operative from 1948. Membership: 27 member countries. Provides technical advice, assistance and training in economic, social and cultural areas to countries of the region.

Warsaw Treaty of Friendship, Co-operation and Mutual Assistance
A treaty of friendship and collaboration by which member states agreed to defend each other against armed aggression in Europe. Signed May 1955, treaty extended July 1975 and again April 1985. Dissolved March 1991. Membership: Bulgaria, Czechoslovakia, German Democratic Republic, Hungary, Poland, Romania, USSR.

UNITED NATIONS
An international association of states, founded in 1945, who pledged to maintain international peace and security through the development of international cooperation in social, cultural and humanitarian spheres. In 1990 there were 159 member countries. The principal organs of the UN are the General Assembly, the Security Council, The Economic and Social Council, The Trusteeship Council, the International Court of Justice and the Secretariat. The majority of the UN's work is carried out through 14 major programs and 19 specialized agencies.

Specialized agencies:
FAO – Food and Agriculture Organization (1945)
Aims to raise levels of nutrition and standards of living through improvements in the production and distribution of agricultural products. 158 members.

GATT – General Agreements on Tariffs and Trade (1948)
Establishes a code of conduct in international trade and trade relations, aims to reduce tariff barriers and assist the trade of developing countries. 96 members.

IAEA – International Atomic Energy Agency (1957)
An autonomous, inter-governmental organization intended to enable the development of atomic energy for peaceful purposes. 113 members.

IBRD – International Bank for Reconstruction and Development (World Bank) (1945)
Provides funds and technical assistance to facilitate economic development in poorer member countries. 151 members.

ICAO – International Civil Aviation Organization (1947)
Founded to develop safe and efficient techniques of international air navigation and to assist in the planning and improvement of international air transport. 158 members.

IDA – International Development Association (1960)
World Bank lending agency which concentrates assistance on the governments of the developing countries. 135 members.

IFC – International Finance Corporation (1956)
Affiliated to the World Bank, the IFC encourages the growth of private enterprise in less developed countries. 133 members.

MIGA – Multilateral Investment Guarantee Agency (1988)
Affiliated to the World Bank, MIGA aims to encourage the flow of investments to the developing countries by the mitigation of noncommercial barriers to investment, especially political ones.

IFAD – International Fund for Agricultural Development (1976)
Established to fund rural development programs for the world's poorest peoples. 130 members.

ILO – International Labor Organization (1919)
Autonomous, inter-governmental agency aiming to raise living standards and improve labor conditions and relations and so establish lasting peace through a foundation of social justice. 150 members.

IMO – International Maritime Organization (1948)
Facilitates intergovernmental cooperation on matters of safety at sea and control of pollution by shipping. Also drafts international maritime conventions. 131 members.

IMF – International Monetary Fund (1945)
Promotes international monetary cooperation, facilitates the expansion and balanced growth of international trade, stabilizes exchange rates and aims to alleviate disequilibrium in members' international balance ofpayments. 151 members.

ITU – International Telecommunication Union (1947)
Encourages world cooperation in the use of telecommunications, promotes technical development and harmonizes national policies in the field. Also allots radio frequencies. 154 members.

UNESCO – United Nations Educational, Scientific and Cultural Organization (1946)
Promotes intergovernmental collaboration in order to further universal respect for justice, the rule of law and human rights and fundamental freedoms. Fosters the free flow of information and encourages the interchange of ideas and cultural achievements. 158 members.

UNIDO – United Nations Industrial Development Organization (1986)
Promotes industrial development in the developing countries.

UPU – Universal Postal Union (1875, became specialized agency in 1948)
Promotes the collaboration of international postal services. 159 members.

WHO – World Health Organization (1948)
Aims to enable all peoples to attain the highest possible level of health. Establishes programs to research and fight disease, promotes international health standards and directs and coordinates international health work. 166 members.

WIPO – World Intellectual Property Organization (1974)
Promotes the protection of copyright throughout the world in the arts, science and industry. 121 members.

WMO – World Meterological Organization (1951)
Aims to improve the exchange of weather information and its applications. 155 members.

FURTHER READING

General

Europa Year Book Europa, London (annual)

Statesman's Year Book Macmillan, London (annual)

Europe

Bark,DL and Gress, DR *A History of West Germany, 1945–88* (Oxford, 1989)

Blouet, B *Malta: An Island Republic* (London, 1981)

Braudel, F *The Identity of France* (Vols I and II) (London, 1988–90)

Crampton, RJ *A Short History of Modern Bulgaria* (Cambridge, 1987)

Danstrup, J *History of Denmark* (Copenhagen, 1949)

Ferreira, HG and Marshall, MW *Portugal's Revolution: Ten Years On* (Cambridge, 1986)

Gilbey, T *Nationalism and Communism in Romania: the Rise and Fall of Ceausescu's Personal Dictatorship* (Oxford, 1990)

Havel, V *Disturbing the Peace* (London, 1990)

Imber, W *Norway* (Oslo, 1980)

Kalvoda, J *The Genesis of Czechoslovakia* (New York, 1986)

Klinge, M *A Brief History of Finland* (Helsinki, 1987)

Krisch, H *The German Democratic Republic* (Boulder, 1985)

Kyle, K *Cyprus* (London, 1984)

Lee, JJ *Ireland 1912–1985: Politics and Society* (Cambridge, 1989)

Lydall, H *Yugoslavia in Crisis* (Oxford, 1989)

Marsh, D *The New Germany: at the Crossroads* (London, 1990)

Misiurias, R-J and Taagepera, R *The Baltic States: Years of Dependence 1940–1980* (Farnborough, 1983)

Morgan, KO *The People's Peace: British History 1945–89* (Oxford, 1990)

Pollo, S and Arben, P *The History of Albania* (London, 1981)

Preston, P *The Triumph of Democracy in Spain* (London and New York, 1986)

Scott, FD *Sweden:The Nation's History* (University of Minnesota Press, 1983)

Spotts, F and Wieser, T *Italy: A Difficult Democracy* (Cambridge, 1986)

Sugar, PF (ed) *A History of Hungary* (London, 1991)

Sully, MA *A Contemporary History of Austria* (London, 1990)

Walesa, L *A Path of Hope* (London, 1989)

Whyte, J *Interpreting Northern Ireland* (Oxford, 1990)

Wildblood, R *What makes Switzerland Tick?* (London, 1988)

Woodhouse, CM *Modern Greece: A Short History* (London, 1977)

The Americas

Anderson, TP *Politics in Central America* (Boston, Mass, 1988)

Barck, OT Jr and Blake, NM *Since 1900: A History of the United States* (New York, 1974)

Bazant, J *A Concise History of Mexico* (Cambridge, 1977)

Beckles, H *A History of Barbados: from Amerindian Settlement to Nation State* (Cambridge, 1990)

Burns, EB *A History of Brazil* (Columbia University Press, 1980)

Cueva, A *The Process of Political Domination in Ecuador* (London, 1982)

Ewell, J *Venezuela: A Century of Change* (London, 1984)

Falcoff, M et al *Chile:Prospects for Democracy* (New York, 1988)

Ferguson, J *Grenada: Revolution in Reverse* (London, 1991)

Fernandez, J *Belize: Case Study for Democracy in Central America* (Aldershot, 1989)

Granatstein, JL *Twentieth Century Canada* (Toronto, 1983)

Klein, H *Bolivia: The Evolution of a Multi-ethnic Society* (Oxford, 1982)

Payne, AJ *Politics in Jamaica* (London and New York, 1988)

Rock, D *Argentina 1516–1982* (London, 1986)

Ruttin, P *Capitalism and Socialism in Cuba: a Study of Dependency, Development and Underdevelopment* (London, 1990)

Sahota, GS *Poverty Theory and Policy: a Study of Panama* (Johns Hopkins University Press, 1990)

Sanders, A *The Powerless People* (London, 1987)

Schlesinger, J *America at Century's End* (Columbia University Press, 1989)

Thorpe, R and Bertram, G *Peru 1890–1977: Growth and Policy in an Open Economy* (London, 1978)

Walker, TW *Nicaragua: The First Five Years* (New York, 1985)

Weinstein, M *Uruguay: Democracy at the Crossroads* (Boulder, 1988)

Africa

Achebe, C *The Trouble with Nigeria* (London, 1984)

Ageron, C-R *A History of Modern Algeria* (London, 1988)

Arnold, G *Modern Kenya* (London, 1982)

Bardill, JE and Cobbe, JH *Lesotho: Dilemmas of Dependence in South Africa* (London, 1986)

Blundy, D and Lycett, A *Qadhafi and the Libyan Revolution* (London, 1987)

Burdette, M *Zambia: Between Two Worlds* (Boulder, 1988)

Gellar, S *Senegal* (Boulder, 1982)

Herbstein, D and Evenston, J *The Devils are Among Us: the War for Namibia* (London, 1989)

Hood, M (ed) *Tanzania and Nyerere* (London, 1988)

Jørgensen, JJ *Uganda: A Modern History* (London, 1981)

Keller, EJ *Revolutionary Ethiopia: From Empire to People's Republic* (Indiana University Press, 1989)

Mondlane, E *The Struggle for Mozambique* (London, 1983)

Ray, DI *Ghana: Politics, Economics and Society* (London, 1986)

Somerville, K *Angola: Politics, Economics and Society* (London and Boulder, 1986)

Sparks, A *The Mind of South Africa* (London, 1990)

Storeman, G *Zimbabwe: Politics, Economics and Society* (London, 1988)

Thompson, L *A History of South Africa* (Yale University Press, 1990)

Thompson, P and Adloff, R *Historical Dictionary of the People's Republic of the Congo* (Metuchen, 1984)

Vatikiotis, PJ *History of Modern Egypt: from Muhammad Ali to Mubarak* (London, 1991)

Wilson, CM *Liberia: Black Africa in Microcosm* (New York, 1971)

Woodward, P *Sudan 1898–1989: the Unstable State* (London, 1991)

Young, C and Turner, T *The Rise and Decline of the Zairian State* (University of Wisconsin Press, 1985)

Zartman, IW *The Political Economy of Nigeria* (New York, 1983)

West and South Asia

Al-Khalil, S *Republic of Fear: the Politics of Modern Iraq* (University of California Press, 1989)

Arney, G *Afghanistan* (London, 1990)

Gilmartin, D *Empire and Islam: Punjab and the making of Pakistan* (London, 1988)

Gubser, P *Jordan* (Boulder, 1982)

Harkabi, Y *Israel's Fateful Decisions* (London, 1989)

Kulke, H & Rothermund, D *A History of India* (London, 1990)

Mackenzie, K *Turkey in Transition: The West's Neglected Ally* (London, 1984)

O'Donnell CR *Bangladesh; Biography of a Muslim Nation* (Boulder, 1986)

Rahnema, A and Nomani, F *The Secular Miracle: Religion, Politics and Economic Activity* (London, 1990)

Safran, N *Saudi Arabia: the Ceaseless Quest for Security* (Harvard University Press, 1985)

Seale, P *Asad of Syria: The Struggle for the Middle East* (London, 1989)

Shehadi, N and Mills, DH *Lebanon: A History of Conflict and Consensus* (London, 1988)

Wilson, J *The Break-up of Sri Lanka: The Sinhalese-Tamil Conflict* (London, 1988)

East Asia and Oceania

Ablin, DA and Hood, M (eds) *The Cambodian Agony* (London and New York, 1987)

Beasley, WT *The Rise of Modern Japan* (London, 1990)

Cotterell, A *China: A Concise Cultural History* (London, 1989)

Drysdale, J *Singapore: Struggle for Success* (Singapore, 1984)

Galli, A *Taiwan ROC: A Chinese Challenge to the World* (London, 1987)

Girling, JIS *Thailand: Society and Politics* (Cornell University Press, 1981)

Hastings, M *The Korean War* (London, 1987)

Karnow, S *In Our Image: America's Empire in the Philippines* (New York, 1989)

Kihl, YW *Politics and Policies in Divided Korea* (Boulder, 1984)

Oxford History of Australia: Vol V (1942-88) (Oxford, 1990)

Palmier, L *Understanding Indonesia* (London, 1986)

Post, K *Revolution, Socialism and Nationalism in Vietnam: Vol I* (Aldershot, 1989)

Sinclair, K *A History of New Zealand* (London, 1980)

Spence, JD *The Search for Modern China* (London, 1989)

Taylor, RH *The State in Burma* (London, 1988)

Zasloff, JJ and Unger, L (eds) *Laos: Beyond the Revolution* (London, 1991)

ACKNOWLEDGEMENTS

Picture Credits

1 Nehru with Mahatma Ghandi, 6th July 1946, TPS

2–3 Football fans, Italy, ASp/David Cannon

6 Beirut, Lebanon, M/Chris Steele-Perkins

8–9 Satellite image of the earth, SPL

9c East German border guard, M/Steve McCurry

81c A young Apache at an American Indian convention, 1990, M/David Hurn

129c South Africans burn their pass books, M/Ian Berry

173c A Koran class in Istanbul, M/Abbas

207c American soldier and Vietnamese civilian, M/Philip Jones-Griffiths

11 Statoil **13** SP/Mark Edwards **15** FSP/Toni Sica **16** K **17** I/Norman Lomax **18** SGA/Adam Woolfitt **19** TPS **21** HDC **22–3** M/Gurary **23t** M/Pinkhassov **24** M/Burt Glinn **25** M/Elliot Erwitt **27** TPS **28** M/Bruno Barbey **29** M/Peter Marlow **31** M/Kondelka **33** M/Erich Lessing **35** M/James Nachtwey **37** M/Steve McCurry **41** M/Bruno Barbey **43** TPS **44** M/Marc Riboud **44–5** ASp/David Cannon **47** M/Robert Cappa **48** M/Ernst Haas **49** M/Harry Gruyaert **51** M/Bruno Barbey **53** M/Henri-Cartier Bresson **54t** M/Marc Riboud **54b** ASp **57** MEPL **58** HDC **59** M/Peter Marlow **60–1** HDC **63** M/Ian Berry **64–5** M/Chris Steele-Perkins **66–7** FSP **68** Peter Newarks Historical Pictures **69** Weimar Archives **70–1** HDC **72** M/Thomas Hoepker **73** M/James Nachtwey **75** TPS **77** M/A. Venzago **80–1** SPL **83** Peter Newarks Military Pictures **84** ASp/Mike Powell **84–5** FSP **87** HDC **88** HDC **89t** Z **89b** M/Marc Riboud **90** ASp/T. Defrisco **91** SPL/NASA **93** Museo Nacional de Historia, Mexico City/David Alfaro Sequeiros **96–7** M/Burt Glinn **97** M/Burt Glinn **102–3** M/Cornell Capa **105** Z **106–7** SAP/Tony Morrison **108** M/Bruno Barbey **111** SAP/Tony Morrison **113** M/Bruce Davidson **115** M/Raymond Depardon **117** SAP/K. Morrison **119** SAP/Marion Morrison **121** M/S. Salgado **123** M/Susan Meiselas **125** M/A. Venzago **126** M/Burt Glinn **128–9** SPL **131** M/Abbas **135** FSP/G. Noel **137** HDC **138** SGA/Tor Eigeland **141** M/S. Salgado **143** HDC **149** Tx/E. Schuurmans **151** M/Chris Steele-Perkins **153** HDC **154** M/Abbas **155** M/G. Mendel **161** M/Don McCullin **162** M/Bruno Barbey **171** M/Abbas **172–3** SPL **174–5** UB **177** FSP **179** M/Chris Steele-Perkins **181** M/Leonard Freed **185** RGS **188** M/Stuart Franklin **191** M/Abbas **193** M/Steve McCurry **195** M/Abbas **197** India Office Library **198** TPS **198–9** F/Maggie Murray **200** M/Raghu Rai **201** SP/Mark Edwards **206–7** SPL **209** Peter Newarks Military Pictures **210–1** M/Marc Riboud **211** M/Raghu Rai **213t** M/Ian Berry **213b** M/Ian Berry **217** HDC **218–9** M/Bruno Barbey **219** M/Richard Kalvar **221** As/J. Alex Langley **223** As/Geoff Tompkinson **227** CP/Terence Spencer **229** M/Philip Jones-Griffiths **231** SGA/Nik Wheeler **233** HDC **235** United Nations **239** M/James Nachtwey **241** FSP/Gamma/Liaison **242–3** A/R. Bryant **243** FSP/Patrick Riviere **245** ASp/Russell Cheyne

Abbreviations used

A	Arcaid, Surrey, UK
As	Aspect Picture Library Ltd, London
ASp	Allsport UK Limited, London
CP	Camera Press Ltd, London
F	Format Partners Photo Library, London
FSP	Frank Spooner Pictures, London
HDC	The Hulton Deutsch Collection, London
I	Impact Photos, London
K	The Kobal Collection, London
M	Magnum Photos Ltd, London
MEPL	Mary Evans Picture Library, London
NASA	National Aeronautics and Space Administration, USA
RGS	Royal Geographic Society, London
SAP	South American Pictures, Suffolk, UK
SGA	Susan Griggs Agency, London
SP	Still Pictures, London
SPL	Science Photo Library, London
Tx	Tropix, Liverpool, UK
TPS	Topham Picture Source, Kent, UK
UB	Ullstein Bilderdienst
Z	Zefa Picture Library (UK) Ltd, London

t: top b: bottom c: center

Editorial and research assistance
Ann Furtado, Sue Phillips, Lin Thomas

Photographs
Thérèse Maitland, Joanne Rapley

Cartography
Maps drafted by Lovell Johns, Oxford; Alan Mais, Hornchurch; and Euromap Ltd, Pangbourne

Typesetting
Brian Blackmore, Niki Whale

Production
Stephen Elliott

Origination
Eray Scan Ltd, Singapore
Lithocraft, Coventry, UK

Index
Fiona Barr

INDEX